T0321069

Statistical Shape Analysis

Statistical Shape Analysis
with Applications in R

Second Edition

Ian L. Dryden
The University of Nottingham, UK

Kanti V. Mardia
University of Leeds and University of Oxford, UK

Library of Congress Cataloging-in-Publication Data

Names: Dryden, I. L. (Ian L.), author. | Mardia, K. V., author.
Title: Statistical shape analysis with applications in R / Ian L. Dryden and Kanti V. Mardia.
Other titles: Statistical shape analysis | Wiley series in probability and statistics.
Description: Second edition. | Chichester, UK ; Hoboken, NJ : John Wiley & Sons, 2016. | Series: Wiley
    series in probability and statistics | Originally published as: Statistical shape analysis, 1998 |
    Includes bibliographical references and index.
Identifiers: LCCN 2016011608 (print) | LCCN 2016016288 (ebook) | ISBN 9780470699621 (cloth) |
    ISBN 9781119072508 (pdf) | ISBN 9781119072515 (epub)
Subjects: LCSH: (1) Shape theory (Topology)–Statistical methods (2) Statistics, multivariate analysis.
Classification: LCC QA612.7 .D79 2016 (print) | LCC QA612.7 (ebook) | DDC 514/.24–dc23
LC record available at https://lccn.loc.gov/2016011608

A catalogue record for this book is available from the British Library.

Cover image: Shape analysis of two proteins 1cyd (left) and 1a27 (right) from the Protein Data Bank

ISBN: 9780470699621

Set in 10/12pt TimesLTStd by Aptara Inc., New Delhi, India

1   2016

*To my wife **Maria** and daughter **Sophia*** (Ian Dryden)

*To my grandsons **Ashwin** and **Sashin*** (Kanti Mardia)

Shapes of all Sort and Sizes, great and small,
That stood along the floor and by the wall;
And some loquacious Vessels were; and some
Listen'd perhaps, but never talk'd at all.

Edward FitzGerald, 3rd edition (1872),
Quatrain 83, Rubaiyat of Omar Khayyam

Contents

Preface

Since the publication of the first edition of this book (Dryden and Mardia 1998) there have been numerous exciting novel developments in the field of statistical shape analysis. Although a book length treatment of the new developments is certainly merited, much of the work that we discussed in the first edition still forms the foundations of new methodology. The shear volume of applications of the methodology has multiplied significantly, and we are frequently amazed by the breadth of applications of the field.

The first edition of the book primarily discussed the topic of landmark shape analysis, which is still the core material of the field. We have updated the material, with a new focus on illustrating the methodology with examples based on the shapes package (Dryden 2015) in R (R Development Core Team 2015). This new focus on R applications and an extension of the material has resulted in the new title 'Statistical Shape Analysis, with Applications in R' for this second edition. There is more emphasis on the joint analysis of size and shape (form) in this edition, treatment of unlabelled size-and-shape and shape analysis, more three-dimensional applications and more discussion of general Riemannian manifolds, providing more context in our discussion of geometry of size and shape spaces. All chapters contain a good deal of new material and we have rearranged some of the ordering of topics for a more coherent treatment. Chapters 6, 13, 14, 16 and 18 are almost entirely new. We have updated the references and give brief descriptions of many of the new and ongoing developments, and we have included some exercises at the end of the book which should be useful when using the book as a class text.

In Chapter 1 we provide an introduction and describe some example datasets that are used in later chapters. Chapter 2 introduces some basic size and shape coordinates, which we feel is an accessible way to understand some of the more elementary ideas. Chapter 3 provides a general informal introduction to Riemannian manifolds to help illustrate some of the geometrical concepts. In Chapter 4 we concentrate on Kendall's shape space and shape distances, and in Chapter 5 the size-and-shape (form) space and distance.

After having provided the geometrical framework in Chapters 2–5, statistical inference is then considered with a focus on the estimation of mean shape or size-and-shape in Chapter 6. Chapter 7 provides a detailed discussion of Procrustes analysis, which is the main technique for registering landmark data. Chapter 8 contains specific two-dimensional methods which exploit the algebraic structure of complex numbers, where rotation and scaling are carried out via multiplication and translation by addition. Chapter 9 contains the main practical inferential methods, based on tangent space approximations. Chapter 10 introduces some shape distributions, primarily for two-dimensional data. Chapter 11 contains shortened material on offset normal shape distributions compared with the first edition, retaining the main results and referring to our original papers for specific details. Chapter 12 discusses size and shape deformations, with a particular focus on thin-plate splines as in the first edition.

In Chapter 13 we have introduced many recent developments in non-parametric shape analysis, with discussion of limit theorems and the bootstrap. Chapter 14 introduces unlabelled shape, where the correspondence between landmarks is unknown and must be estimated, and the topic is of particularly strong interest in bioinformatics. Chapter 15 lays out some distance-based measures, and some techniques based on multidimensional scaling. Chapter 16 provides a brief summary of some recent work on analysing curves, surfaces and volumes. Although this area is extensive in terms of applications and methods, many of the basic concepts are extensions of the simpler methods for landmark data analysis. Chapter 17 is a more minor update of shapes in images, which is a long-standing application area, particularly Bayesian image analysis using deformable templates. Chapter 18 completes the material with discussion of a wide variety of recent methods, including statistics on other manifolds and the broad field of Object Data Analysis.

There are many other books on the topic of shape analysis which complement our own including Bookstein (1991); Stoyan and Stoyan (1994); Stoyan et al. (1995); Small (1996); Kendall et al. (1999); Lele and Richtsmeier (2001); Grenander and Miller (2007); Bhattacharya and Bhattacharya (2008); Claude (2008); Davies et al. (2008b); da Fontoura Costa and Marcondes Cesar J. (2009); Younes (2010); Zelditch et al. (2012); Brombin and Salmaso (2013); Bookstein (2014) and Patrangenaru and Ellingson (2015). A brief discussion of other books and reviews is given in Section 18.7.

Our own work has been influenced by the long-running series of Leeds Annual Statistical Research (LASR) Workshops, which have now been taking place for 40 years (Mardia et al. 2015). A strong theme since the 1990s has been statistical shape analysis, and a particularly influential meeting in 1995 had talks by both Kendall and Bookstein among many others (Mardia and Gill 1995), and the proceedings volume was dedicated to both David Kendall and Fred Bookstein.

We are very grateful for the help of numerous colleagues in our work, notably at the University of Leeds and The University of Nottingham. We give our special thanks to Fred Bookstein and John Kent who provided many very insightful comments on the first edition and we are grateful for Fred Bookstein's comments on the current edition. Their challenging comments have always been very helpful indeed. Also, support of a Royal Society Wolfson Research Merit Award WM110140 and EPSRC grant EP/K022547/1 is gratefully acknowledged.

We would be pleased to hear about any typographical or other errors in the text.

Ian Dryden and Kanti Mardia
Nottingham, Leeds and Oxford,
January 2016

Preface to the first edition

Whence and what art thou, execrable shape?
John Milton (1667) *Paradise Lost II*, 681

In a wide variety of disciplines it is of great practical importance to measure, describe and compare the shapes of objects. The field of shape analysis involves methods for the study of the shape of objects where location, rotation and scale information can be removed. In particular, we focus on the situation where the objects are summarized by key points called landmarks. A geometrical approach is favoured throughout and so rather than selecting a few angles or lengths we work with the full geometry of the objects (up to similarity transformations of each object). Statistical shape analysis is concerned with methodology for analysing shapes in the presence of randomness. The objects under study could be sampled at random from a population and the main aims of statistical shape analysis are to estimate population average shapes, to estimate the structure of population shape variability and to carry out inference on population quantities.

Interest in shape analysis at Leeds began with an application in Central Place Theory in Geography. Mardia (1977) investigated the distribution of the shapes of triangles generated by certain point processes, and in particular considered whether towns in a plain are spread regularly with equal distances between neighbouring towns. Our joint interest in statistical shape analysis began in 1986, with an approach from Paul O'Higgins and David Johnson in the Department of Anatomy at Leeds, asking for advice about the analysis of the shape of some mouse vertebrae. Some of their data are used in examples in the book.

In 1986 the journal *Statistical Science* began, and the thought-provoking article by Fred Bookstein was published in Volume 1. David Kendall was a discussant of the paper and it had become clear that the elegant and deep mathematical work in shape theory from his landmark paper (Kendall 1984) was of great relevance to the practical applications in Bookstein's paper. The pioneering work of these two authors provides the foundation for the work that we present here. The penultimate chapter on shape in image analysis is rather different and is inspired by Grenander and his co-workers.

This text aims to introduce statisticians and applied researchers to the field of statistical shape analysis. Some maturity in Statistics and Mathematics is assumed, in order to fully appreciate the work, especially in Chapters 4, 6, 8 and 9. However, we believe that interested researchers in various disciplines including biology, computer science and image analysis will also benefit, with Chapters 1–3, 5, 7, 8, 10 and 11 being of most interest.

As shape analysis is a new area we have given many definitions to help the reader. Also, important points that are not covered by definitions or results have been highlighted in various places. Throughout the text we have attempted to assist the applied researcher with practical advice, especially in Chapter 2 on size measures and simple shape coordinates, in Chapter 3 on two-dimensional Procrustes analysis, in Section 5.5.3 on principal component analysis, Section 6.9 on choice of models, Section 7.3.3 on analysis with Bookstein coordinates, Section 8.8 on size-and-shape versus shape and Section 9.1 on higher dimensional work. We are aware of current discussions about the advantages and disadvantages of superimposition type approaches versus distance-based methods, and the reader is referred to Section 12.2.5 for some discussion.

Chapter 1 provides an introduction to the topic of shape analysis and introduces the practical applications that are used throughout the text to illustrate the work. Chapter 2 provides some preliminary material on simple measures of size and shape, in order to familiarize the reader with the topic. In Chapter 3 we outline the key concepts of shape distance, mean shape and shape variability for two-dimensional data using Procrustes analysis. Complex arithmetic leads to neat solutions. Procrustes methods are covered in Chapters 3–5. We have brought forward some of the essential elements of two-dimensional Procrustes methods into Chapter 3 in order to introduce the more in-depth coverage of Chapters 4 and 5, again with a view to helping the reader. In Chapter 4 we introduce the shape space. Various distances in the shape space are described, together with some further choices of shape coordinates. Chapter 5 provides further details on the Procrustes analysis of shape suitable for two and higher dimensions. We also include further discussion of principal component analysis for shape.

Chapter 6 introduces some suitable distributions for shape analysis in two dimensions, notably the complex Bingham distribution, the complex Watson distribution and the various offset normal shape distributions. The offset normal distributions are referred to as 'Mardia–Dryden' distributions in the literature. Chapter 7 develops some inference procedures for shape analysis, where variations are considered to be small. Three approaches are considered: tangent space methods, approximate distributions of Procrustes statistics and edge superimposition procedures. The two sample tests for mean shape difference are particularly useful.

Chapter 8 discusses size-and-shape analysis – the situation where invariance is with respect to location and rotation, but not scale. We discuss allometry which involves studying the relationship of shape and size. The geometry of the size-and-shape space is described and some size-and-shape distributions are discussed. Chapter 9 involves the extension of the distributional results into higher than two dimensions, which is a more difficult situation to deal with than the planar case.

Chapter 10 considers methods for describing the shape change between objects. A particularly useful tool is the thin-plate spline deformation used by Bookstein (1989) in shape analysis. Pictures can be easily drawn for describing shape differences in the spirit of D'Arcy Thompson (1917). We describe some of the historical developments and some recent work using derivative information and kriging. The method of relative warps is also described, which provides an alternative to principal component analysis emphasizing large or small scale shape variability.

Chapter 11 is fundamentally different from the rest of the book. Shape plays an important part in high-level image analysis. We discuss various prior modelling procedures in Bayesian image analysis, where it is often convenient to model the similarity transformations and the shape parameters separately. Some recent work on image warping using deformations is also described.

Finally, Chapter 12 involves a brief description of alternative methods and issues in shape analysis, including consistency, distance-based methods, more general shape spaces, affine shape, robust methods, smoothing, unlabelled shape, probabilistic issues and landmark-free methods.

We have attempted to present the essential ingredients of the statistical shape analysis of landmark data. Other books involving shape analysis of landmarks are Bookstein (1991), Stoyan and Stoyan (1994, Part II) (a broad view including non-landmark methods) and Small (1996) (a more mathematical treatment which appeared while our text was at the manuscript stage).

In the last few years the Leeds Annual Statistics Research workshop has discussed various ideas and issues in statistical shape analysis, with participants from a wide variety of fields. The edited volumes of the last three workshops (Mardia and Gill 1995; Mardia et al. 1996c, 1997a) contain many topical papers which have made an impact on the subject.

If there are any errors or obscurities in the text, then we would be grateful to receive comments about them.

Real examples are used throughout the text, taken from biology, medicine, image analysis and other fields. Some of the datasets are available from the authors on the internet and we reprint three of the datasets in Appendix B.

Ian Dryden and Kanti Mardia
Leeds, December 1997

Acknowledgements for the first edition

We are very grateful to John Kent and Fred Bookstein for their insightful comments on a late draft of the book. We are also grateful to colleagues for many useful discussions and for sending preprints: Yali Amit, Bidyut Chaudhuri, Colin Goodall, Tim Hainsworth, Val Johnson, David Johnson, Peter Jupp, David Kendall, Wilfrid Kendall, Huiling Le, Paul O'Higgins, Dietrich Stoyan, Charles Taylor and Alistair Walder.

Various aspects of our work in shape analysis were developed with research fellows and students at Leeds. In particular we are grateful for the help of current and former Postdoctoral Research Fellows Kevin de Souza, Merrilee Hurn, Delman Lee, Richard Morris, Wei Qian, Sophia Rabe-Hesketh, Druti Shah, Lisa Walder, Janet West and Postgraduate Students Cath Anderson, Mohammad Faghihi, Bree James, Jayne Kirkbride, John Little, Ian Moreton, Mark Scarr, Catherine Thomson and Gary Walker.

Comments from those attending the Shape Analysis course given by ILD at the University of Chicago were very valuable (Xiao-Li Meng, Chih-Rung Chen, Ernesto Fontenla and Scott Hadley). The shape workshop at Duke University in March 1997, organized by Val Johnson, was also very helpful.

The use of computer and other facilities at the University of Leeds and the University of Chicago is gratefully acknowledged.

We are very grateful to the following for providing and helping with the data: Cath Anderson, Fred Bookstein, Chris Glasbey, Graham Horgan, Nina Jablonski, David Johnson, John Marchant, Terry McAndrew, Paul O'Higgins, and Robin Tillett.

The authors would like to thank their wives Maria Dryden and Pavan Mardia for their support and encouragement.

Ian Dryden and Kanti Mardia
Leeds, December 1997

1

Introduction

1.1 Definition and motivation

Objects are everywhere, natural and man-made. Geometrical data from objects are routinely collected all around us, from sophisticated medical scans in hospitals to ubiquitous smart-phone camera images. Decisions about objects are often made using their sizes and shapes in geometrical data, for example disease diagnosis, face recognition and protein identification. Hence, developing methods for the analysis of size and shape is of wide, growing importance. Locating points on objects is often straightforward and we initially consider analysing such data, before extending to curved outlines, smooth surfaces and full volumes.

Size and shape analysis is of great interest in a wide variety of disciplines. Some specific applications follow in Section 1.4 from biology, chemistry, medicine, image analysis, archaeology, bioinformatics, geology, particle science, genetics, geography, law, pharmacy and physiotherapy. As many of the earliest applications of shape analysis were in biology we concentrate initially on biological examples and terminology, but the domain of applications is in fact very broad indeed.

The word 'shape' is very commonly used in everyday language, usually referring to the appearance of an object. Following Kendall (1977) the definition of shape that we consider is intuitive.

Definition 1.1 **Shape** *is all the geometrical information that remains when location, scale and rotational effects are removed from an object.*

An object's shape is invariant under the Euclidean similarity transformations of translation, scaling and rotation. For example, the shape of a human skull consists of all the geometrical properties of the skull that are unchanged when it is translated, rescaled or rotated in an arbitrary coordinate system. Two objects have the same shape

Statistical Shape Analysis, with Applications in R, Second Edition. Ian L. Dryden and Kanti V. Mardia.
© 2016 John Wiley & Sons, Ltd. Published 2016 by John Wiley & Sons, Ltd.

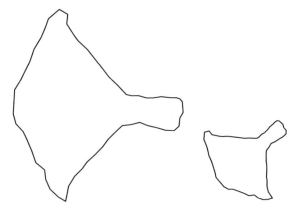

Figure 1.1   Two outlines of the same second thoracic (T2) vertebra of a mouse, which have different locations, rotations and scales but the same shape.

if they can be translated, rescaled and rotated to each other so that they match exactly, that is if the objects are similar. In Figure 1.1 the two mouse vertebrae outlines have the same shape. In practice we are interested in comparing objects with different shapes and so we require a way of measuring shape, some notion of distance between two shapes and methods for the statistical analysis of shape.

Sometimes we are also interested in retaining scale information (size) as well as the shape of the object, and so the joint analysis of size and shape (or form) is also very important.

Definition 1.2   Size-and-shape *is all the geometrical information that remains when location and rotational effects are removed from an object.*

Two objects have the same size-and-shape if they can be translated and rotated to each other so that they match exactly, that is if the objects are rigid-body transformations of each other. 'Size-and-shape' is also frequently denoted as **form** and we use the terms equivalently throughout the text.

A common theme throughout the text is the geometrical transformation of objects. The terms superimposition, superposition, registration, transformation, pose and matching are often used equivalently for operations which involve transforming objects, either with respect to each other or into a specified reference frame.

An early writing on shape was by Galileo (1638), who observed that bones in larger animals are not purely scaled up versions of those in smaller animals; there is a shape difference too. A bone has to become proportionally thicker so that it does not break under the increased weight of the heavier animal, see Figure 1.2. The field of geometrical shape analysis was initially developed from a biological point of view by Thompson (1917), who also discussed this application.

How should a scientist wishing to investigate a shape change proceed? Even describing an object's shape is difficult. In everyday conversation an object's shape

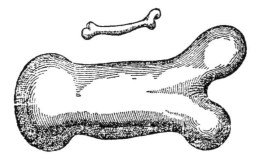

Figure 1.2   From Galileo (1638) illustrating the differences in shapes of the bones of small and large animals.

is usually described by naming a second more familiar shape which it looks like, for example a map of Italy is 'boot shaped'. This leads to very subjective descriptions that are unsuitable for most applications. A practical way forward is to locate a finite set of points on each object, which summarize the key geometrical information.

1.2   Landmarks

Initially we will describe a shape by locating a finite number of points on each specimen which are called landmarks.

Definition 1.3   A **landmark** *is a point of correspondence on each object that matches between and within populations.*

There are three basic types of landmarks in our applications: scientific, mathematical and pseudo-landmarks. In the literature there have been various synonyms for landmarks, including vertices, anchor points, control points, sites, profile points, 'sampling' points, design points, key points, facets, nodes, model points, markers, fiducial markers, markers, and so on.

A **scientific landmark** is a point assigned by an expert that corresponds between objects in some scientifically meaningful way, for example the corner of an eye or the meeting of two sutures on a skull. In biological applications such landmarks are also known as **anatomical landmarks** and they designate parts of an organism that correspond in terms of biological derivation, and these parts are called homologous (e.g. see Jardine 1969). In Figure 1.3 we see some anatomical landmarks located on the skull of a macaque monkey, viewed from the side. This application is described further in Section 1.4.3. Another example of a scientific landmark is a carbon C_α atom of an amino acid on a protein backbone, as seen in Section 1.4.9.

Mathematical landmarks are points located on an object according to some mathematical or geometrical property of the figure, for example at a point of high curvature or at an extreme point. The use of mathematical landmarks is particularly useful in automated recognition and analysis.

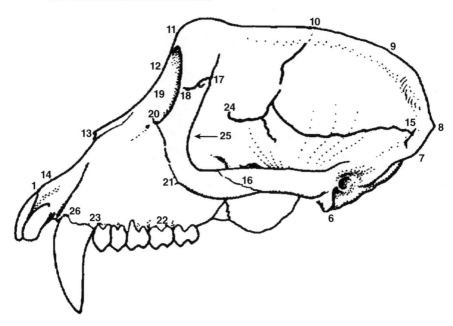

Figure 1.3   Anatomical landmarks located on the side view of a macaque monkey skull.

Pseudo-landmarks are constructed points on an object, located either around the outline or in between scientific or mathematical landmarks. For example, Lohmann (1983) took equally spaced points on the outlines of micro-fossils. In Figure 1.4 we see six mathematical landmarks at points of high curvature and seven pseudo-landmarks marked on the outline inbetween each pair of landmarks on a second thoracic (T2) mouse vertebra. Continuous curves can be approximated by a large number of pseudo-landmarks along the curve. Hence, continuous data can also be studied by landmark methods, although one needs to work with discrete approximations and the choice of spacing of the pseudo-landmarks is crucial. Examples of such approaches include the analysis of hand shapes (Grenander *et al.* 1991; Mardia *et al.* 1991; Cootes *et al.* 1992), resistors (Cootes *et al.* 1992, 1994), mitochondrial outlines (Grenander and Miller 1994), carotid arteries (Cheng *et al.* 2014; Sangalli *et al.* 2014) and mouse vertebrae (Cheng *et al.* 2016). Also, pseudo-landmarks are useful in matching surfaces, when points can be located on a grid over each surface, for example the cortical surface of the brain (Brignell *et al.* 2010) or the surface of the hippocampus (Kurtek *et al.* 2011).

Bookstein (1991) also demarks landmarks into three further types, which are of particular use in biology. **Type I landmarks** occur at the joins of tissues/bones; **type II landmarks** are defined by local properties such as maximal curvatures, and **type III landmarks** occur at extremal points or constructed landmarks, such as maximal diameters and centroids.

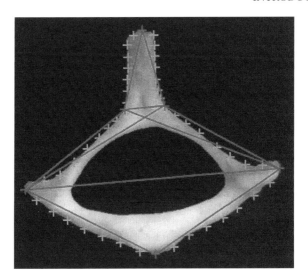

Figure 1.4   Image of a T2 mouse vertebra with six mathematical landmarks on the outline joined by lines (dark +) and 42 pseudo-landmarks (light +). Source: Dryden & Mardia 1998. Reproduced with permission from John Wiley & Sons. For a colour version of this figure, see the colour plate section.

A further type of landmark is the **semi-landmark** which is a point located on a curve and allowed to slip a small distance in a direction tangent to another corresponding curve (Bookstein 1996a,c; Green 1996; Gunz *et al.* 2005). The term 'semi-' is used because the landmark lies in a lower number of dimensions than other types of landmarks, for example along a one-dimensional (1D) curve in a two-dimensional (2D) image (see Section 16.3).

A further situation that may arise is the combination of landmarks and geometrical curves. For example, the pupil of the eye may be represented by a landmark at the centre surrounded by a circle, with the radius as an additional parameter. Yuille (1991) and Phillips and Smith (1993, 1994) considered such representations for analysing images of the human face.

Definition 1.4   *A* **label** *is a name or number associated with a landmark, and identifies which pairs of landmarks correspond when comparing two objects. Such landmarks are called* **labelled landmarks**.

The landmark with, say, label 1 on one specimen corresponds in some meaningful way with landmark 1 on another specimen. A labelling is usually known and given as part of the dataset. For example, in labelling the anatomical landmarks on a skull the labelling follows from the definition of the points. When we refer to just 'shape' of landmarks we implicitly mean the shape of labelled landmarks, that is **labelled shape**.

Unlabelled landmarks are those where no labelling correspondence is given between points on different specimens. It may make sense to try to estimate a correspondence between landmarks, although there is usually some uncertainty involved. This approach is common in bioinformatics for example, as seen in Chapter 14.

In some applications there is no natural labelling, and one must treat all permutations of labels as equivalent. The **unlabelled shape** of an object is the geometrical information that is invariant under permutations of the labels, and translation, rotation and scale.

Example 1.1   Consider the simple example in Figure 1.5. The six triangles (A, B, C, D, E and F) are constructed from triples of labelled points (1,2,3). Triangles A and

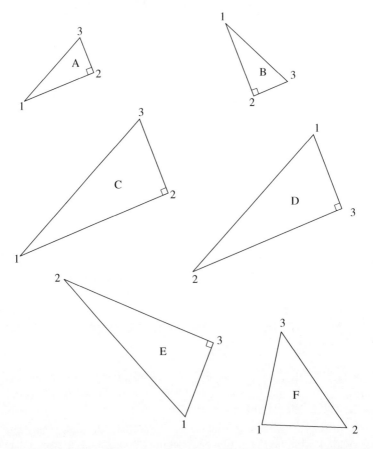

Figure 1.5   Six labelled triangles: A and B have the same size and labelled shape; C has the same labelled shape as A and B (but larger size); D has a different labelled shape but its labels can be permuted to give the same unlabelled shape as A, B and C; triangle E can be reflected to have the same labelled shape as D; triangle F has a different shape from A, B, C, D and E.

B have the same size and the same labelled shape because they can be translated and rotated to be coincident. Triangle C has the same labelled shape as A and B (but has a larger size) because it can be translated, rotated and rescaled to be coincident with A and B. Triangle D has a different labelled shape but, if ignoring the labelling, it has the same unlabelled shape as A, B and C. Triangle E has a different shape to D but it can be reflected and translated to be coincident, and so D and E have the same reflection shape. Triangle F has a different shape from all the rest.   □

In the majority of this book the methodology is appropriate for landmark data or other point set data. Following Kendall (1984) our notation will be that there are k landmarks in m dimensions, where we usually have $k \geq 3$ and $m = 2$ or $m = 3$. Extensions to size and shape analysis methods for outline data, surface data and volume data are then considered in the latter chapters of the book, and many of the basic ideas from landmark shape analysis are very helpful for studying these more complex applications.

1.3   The `shapes` package in R

The statistical package and programming language R is an extremely powerful and wide ranging environment for carrying out statistical analysis (Ihaka and Gentleman 1996; R Development Core Team 2015). The progam is available for free download from `http://cran.r-project.org` and is a very widely used and popular platform for carrying out modern statistical analysis. R is continually updated and enhanced by a dedicated and enthusiastic team of developers. R has thousands of contributed packages available, including the `shapes` package (Dryden 2015), which includes many of the methods and datasets from this book. There are numerous introductory texts on using R, including Crawley (2007). For an excellent, comprehensive summary of a wide range of statistical analysis in R, see Venables and Ripley (2002).

We shall make use of the `shapes` package in R (Dryden 2015) throughout the text. Although it is not necessary to follow the R commands, we believe it may be helpful for many readers. To join in interactively the reader should ensure that they have downloaded and installed the base version of R onto their machine. Installation instructions for specific operating systems are given at `http://cran .r-project.org`. After successful installation of the base system the reader should install the `shapes` package. The reader will then be able to repeat many of the examples in the book by typing in the displayed commands.

The first command to issue is:

```
library(shapes)
```

which makes the package available for use. In order to obtain a quick listing of the commands type:

```
library(help=shapes)
```

Also, at any stage it is extremely useful to use the 'help' system, by typing:

```
help(commandname)
```

where commandname is the name of a command from the shapes package. For example,

```
help(plotshapes)
```

gives information about a basic plotting function for landmark data.

1.4   Practical applications

We now describe several specific applications that will be used throughout the text to illustrate the methodology. Some typical tasks are to study how shape changes during growth; how shape changes during evolution; how shape is related to size; how shape is affected by disease; how shape is related to other covariates such as sex, age or environmental conditions; how to discriminate and classify using shape; and how to describe shape variability. Various methodologies of multivariate analysis have been used to answer such questions over the last 75 years or so. Many of the questions in traditional areas such as biology are the same as they have always been and many of the techniques of shape analysis are closely related to those in multivariate analysis. One of the practical problems is that small sample sizes are often available with a large number of variables, and so high dimension, low sample size issues (large p, small n) are prevalent (e.g. Hall *et al.* 2005). We shall describe many new techniques that are not part of the general multivariate toolkit. As well as traditional biological applications many new problems can be tackled with statistical size and shape analysis.

1.4.1   Biology: Mouse vertebrae

In an experiment to assess the effects of selection for body weight on the shape of mouse vertebrae, three groups of mice were obtained: Large, Small and Control. The Large group contains mice selected at each generation according to large body weight, the Small group was selected for small body weight and the Control group contains unselected mice. The bones form part of a much larger study and these bones are from replicate E of the study (Falconer 1973; Truslove 1976; Johnson *et al.* 1985, 1988; Mardia and Dryden 1989b).

We consider the second thoracic vertebra T2. There are 30 Control, 23 Large and 23 Small bones. The aims are to assess whether there is a difference in size and shape between the three groups and to provide descriptions of any differences. Each vertebra was placed under a microscope and digitized using a video camera to give a grey level image, see Figure 1.4. The outline of the bone is then extracted using

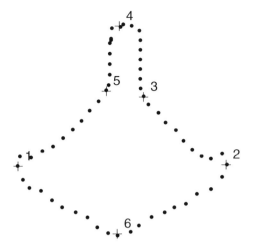

Figure 1.6   Six mathematical landmarks (+) on a second thoracic mouse vertebra, together with 54 pseudo-landmarks around the outline, approximately equally spaced between pairs of landmarks. The landmarks are 1 and 2 at maximum points of approximate curvature function (usually at the widest part of the vertebra rather than on the tips), 3 and 5 at the extreme points of negative curvature at the base of the spinous process, 4 at the tip of the spinous process, and 6 at the maximal curvature point on the opposite side of the bone from 4.

standard image processing techniques (for further details see Johnson *et al.* 1985) to give a stream of about 300 coordinates around the outline. Six landmarks were taken from the outline using a semi-automatic procedure described by Mardia (1989a) and Dryden (1989, Chapter 5), where an approximate curvature function of the smoothed outline is derived and the mathematical landmarks are placed at points of extreme curvature as measured by this function. In Figure 1.6 we see the six landmarks and also in between each pair of landmarks, nine equally spaced pseudo-landmarks are placed.

The dataset is available in the R package shapes and the three groups can be accessed by typing:

```
library(shapes)
data(mice)
```

The dataset is stored as a list with three components: mice$x is an array of the coordinates in two dimensions of the six landmarks for each bone; mice$group is a vector of group labels; and mice$outlines is an array of 60 points on each outline in two dimensions, containing the landmarks and pseudo-landmarks. To print the $k \times m \times n$ array of landmarks in R, type mice$x and to print the group labels

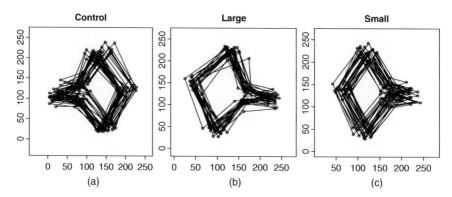

Figure 1.7   The three groups of T2 mouse landmarks, with k = 6 landmarks per bone: (a) 30 Control; (b) 23 Large; and (c) 23 Small mice.

type mice$group ('c' for Control, 'l' for Large and 's' for Small). In order to plot the landmark data we can use:

```
par(mfrow=c(1,3))
joins<-c(1,6,2:5,1)
plotshapes(mice$x[,,mice$group=="c"],joinline=joins)
title("Control")
plotshapes(mice$x[,,mice$group=="l"],joinline=joins)
title("Large")
plotshapes(mice$x[,,mice$group=="s"],joinline=joins)
title("Small")
```

Here the plotshapes function plots 2D (x, y) coordinates of each object, and lines are drawn between the landmarks given in the joinline option. Here lines are drawn from landmark 1 to 6 to 2 to 3 to 4 to 5 and finally back to 1 on each object. The plot is given in Figure 1.7.

In order to plot the outline data we can use:

```
par(mfrow=c(1,3))
joins<-c(1:60,1)
plotshapes(mice$outlines[,,mice$group=="c"],joinline=joins,col=2)
title("Control")
plotshapes(mice$outlines[,,mice$group=="l"],joinline=joins,col=2)
title("Large")
plotshapes(mice$outlines[,,mice$group=="s"],joinline=joins,col=2)
title("Small")
```

and the result is given in Figure 1.8. Here the points are drawn in red (col=2) and the lines are drawn to connect points 1 through 60 and back to 1.

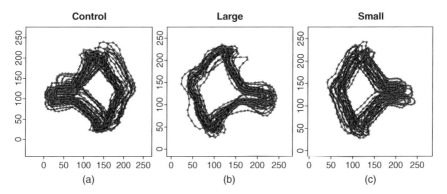

Figure 1.8   The three groups of T2 vertebra outlines, with 60 points per bone: (a) 30 Control; (b) 23 Large; and (c) 23 Small mice.

Note that the coordinates in `mice$x` are also available in the `shapes` package individually by group: `qcet2.dat`, `qlet2.dat`, and `qset2.dat`, which can be useful for short-cuts in coding.

It is of interest to examine size and shape differences in the three groups, and how shape is related to size.

1.4.2   Image analysis: Postcode recognition

A random sample of handwritten British postcodes was collected and digitized (Anderson 1997), and an example digit '3' is shown in Figure 1.9. It is of interest to classify each of the handwritten characters so that mail can be automatically sorted. The problem is a classic one in image analysis and many methods have been suggested, with varying degrees of success (e.g. see Hull 1990). The location and size of the characters are not so important for recognition but orientation information may be crucial, for example an 'M' must not be confused with a 'W'. Some successful attempts at reading handwritten numbers include Simard *et al.* (1993); Hastie and Tibshirani (1994); and Hastie and Simard (1998). A survey of relevant work is given by Plamondon and Srihari (2000), and a related topic is hand-drawn gesture recognition (e.g. see Mardia *et al.* 1993).

Anderson (1997) obtained mathematical landmarks and pseudo-landmarks on the digital images by hand, and in particular for the digit 3 there were 13 landmarks, as shown in Figure 1.9. It is of interest to examine the average shape and variability in shape in the data, which can then be used as a prior model for digit recognition from images of handwritten postcodes.

The landmark data are given the dataset `digit3.dat` in the `shapes` package. The data are displayed in Figure 1.10 using the R command:

```
plotshapes(digit3.dat,joinline=1:13)
```

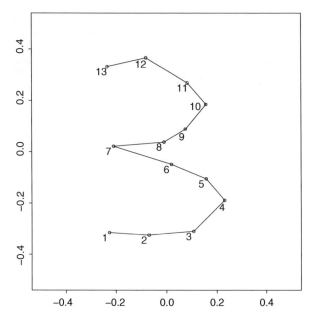

Figure 1.9   A handwritten digit '3' from the postcode dataset, with 13 labelled mathematical landmarks. Landmark 1 is at the extreme bottom left, 4 is at the maximum curvature of the bottom arc, 7 is at the extreme end of the central protrusion, 10 is at the maximum curvature of the top arc and 13 is the extreme top left point. Landmarks 2, 3, 5, 6, 8, 9, 11 and 12 are pseudo-landmarks at approximately equal intervals between the mathematical landmarks.

1.4.3   Biology: Macaque skulls

In an investigation into sex differences in the crania of a species of macaque *Macaca fascicularis* (a type of monkey), random samples of 9 male and 9 female skulls were obtained by Paul O'Higgins (Hull-York Medical School) (Dryden and Mardia 1993). A subset of seven anatomical landmarks was located on each cranium and the three-dimensional (3D) coordinates of each point were recorded.

It is of interest to assess whether there are any size and shape differences between the sexes. If there are any differences, then a description of the differences is required. An artist's impression of the 3D skull with the anatomical landmarks is given in Figure 1.11.

The data are obtained by typing:

```
library(shapes)
data(macaques)
```

The 3D landmarks are available in the $7 \times 3 \times 18$ dimensional array macaques$x, and the genders are in macaques$group.

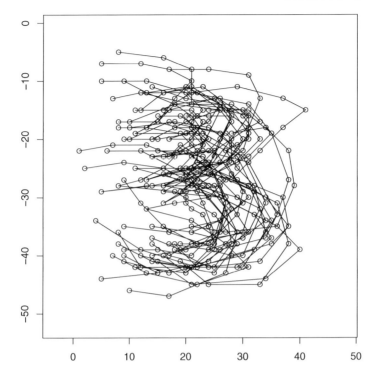

Figure 1.10   The thirty digit 3 configurations, each with 13 landmarks.

We plot the data in Figure 1.12 using the command `shapes3d` as follows:

```
joins<-c(1,2,5,2,3,4,1,6,5,3,7,6,4,7)
colpts<-rep(1:7,times=18)
shapes3d(macaques$x,col=colpts,joinline=joins)
```

The command `shapes3d` uses the `rgl` library in R, which in turn uses OpenGL graphics. The 3D plots can be easily rotated and moved in the graphics window by clicking and moving the mouse, thus giving a good idea of the 3D geometry of the configuration.

Note that the coordinates in `macaques$x` are also available in the `shapes` package individually by group: `macf.dat`, and `macm.dat`.

1.4.4   Chemistry: Steroid molecules

Dryden *et al.* (2007) and Czogiel *et al.* (2011) analyse a dataset of steroids, which are small molecules with a wide variety of uses. The dataset consists of between 42 and 61 atoms for each of 31 steroid molecules. The three-dimensional coordinates, atom type, van der Waals radius and partial charges of each atom are given. The collection of steroids has been considered by a number of authors, including Wagener *et al.*

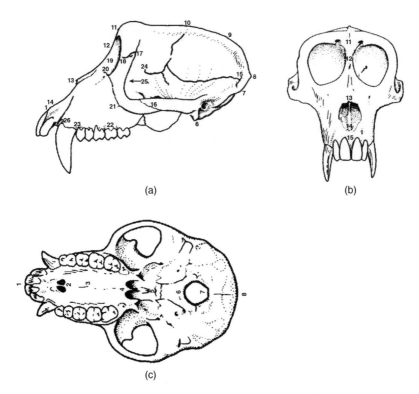

Figure 1.11   A 3D macaque skull: (a) side view; (b) frontal view; and (c) bottom view. A total of 26 landmarks are displayed on the skull and a subset of 7 was taken for the analysis. The seven chosen landmarks are: 1, prosthion; *7,* opisthion; *10,* bregma; *12,* nasion; *15,* asterion; *16,* midpoint of zyg/temp suture; *and 17,* interfrontomalare.

(1995). This particular version of the data was constructed by Jonathan Hirst and James Melville (School of Chemistry, University of Nottingham). The steroids have different binding affinities to the corticosteroid binding globulin (CBG) receptor, and so each molecule has an activity class of either '1' high, '2' intermediate or '3' low binding affinity. It is of interest to examine how the shape ('steric') properties of the molecules are related to activity class. This dataset is quite challenging in that the molecules have different numbers of atoms, and the correspondence between atoms (labelling) is not given. However, the 17 carbon atoms in the three cyclohexane rings and one cyclopentane ring are common to all the steroids, and these do correspond in a sensible way. The carbon rings are plotted in Figure 1.13 using the following commands.

```
data(steroids)
joins<-c(1:6,1,6,5,4,7:10,5,4,7,11:14,8,14:17,13)
shapes3d(steroids$x[,,],col=rep(1:17,times=31),joinline=joins)
```

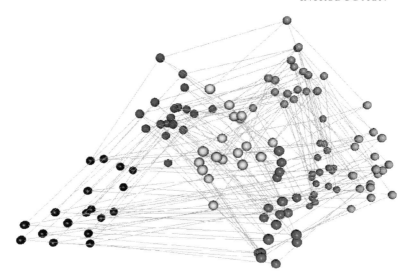

Figure 1.12   The macaque skull data with seven landmarks from 18 individuals, with each landmark displayed by a different colour. For a colour version of this figure, see the colour plate section.

1.4.5   Medicine: Schizophrenia magnetic resonance images

Bookstein (1996b) considers 13 landmarks taken on near midsagittal 2D slices from magnetic resonance (MR) brain scans of 14 control volunteers and 14 schizophrenia patients. It is of interest to study any shape differences in the brain between the two groups, either in average shape or in shape variability. If shape differences between the two groups can be established, then this should enable researchers to gain an increased understanding about the condition. In Figure 1.14 we see the 13 landmarks on a 2D slice from a scan of a schizophrenia patient. The landmarks are: 1, splenium, posteriormost point on corpus callosum; 2, genu, anteriormost point on corpus

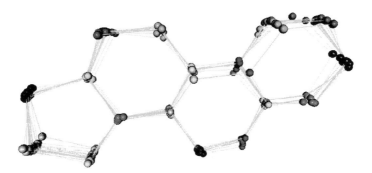

Figure 1.13   The first 17 carbon atoms in the 31 steroid molecules. For a colour version of this figure, see the colour plate section.

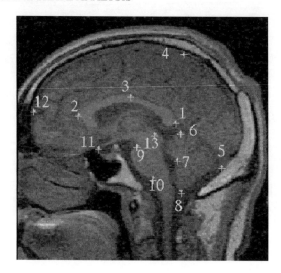

Figure 1.14   The 13 landmarks on a near midsagittal section from a brain scan of a schizophrenia patient. The landmark positions are approximately located at each cross (+). Source: Adapted from Bookstein 1996b. Reproduced with permission from Springer Science+Business Media.

callosum; 3, top of corpus callosum, uppermost point on arch of callosum (all three landmarks registered to the diameter of the callosum); 4, top of head, a point relaxed from a standard landmark along the apparent margin of the dura; 5, tentorium of cerebellum at dura; 6, top of cerebellum; 7, tip of fourth ventricle; 8, bottom of cerebellum; 9, top of pons, anterior margin; 10, bottom of pons, anterior margin; 11, optic chiasm; 12, frontal pole, extension of a line from 1 through 2 until it intersects the dura; and 13, superior colliculus.

The data are plotted in Figure 1.15 using the following commands.

```
data(schizophrenia)
plotshapes(schizophrenia$x,symbol=as.integer(schizophrenia$group))
```

1.4.6   Medicine and law: Fetal alcohol spectrum disorder

Another important application of shape analysis is in the assessment of fetal alcohol spectrum disorders (FASDs). An MR image from the corpus callosum of a prisoner with a landmark (Rostrum) and 39 semi-landmarks is displayed in Figure 1.16. The shape of the corpus callosum has been used in court cases in expert witness testimony to help assess whether or not a defendant had been affected by FASD. Statistical shape analysis has been used successfully to help waive the death penalty for many defendants. For further details see Mardia *et al.* (2013a) and Section 7.10.

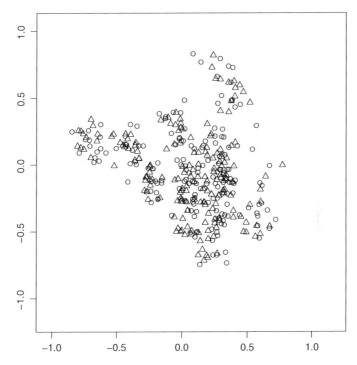

Figure 1.15   The dataset of 13 landmarks per individual from the schizophrenia study, with circles for controls and triangles for patients.

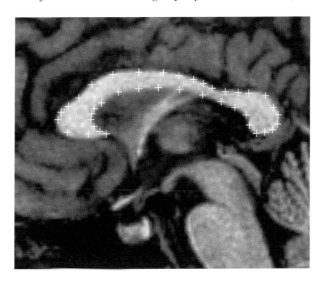

Figure 1.16   A landmark and 39 semi-landmarks on the outline of the corpus callosum from an MR image of a prisoner. Source: Mardia et al. 2013a. Reproduced with permission from John Wiley & Sons.

1.4.7   Pharmacy: DNA molecules

Molecular dynamics simulations are a widely used and powerful method of gaining an understanding of the properties of molecules, particularly biological molecules such as DNA. The simulations are undertaken with a computer package, for example AMBER (Salomon-Ferrer *et al.* 2013), and involve a deterministic model being specified for the molecule. The model consists of point masses (atoms) connected by springs (bonds) moving in an environment of water molecules, also treated as point masses and springs. At each time step the equations of motion are solved to provide the next position of the configuration in space. The simulations are very time-consuming to run – for example several weeks of computer time may be needed to generate a few nanoseconds of data.

We consider the statistical modelling of a specific DNA molecule configuration in water. In particular, we concentrate on the simple case of 22 phosphorous atoms, where the $k = 22$ atom locations are recorded in angstroms (in $m = 3$ dimensions) and are observed over a period of time. The temporal data are highly correlated. There are many questions of interest, for example, can we describe the main features of geometric variability, can we estimate the full configuration space of the molecule, and can we simulate the molecule over much longer time scales using fast statistical techniques. In the shape package there is a very small set of DNA data with $n = 30$ observations, which can be obtained using:

```
data(dna.dat)
```

A plot of the data is given in Figure 1.17.

Figure 1.17   A small dataset of 22 phosphorous atoms from a DNA molecule at $n = 30$ time points. For a colour version of this figure, see the colour plate section.

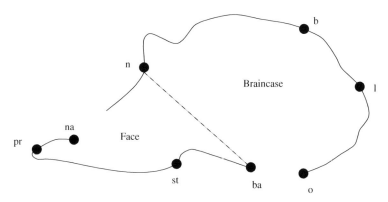

Figure 1.18   Eight landmarks on the midline section of the ape cranium. The face region is taken to be comprised of landmarks: 7, nasion (n); 4, basion (ba); 5, staphylion (st); 1, prosthion (pr); and 6, nariale (na). The braincase region is taken to be comprised of landmarks: 7, 4 and 8, bregma (b); 2, lambda (l); and 3, opisthion (o).

1.4.8   Biology: Great ape skulls

In an investigation to assess the cranial differences between the sexes of ape, 29 male and 30 female adult gorillas (*Gorilla gorilla*), 28 male and 26 female adult chimpanzees (*Pan*), and 30 male and 24 female adult orangutans (*Pongo*) were studied. The data are described in detail by O'Higgins (1989) and O'Higgins and Dryden (1993). Eight landmarks are chosen in the midline plane of each skull as shown in Figure 1.18. The landmarks are anatomical landmarks and are located by an expert biologist.

The dataset is available in the R package `shapes` and can be accessed by typing:

```
data(apes)
```

The dataset is stored as a list with two components: `apes$x` is an array of coordinates in eight landmarks in two dimensions for each skull; and `apes$group` is a vector of group labels. To print the $k \times m \times n$ array of landmarks in R, type `apes$x` and to see the group labels type `apes$group` ('gorf' for female gorillas, 'gorm' for male gorillas, 'panf' for female chimpanzees, 'panm' for male chimpanzees, 'pongof' for female orangutans, and 'pongom' for male orangutans). In order to plot the landmark data we can use:

```
par(mfcol=c(2,3))
plotshapes(apes$x[,,apes$group=="gorf"],col=1)
title("Female Gorillas")
plotshapes(apes$x[,,apes$group=="gorm"],col=2)
title("Male Gorillas")
```

```
plotshapes(apes$x[,,apes$group=="panf"],col=3)
title("Female Chimpanzees")
plotshapes(apes$x[,,apes$group=="panm"],col=4)
title("Male Chimpanzees")
plotshapes(apes$x[,,apes$group=="pongof"],col=5)
title("Female Orang Utans")
plotshapes(apes$x[,,apes$group=="pongom"],col=6)
title("Male Orang Utans")
```

Again the plotshapes function plots 2D (x, y) coordinates of each object, and here we have drawn each group using a different colour. The plot is given in Figure 1.19.

It is of interest to assess whether there is a size difference between the sexes and whether there are any shape differences between the sexes in the face and braincase regions. A biologist would also be interested in geometrical descriptions of the shape difference, and how shape relates to size and other covariates.

Note that the coordinates in apes$x also available in the shapes package individually by group: gorf.dat, gorm.dat, panf.dat, panm.dat, pongof.dat and pongom.dat.

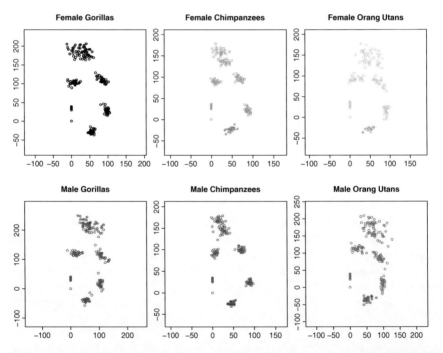

Figure 1.19   The six groups of great ape skull landmarks: (left column) female and male gorillas; (middle column) female and male chimpanzees; and (right column) female and male orangutans. For a colour version of this figure, see the colour plate section.

1.4.9   Bioinformatics: Protein matching

A protein is a sequence of amino acids, of which there are 20 types, and each amino acid has a one-letter code: $A, C, D, E, F, G, H, I, K, L, M, N, P, Q, R, S, T, V, W, Y$. A protein structure is the 3D configuration of atoms determined by the sequence and the 3D size and shape of the protein structure determines its function. The structure is summarized using key atoms, such as α-carbon atoms in each amino acid, and other secondary structure features such as α-helices and β-sheets (e.g. Mardia 2013b). Predicting the 3D structure of the protein from the amino acid sequence (protein folding) is one of the grand challenges in science.

Green and Mardia (2004, 2006) considered matching two proteins from the PDB data bank http://www.rcsb.org/pdb (Berman *et al.* 2000). The particular pair of proteins that were compared had PDB codes 1a27 (Human Type I 17Beta-Hydroxysteroid Dehydrogenase) (Mazza 1997) with 63 points; and 1cyd (Mouse lung carbonyl reductase) (Tanaka *et al.* 1996) with 40 points. The points are the active sites of the proteins and consist of the coordinates of the centres of gravity of the amino acids that make up the nicotinamide adenine dinucleotide phosphate (NADP) binding sites of two proteins. The 20 amino acid types of each active site are labelled in one of four groups as follows:

- Group 1: Hydrophobic. A, F, I, L, M, P, V.

- Group 2: Charged. D, E, K, R.

- Group 3: Polar. C, H, N, Q, S, T, W, Y.

- Group 4: Glycine. G.

Representations of the full proteins are given in Figure 1.20.

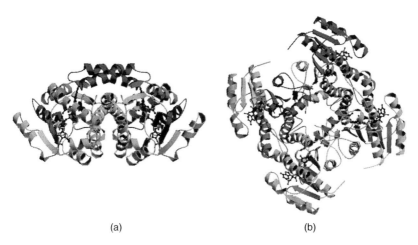

(a)   (b)

Figure 1.20   The proteins 1a27 (a) and 1cyd (b) from the PDB databank (Tanaka et al. *1996; Mazza 1997; Berman* et al. *2000). For a colour version of this figure, see the colour plate section.*

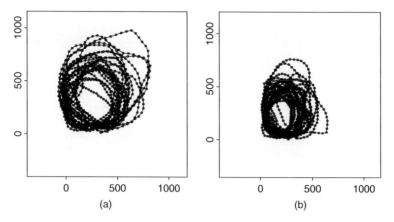

Figure 1.21   The sand particle outlines: (a) sea sand; and (b) river sand.

It is of interest to align the two molecules in order to find common similar geometrical parts. The size and shape of the proteins are of interest, and scale invariance is not required here. In this application the landmarks are unlabelled, and estimation of the correspondence between subsets of the active sites of each protein is required.

1.4.10   Particle science: Sand grains

A dataset of the curved outlines of sand grain profiles is available. There are 24 sea sand and 25 river sand grain profiles in two dimensions. The original data were provided by Dietrich Stoyan (Stoyan and Stoyan 1994; Stoyan 1997). On each particle outline there are 50 points, which were were extracted at approximately equal arc lengths by the method described in Kent *et al.* (2000, section 8.1). The sea particles are from the Baltic Sea and the river particles from the Caucasian River Selenchuk. It is of interest to describe the differences in shape variability of the sand grains between the two groups. Here we are interested in the shapes of the continuous closed curves. The points on each outline are unlabelled – there is no natural correspondence between points. The data are displayed in Figure 1.21.

```
data(sand)
plotshapes(sand$x[,,sand$group=="sea"],
    sand$x[,,sand$group=="river"],joinline=c(1:50,1))
```

1.4.11   Biology: Rat skull growth

We consider a well-known dataset of landmarks located on X-rays of rat skulls as they grow. The data are described in Bookstein (1991) and studied by several other authors including Goodall and Lange (1989); Monteiro (1999); Le and Kume (2000a); Kent *et al.* (2001); and Kenobi *et al.* (2010). The rats were carefully X-rayed at ages 7,

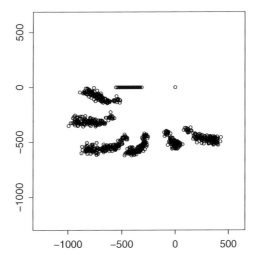

Figure 1.22   The eight landmarks on the 18 rat skulls, observed at eight time points.

14, 21, 30, 40, 60, 90 and 150 days, and there are 18 rats with complete sets of eight landmarks in two dimensions at each age. The X-rays were taken of the skulls of the same rats recorded throughout their lifetimes, and it is of interest to describe the size and shape changes as the rats undergo growth.

The data for the 18 rats with complete data are displayed in Figure 1.22 and are available in the `shapes` package using:

```
data(rats)
```

with the landmark coordinates in `rats$x`, identification number in `rats$no` and the time in `rats$time`.

1.4.12   Biology: Sooty mangabeys

Twelve landmarks are taken from the midline of the skulls of a type of monkey, sooty mangabey (*Cercocebus atys*), in a study described by O'Higgins and Dryden (1992), see Figure 1.23. The specimens ranged from young juveniles to an adult female and an adult male. The objective is to describe the size and shape differences in the individuals in the series from the young juveniles to the older juveniles, and then to the adults. A further problem is to examine whether the individuals can be modelled by a regression line in shape space.

The data are available in the `shapes package` using:

```
data(sooty)
```

The data is an array of size $10 \times 2 \times 7$, with the first two configurations being centred, rotated coordinates for the smallest juvenile and the male adult, and then the last five observations contain the original data for the three juveniles, then the female adult and the male adult.

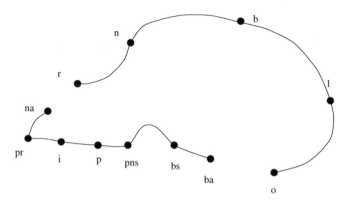

Figure 1.23   The 12 landmarks on the midline of the skull of a juvenile sooty mangabey. The chosen landmarks are nasion *(n),* rhinion *(r),* nariale *(na),* prosthion *(pr),* incisive canal *(i),* palatine junction *(p),* posterior nasal spine *(pns),* basisphenoid *(bs),* basion *(ba),* opisthion *(o),* lambda *(l) and* bregma *(b).*

1.4.13   Physiotherapy: Human movement data

Kume *et al.* (2007) consider an application in the study of human movement data, consisting of $k = 4$ landmarks (lower back, shoulder, wrist and index finger) moving in time. The landmark coordinates are obtained by recording the 3D locations of small reflective markers using a system of seven video cameras. Each individual in the study sat at a table and was asked to move his or her index finger towards a target point which was positioned straight, or to the right or or to the left at different angles. The data were collected by Dr James Richardson, Université Paris Sud, France. We are interested in the shapes of the landmark configurations, and we concentrate on the shapes of the configurations in the plane of the table and this subset was considered by Kume *et al.* (2007). We shall concentrate on a subset of the data where five curves are available, which are labelled a, b, c, d, e. The dataset consists of the projected view of the movements in the plane of the table at which the subject is sitting. For each of the individuals we have 10 equally spaced time points (after initially linearly transforming the times to $[0, 1]$ for each curve). The data are available as a four-dimensional array in the shapes package using:

```
data(humanmove)
```

The array is of dimension $4 \times 2 \times 10 \times 5$ and represents the coordinates of the $k = 4$ landmarks in $m = 2$ dimensions with 10 time points for each of the 5 movements.

1.4.14   Genetics: Electrophoretic gels

A technique for the identification of proteins involves the comparison of electrophoretic gel images (Horgan *et al.* 1992). Two examples of such images are gel

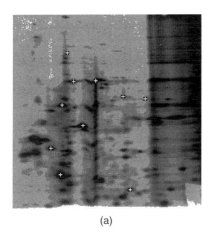
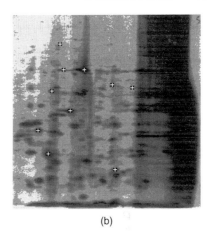

(a)   (b)

*Figure 1.24   The electrophoretic gel images from (a) gel A and (b) gel B. The invari-
ant spots are marked with a '+' in both images. Source: Adapted from Horgan et al.
1992.*

A and gel B shown in Figure 1.24. The images were obtained from particular strains
of parasites which carry malaria. The objective is to use the gel image to identify the
strain of parasite.

In each gel there are a number of black spots, where each spot can be one of two
types – invariant or variant. The invariant spots are present for all parasites and the
arrangement of variant spots enables identification of the parasite. A problem with
the technique when used in the field is that the gels are prone to deformations and so
the gel images first need to be 'registered' (transformed) so that direct comparisons
can be made.

In this application, 10 invariant spots have been picked out by an expert, as shown
in Figure 1.25. The spots are available in the shapes package using:

```
data(gels)
```

The invariant spots are used to match gel A to gel B, either by a similarity transformation or by a more complicated transformation. A question of interest is: can a
matching procedure be made resistant to some outlier points, for example mislabelled
points?

1.4.15   Medicine: Cortical surface shape

In a study of the shape of cortical surfaces of schizophrenia patients and normal volunteers, some structural MR images were taken of the brain (Brignell *et al.* 2010).
The dataset consists of $n = 68$ 3D MR images of the brain from 29 male healthy
controls, 25 male schizophrenia patients, 9 female healthy controls, and 5 female

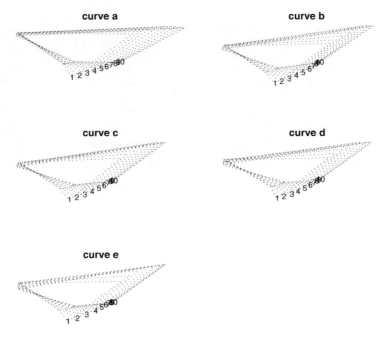

Figure 1.25   The five series of configurations projected into the plane of the table. Each series consists of 10 quadrilaterals observed at equal fractions of the time taken to carry out the pointing movement. Source: Kume et al. 2007. Reproduced with permission of Oxford University Press. For a colour version of this figure, see the colour plate section.

schizophrenia patients. The MR images are proton density weighted images and were collected by Sean Flynn at the University of British Columbia, Canada. Each volunteer's image consists of $256 \times 256 \times 256$ voxels (3D pixels of size 1 mm^3). The cortical surface was extracted from each dataset using image processing techniques [specifically the brain extraction tool of Smith (2002)]. Finally 62 501 points were located on the surface of the cortex above a transverse plane passing through the anterior and posterior commisures, with the brain midline plane being located in a sagittal plane. We see an example of the cortical surface points of one of the subjects in Figure 1.26. It is of interest to describe the cortical surface shape, and whether there are any differences in shape between the schizophrenia group and the control group.

A subset of the data are available in the shapes package:

```
data(cortical)
```

which contains the 250 points on the outline intersecting the axial slice containing the commisures (the bottom-most axial cross-section in Figure 1.26).

Figure 1.26   A set of 62 501 cortical surface points in three dimensions. The colouring indicates the ordering of the points. For a colour version of this figure, see the colour plate section.

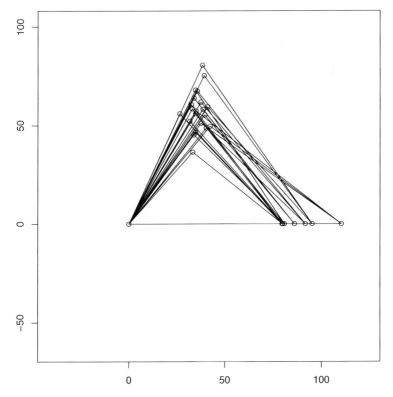

Figure 1.27   Landmarks from 21 mean outlines of microfossils.

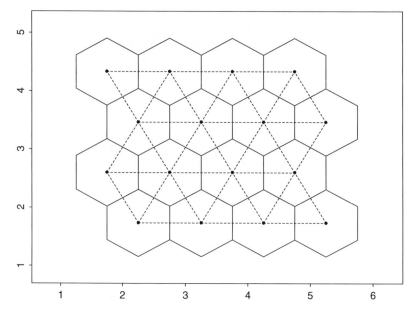

Figure 1.28   The Voronoi polygons (unbroken lines) and Delaunay triangulation (broken lines) for a completely regular configuration, that is ideal central places.

1.4.16   Geology: Microfossils

The microfossil *Globorotalia truncatulinoides* is a microscopic planktonic found in the ooze on the ocean bed. Lohmann (1983) published 21 mean outlines of the microfossil which were based on random samples of organisms taken at different latitudes in the South Indian Ocean. Figure 1.27 shows the three mathematical landmarks selected on each of the outlines, where one landmark has been placed at the origin. The coordinates of the landmarks are extracted from Figure 7 of Bookstein (1986). It is of interest to examine whether the size of the organisms is related to the shape, and whether size or shape are related to the covariate of latitude. A more basic problem would be to obtain an estimate of the average shape of the fossils and to describe the structure of the shape variability.

The data are available in the shapes package using:

```
data(shells)
```

1.4.17   Geography: Central Place Theory

Central Place Theory was postulated by Christaller (1933) and is the situation where towns are distributed on a regular hexagonal lattice over a homogeneous area (with towns at hexagon centres, see Figure 1.28). Mardia *et al.* (1977) consider this hypothesis for a map of 44 places in 6 counties in Iowa, namely Union, Ringgold, Clarke, Decatur, Lucas and Wayne.

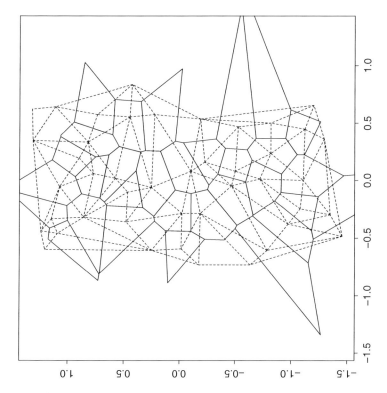

Figure 1.29   The Voronoi polygons (unbroken lines) and Delaunay triangles (broken lines) for the Iowa towns. The Voronoi polygons at the edges are not shown fully.

In order to examine whether Central Place Theory holds, one could examine the shapes of the triangles formed by a town and its neighbours to see if they are more equilateral than expected under a randomness hypothesis. A convenient triangulation of the towns is a Delaunay triangulation (Mardia *et al.* 1977; Green and Sibson 1978; Okabe *et al.* 2000). In Figure 1.28 we see that Voronoi polygons for ideal central places would be hexagons and Delaunay triangles would be equilateral triangles. In Figure 1.29 we see a Delaunay triangulation and the Voronoi polygons for the Iowa data. An important question to ask is: are the Delaunay triangles more equilateral than expected by chance?

The points here are unlabelled (there is no correspondence in the vertices of the triangles). Also the triangles are correlated due to neighbouring triangles sharing points. Kendall (1983, 1989) also studied shape in Delaunay triangulations, in order to investigate the Central Place Theory hypothesis, following Mardia *et al.* (1977).

1.4.18   Archaeology: Alignments of standing stones

Consider a map of the 52 megalithic sites that form the 'Old Stones of Land's End' in Cornwall, UK, given in Figure 1.30. It was proposed by Alfred Watkins, in the early

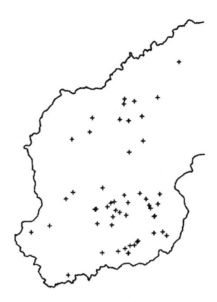

Figure 1.30   The map of 52 megalithic sites (+) that form the 'Old Stones of Land's End' in Cornwall. Source: Stoyan, Kendall & Mecke 1995. Reproduced with permission from John Wiley & Sons.

1920s, that these and other megalithic sites were placed in deliberate straight lines, called ley lines. One approach is to consider the shapes of all possible triangles and to see if there are more 'flat' triangles (triangles with the largest angle close to 180°) than expected under a randomness hypothesis. The points are unlabelled, and in this dataset there are $\binom{52}{3}$ triangles in two dimensions.

This dataset is particularly important in the history of shape analysis because it motivated D.G. Kendall's pioneering work. Analysis of these data is considered by Broadbent (1980), Kendall and Kendall (1980), Small (1988) and Stoyan *et al.* (1995) among others.

2

Size measures and shape coordinates

In this chapter we provide some brief historical background to traditional and geometrical methods of shape analysis. We then define what is meant by a configuration of landmarks and the configuration space, discuss measures of the size, and then consider some simple coordinate systems for representing shapes.

2.1 History

Shape analysis has a long history, especially in biology. In order to analyse size or shape, a biologist traditionally selects ratios of distances between landmarks or angles, and then carries out multivariate analysis (e.g. Rao 1948). This approach has been called 'multivariate morphometrics' in biology and a review is given by Reyment *et al.* (1984, p. 120). Early researchers using such methods include Pearson (1926), who studied a measure of similarity between skulls based on many lengths between landmarks. In fact, various studies have been cited in the journal *Biometrika*, starting from studies in volume 1 itself (Fawcett and Lee 1902). The data were usually distances between landmarks (such as lengths and widths) or angles between landmarks, rather than landmarks themselves. The definitions of landmarks on skulls follow the Frankfurt Concordat (*Frankfurter Verständigung*) of the 13th Congress of the German Anthropological Society which met in Frankfurt-am-Main, 14–17 August 1882 (e.g. see Trevor 1950).

Another large area of work in biology has been in the study of allometry, that is differences in shape associated with size (Hopkins 1966; Sprent 1972). The notion of allometry was introduced by Huxley (1924, 1932) and often involves fitting

Statistical Shape Analysis, with Applications in R, Second Edition. Ian L. Dryden and Kanti V. Mardia.
© 2016 John Wiley & Sons, Ltd. Published 2016 by John Wiley & Sons, Ltd.

simple non-linear equations to length measurements. We consider a discussion of some related distance-based methods later in Chapter 15.

In the studies of multivariate morphometrics one deals exclusively with positive variables (lengths, angles and ratios of lengths). However, considering only distances and angles can be inferior to using the actual coordinates of the landmarks, because the geometry is often thrown away when using the former. Ratios of distances can easily be calculated from coordinates whereas the converse is not generally true. However, if enough distances are taken, then a configuration can be reconstructed up to a reflection and analysis can be carried out using multidimensional scaling (MDS) or Euclidean distance matrix analysis (see Kent 1994; Lele and Richtsmeier 2001; Dryden *et al.* 2008a), see Chapter 15.

Since the 1980s there have been many key developments in shape analysis that allow us to work on the landmark coordinates directly. Also, the advances in technology of measuring landmarks have been helpful, for example landmarks from digitized objects. Of course if there were no constraints from registration, then we could use standard multivariate analysis in Euclidean space, but in general the statistical methodology for shape is inherently non-Euclidean.

The main idea of the geometrical approach to shape analysis is that, rather than working with quantities derived from organisms, one works with the complete geometrical object itself (up to similarity transformations). As pointed out by Bookstein (1978) the approach is very much in the spirit of Thompson (1917) who considered the geometric transformations of one species to another (e.g. see Figure 2.1). D'Arcy Thompson's key ideas will be discussed in more detail in Chapter 12, but the important point to note is that he worked with geometrical pictures of organisms rather than derived quantities.

Throughout the text it will be observed that pictures of the objects under study can always be easily constructed and it is this fact that embodies the geometrical approach to shape analysis. In many applications the statistical goal is inference, for example testing for shape difference. However, the scientific goal is to depict or describe the size and shape changes in a study, and this is a major strength of the geometrical methods that we describe.

We shall consider a shape space obtained directly from the landmark coordinates, which retains the geometry of a point configuration. This approach to shape analysis has been called 'geometric shape analysis' or 'geometric morphometrics' by various authors and the subject progressed rapidly around the late 1970s/early 1980s. A glossary of geometrical morphometrics is given by Slice *et al.* (1996).

Following Thompson (1917) some of the earlier work in geometrical shape analysis included that by Medawar (1944) and Sneath (1967). However, the research area became firmly established with the pioneering work of D.G. Kendall and F.L. Bookstein, who independently developed many of the key ideas, in very different styles. Also, some very important mathematical work in the area was also given by Ziezold (1977). Bookstein (1994a) has summarized the history of geometrical shape analysis, mainly through applications in biology. Kendall (1989) has reviewed shape theory and its development from a different, more theoretical, viewpoint, with applications in archaeology, astronomy and geography. Kendall (1989, 1995) provides

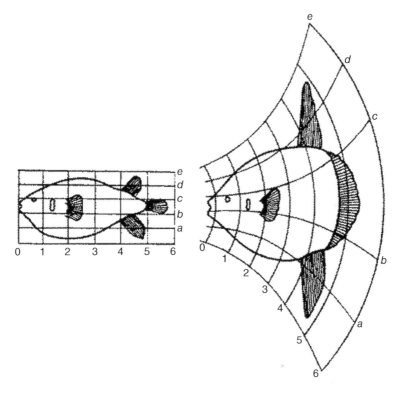

Figure 2.1   Thompson (1917)'s example of a species of fish Diodon *being geometrically transformed into another species* Orthagoriscus. *Source: Thompson 1917. Reproduced with permission of Cambridge University Press.*

some historical remarks on his development of shape theory, and some of the contributions of his colleagues. The first articles on the subject were by Kendall (1977), Ziezold (1977) and Bookstein (1978). Some key early papers in the field include those by Kendall (1984, 1989), Bookstein (1986), Goodall (1991), Le and Kendall (1993) and Kent (1994). Also, development in non-i.i.d. (independent and identically distributed) distribution theory for shape started with Mardia and Dryden (1989a).

2.2   Size

2.2.1   Configuration space

Definition 2.1   *The* **configuration** *is the set of landmarks on a particular object. The* **configuration matrix** X *is the* $k \times m$ *matrix of Cartesian coordinates of the* k *landmarks in* m *dimensions. The* **configuration space** *is the space of all landmark coordinates.*

In our applications we have $k \geq 3$ landmarks in $m = 2$ or $m = 3$ dimensions and the configuration space is typically $\mathbb{M}^{k \times m}$ the space of real $k \times m$ matrices (or equivalently $\mathbb{R}^{km}$) with possibly some special cases removed, such as coincident points.

2.2.2   Centroid size

Before defining shape we should define what we mean by size, so that it can be removed from a configuration. Consider X to be a $k \times m$ matrix of the Cartesian coordinates of k landmarks in m real dimensions, that is the configuration matrix of the object.

Definition 2.2   *A **size measure** $g(X)$ is any positive real valued function of the configuration matrix such that*

$$g(aX) = ag(X) \tag{2.1}$$

for any positive scalar $a > 0$.

Definition 2.3   *The **centroid size** is given by:*

$$S(X) = \|CX\| = \sqrt{\sum_{i=1}^{k} \sum_{j=1}^{m} (X_{ij} - \bar{X}_j)^2}, \quad X \in \mathbb{R}^{km}, \tag{2.2}$$

where X_{ij} is the (i,j)th entry of X, the arithmetic mean in the jth dimension is $\bar{X}_j = \frac{1}{k} \sum_{i=1}^{k} X_{ij}$,

$$C = I_k - \frac{1}{k} 1_k 1_k^{\mathrm{T}} \tag{2.3}$$

is the centring matrix,

$$\|X\| = \sqrt{\mathrm{trace}(X^{\mathrm{T}} X)}$$

is the Euclidean norm, I_k is the $k \times k$ identity matrix (diagonal matrix with ones on the diagonal), and 1_k is the $k \times 1$ vector of ones.

Obviously $S(aX) = aS(X)$ for $a > 0$, thus satisfying Equation (2.1). The centroid size $S(X)$ is the square root of the sum of squared Euclidean distances from each landmark to the centroid, namely

$$S(X) = \sqrt{\sum_{i=1}^{k} \|(X)_i - \bar{X}\|^2},$$

where $(X)_i$ is the ith row of $X(i = 1, \ldots, k)$ and $\bar{X} = (\bar{X}_1, \ldots, \bar{X}_m)$ is the centroid. This measure of size will be used throughout the book. In fact, the centroid size is the most commonly used size measure in geometrical shape analysis (e.g. Kendall 1984;

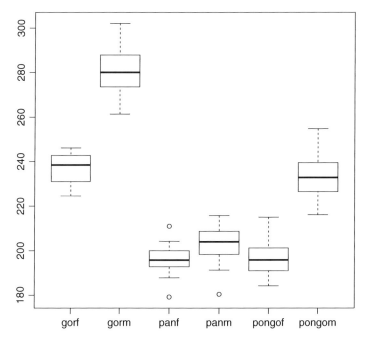

Figure 2.2   Boxplots of the centroid sizes for the apes' data by each group.

Bookstein 1986; Goodall 1991; Dryden and Mardia 1992). The centroid size could also be used in a normalized form, for example $S(X)/\sqrt{k}$ or $S(X)/(km)^{1/2}$, and this would be particularly appropriate when comparing configurations with different numbers of landmarks. The squared centroid size can also be interpreted as $(2k)^{-1}$ times the sum of the squared inter-landmark distances since

$$\sum_{j=1}^{m}\sum_{i=1}^{k}\sum_{l=1}^{k}(X_{ij}-X_{lj})^2 = \sum_{j=1}^{m}\sum_{i=1}^{k}\sum_{l=1}^{k}(X_{ij}-\bar{X}_j+\bar{X}_j-X_{lj})^2$$

$$= 2k\sum_{i=1}^{k}\sum_{j=1}^{m}(X_{ij}-\bar{X}_j)^2 = 2kS(X)^2.$$

In Figure 2.2 we display boxplots of the centroid size of the six groups of great apes which were described in Section 1.4.8. The plot is obtained using the R commands:

```
data(apes)
size<-centroid.size(apes$x)
plot(apes$group,size)
```

It is clear that there are sex differences in centroid size for gorillas and orangutans (*pongo*), with the males all being larger than the females. For chimpanzees (*pan*) there is an overlap in the centroid size distributions of the sexes, but the males are significantly larger in mean centroid.size, for example using a t-test (p-value = 0.0002):

```
> t.test(size[apes$group=="panm"],size[apes$group=="panf"],
                              alternative="greater")
        Welch Two Sample t-test

data: size[apes$group == "panm"] and size[apes$group == "panf"]
t = 3.8176, df = 50.704, p-value = 0.0001841
```

Example 2.1   de Souza *et al.* (2001a,b) analysed the sizes of stereolithography (SLA) skull models which had been constructed by a manufacturer using CT scans from a real head. In a test to assess the accuracy of the SLA models, six heads (cadavers) were CT scanned, and the CT scans were used to build SLA models. After the removal of the soft tissues, 37 specific anatomical landmarks were measured on both the SLA models and the corresponding real skulls. Using a paired t-test on the log of the centroid sizes, it was found that the SLA model was significantly larger by 1.5% than the original real skull. In particular, this size error was taken into account when designing custom implants, because even a small size increase would have severe clinical consequences.   □

2.2.3   Other size measures

An alternative size measure is the baseline size, that is the length between landmarks 1 and 2:

$$D_{12}(X) = \|(X)_2 - (X)_1\|. \tag{2.4}$$

The baseline size was used as early as 1907 by Galton (1907) for normalizing faces and its use came into prominence with Bookstein coordinates, described in Section 2.4. This size variable is also useful in calculating size-and-shape distributions (Dryden and Mardia 1992) which we outline in Section 11.5.

Other alternative size measures include the square root of the area of the convex hull for planar configurations or the cube root of the volume of the convex hull for configurations in three dimensions which intuitively describes size, but these measures have not become popular. For triangles with coordinates $(X_1, Y_1), (X_2, Y_2)$ and (X_3, Y_3) the area is simply given by:

$$A = \frac{1}{2} \left| (Y_3 - Y_1)(X_2 - X_1) - (X_3 - X_1)(Y_2 - Y_1) \right|.$$

Note that the volume of a parallelepiped in three dimensions formed by v_1, v_2, v_3 is given by the absolute value of the determinant of the matrix with columns v_1, v_2, v_3. Other size measures for triangles include the radius of the inscribed circle and the circumradius (e.g. Miles 1970), the latter being equal to $S(X)/\sqrt{3}$.

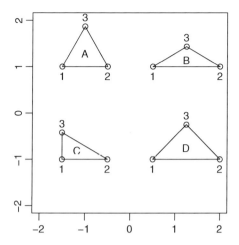

Figure 2.3   Four example triangles. The ranking of the triangles in terms of size differs when different choices of size measure are considered.

In general the choice of size will affect reported conclusions. In particular, if object A has twice the centroid size of object B it does not necessarily follow that object A is twice the size of object B when using a different measure. When reporting size differences one needs to state which size measure was used. The exception is when both objects have the same shape, in which case the ratio of sizes will be the same regardless of the choice of size variable. If comparing sets of collinear points, then very different conclusions about size are reached, with these pathological configurations all having zero area but non-zero centroid size for non-coincident collinear points.

Example 2.2   Consider the triangles in Figure 2.3. The centroid sizes of the triangles to three decimal places are:

$$A : 1.000, \ B : 1.118, \ C : 0.943, \ D : 1.225.$$

The baseline sizes are:

$$A : 1, \ B : 1.5, \ C : 1, \ D : 1.5.$$

The square root of area sizes to three decimal places are:

$$A : 0.658, \ B : 0.570, \ C : 0.537, \ D : 0.750.$$

The relative ranking of triangles in terms of size differs depending on the choice of size measure. In particular, in terms of area A is larger than B, but in terms of centroid size B is larger than A. This example demonstrates that different choices of size measure can lead to different conclusions about size. □

The following R commands give the calculations in the example:

```
> library(shapes)
> x1<-matrix(c(-1.5,-0.5,-1,1,1,sqrt(3)/2+1),3,2)
> x2<-matrix(c(0.5,2,1.25,1,1,1+sqrt(3)/4),3,2)
> x3<-matrix(c(-1.5,-0.5,-1.5,-1,-1,-1+1/sqrt(3)),3,2)
> x4<-matrix(c(0.5,2,1.25,-1,-1,0.75-1),3,2)
> centroid.size(x1)
[1] 1
> centroid.size(x2)
[1] 1.118034
> centroid.size(x3)
[1] 0.942809
> centroid.size(x4)
[1] 1.224745
> #baseline size
> sqrt( (x1[2,1]-x1[1,1])**2+(x1[2,2]-x1[1,2])**2 )
[1] 1
> sqrt( (x2[2,1]-x2[1,1])**2+(x2[2,2]-x2[1,2])**2 )
[1] 1.5
> sqrt( (x3[2,1]-x3[1,1])**2+(x3[2,2]-x3[1,2])**2 )
[1] 1
> sqrt( (x4[2,1]-x4[1,1])**2+(x4[2,2]-x4[1,2])**2 )
[1] 1.5
> #areas
> sqrt(abs(((x1[3,2]-x1[1,2])*(x1[2,1]-x1[1,1])-
                (x1[3,1]-x1[1,1])*(x1[2,2]-x1[1,2]))/2)
[1] 0.658037
> sqrt(abs(((x2[3,2]-x2[1,2])*(x2[2,1]-x2[1,1])-
                (x2[3,1]-x2[1,1])*(x2[2,2]-x2[1,2]))/2)
[1] 0.5698768
> sqrt(abs(((x3[3,2]-x3[1,2])*(x3[2,1]-x3[1,1])-
                (x3[3,1]-x3[1,1])*(x3[2,2]-x3[1,2]))/2)
[1] 0.537285
> sqrt(abs(((x4[3,2]-x4[1,2])*(x4[2,1]-x4[1,1])-
                (x4[3,1]-x4[1,1])*(x4[2,2]-x4[1,2]))/2)
[1] 0.75
```

Mosimann (1970) considers a definition of size based on length measurements which satisfies Equation (2.1), but with vectors of positive measurements (length, width, etc.) in place of the configuration matrix X.

In order to describe an object's shape it is useful to specify a coordinate system. We initially consider some of the most straightforward coordinate systems, which helps to provide an elementary introduction to aspects of shape. A suitable choice of coordinate system for shape is invariant under translation, scaling and rotation of the configuration. Further coordinate systems are discussed later in Section 4.4.

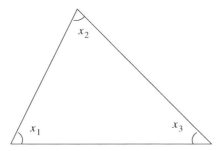

Figure 2.4   A labelled triangle with three internal angles marked. Two of the internal angles could be used to measure the shape of the triangle.

2.3   Traditional shape coordinates

2.3.1   Angles

For $k = 3$ points in $m = 2$ dimensions two internal angles are an obvious choice of coordinates that are invariant under the similarity transformations. For example, x_1 and x_2 could measure the shape of the triangle in Figure 2.4, where $x_1 + x_2 + x_3 = 180°$.

However, it soon becomes apparent that using angles to describe shape can be problematic. For cases such as very flat triangles (three points in a straight line) there are many different arrangements of three points. For example, see Figure 2.5, where the triangles all have $x_1 = 0°, x_3 = 0°, x_2 = 180°$, and yet the configurations have different shapes.

For larger numbers of points ($k > 3$) one could subdivide the configuration into triangles and so $2k - 4$ angles would be needed. For the $m = 3$ dimensional case angles could also be used, but again these suffer from problems in pathological cases. Also, probability distributions of the angles themselves are not easy to work with (see Mardia *et al.* 1977). If the angles of the triangle are x_1, x_2 and x_3, then the

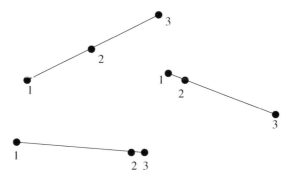

Figure 2.5   Examples of pathological cases where angles are an inappropriate shape measure. The landmarks are at the centre of the discs.

use of $\log(x_1/x_3)$ and $\log(x_2/x_3)$ (where $x_1 + x_2 + x_3 = 180°$) has some potential for analysing triangle shape using compositional data analysis (Aitchison 1986; Pukkila and Rao 1988) and the approach can be adapted to higher dimensions. Mardia *et al.* (1996b) used angles from triangulations of landmark data on photographs of faces, and Mardia *et al.* (1977) used the angles between towns in central place data. In such cases, analysis of shape can proceed as in directional statistics (Mardia and Jupp 2000).

A Ramachandran plot (Ramachandran *et al.* 1963) is a method for visualizing backbone dihedral torsion angles ψ versus ϕ of amino acid residues in protein secondary structures. The two angles describe the torsion angles either side of each α-carbon atom, and provide very useful shape information. This plot contains important angular shape information about the protein backbone and has proved very popular in the proteomics literature. The torsion angles are helpful in the prediction of a protein's 3D folding. Note that each angle is on the circle, S^1, and so the plots are on a torus $S^1 \times S^1$. The plot is also known as a $[\phi, \psi]$ plot or a Ramachandran diagram. Another geometrical description of a protein backbone was given by Zacharias and Knapp (2013) called a (d,θ) plot.

2.3.2   Ratios of lengths

Another traditional approach to representing shape is through ratios of length measurements, or lengths divided by size. A considerable amount of work has been carried out in multivariate morphometrics using distances, ratios, angles, and so on and it is still very commonly used in the biological literature (Reyment *et al.* 1984). There are many situations, such as classification problems, where the techniques can be very powerful. However, sometimes the interpretation of the important linear combinations of ratios of lengths and angles can be difficult. It is often easier to interpret pictures in the original space of the specimens than in some derived multivariate space, so we consider methods where the geometry of the object is retained.

Typical applications of the multivariate morphometrics approach include the classification of species (taxonomy) or the sexing of skulls using lengths, angles or ratios of lengths between landmarks. A very commonly used method is to perform a principal component analysis (PCA) of the multivariate measurements and to interpret each component as a measure of some aspect of size or shape (e.g. Jolicoeur and Mosimann 1960). Quite often the first principal component has approximately equal loadings on each variable and so can be interpreted as an overall measure of size.

Mosimann (1970) gives a rigorous treatment of this subject and provides a theoretical framework for the study of the size and shape of positive length measurements. Mosimann (1970) defines the shape vector to be $l/g(l)$, where $g(l)$ is a size measure. He provides theorems for the independence of population size and shape, including characterizations of various distributions. Further details of the approach can be found in Mosimann (1975a,b, 1988) and Darroch and Mosimann (1985) and the references therein.

2.3.3   Penrose coefficent

Given a set of measurements $d_1, \dots, d_m$ the Penrose (1952) size coefficient is:

$$P_{Size} = C_Q^2 = \frac{1}{m^2} \left(\sum_{i=1}^m d_i \right)^2,$$

and the Penrose shape coefficient is:

$$P_{Shape} = \sum_{i=1}^m \frac{d_i^2}{m} - C_Q^2.$$

The coefficients are related to a coefficient of racial likeness (Penrose, 1952; after Pearson, 1926) given by:

$$CRL = P_{Size} + P_{Shape}.$$

These very traditional shape measures usually only capture part of the shape or size-and-shape information, so we prefer a newer approach where the geometry of the objects under study is retained.

2.4   Bookstein shape coordinates

2.4.1   Planar landmarks

Let $(x_j, y_j), j = 1, \dots, k$, be $k \geq 3$ landmarks in a plane ($m = 2$ dimensions). Bookstein (1986, 1984) suggests removing the similarity transformations by translating, rotating and rescaling such that landmarks 1 and 2 are sent to a fixed position. If landmark 1 is sent to $(0, 0)$ and landmark 2 is sent to $(1, 0)$, then suitable shape variables are the coordinates of the remaining $k - 2$ coordinates after these operations. To preserve symmetry, we consider the coordinate system where the baseline landmarks are sent to $(-\frac{1}{2}, 0)$ and $(\frac{1}{2}, 0)$.

Definition 2.4   **Bookstein coordinates** $(u_j^B, v_j^B)^T, j = 3, \dots, k$, *are the remaining coordinates of an object after translating, rotating and rescaling the baseline to* $(-\frac{1}{2}, 0)$ *and* $(\frac{1}{2}, 0)$ *so that*

$$u_j^B = \{(x_2 - x_1)(x_j - x_1) + (y_2 - y_1)(y_j - y_1)\}/D_{12}^2 - \tfrac{1}{2}, \qquad (2.5a)$$

$$v_j^B = \{(x_2 - x_1)(y_j - y_1) - (y_2 - y_1)(x_j - x_1)\}/D_{12}^2, \qquad (2.5b)$$

where $j = 3, \dots, k$, $D_{12}^2 = (x_2 - x_1)^2 + (y_2 - y_1)^2 > 0$ *and* $-\infty < u_j^B, v_j^B < \infty$.

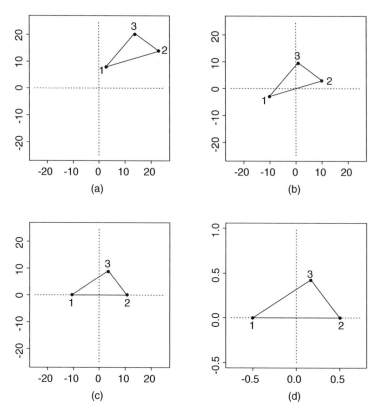

Figure 2.6   The geometrical interpretation of Bookstein coordinates. The original triangle in (a) is transformed by (b) translation, (c) rotation and finally (d) rescaling to give the Bookstein coordinates as the coordinates of the point labelled 3 in plot (d).

A geometrical illustration of these transformations is given in Figure 2.6. If the baseline is taken as $(0,0)$ and $(1,0)$, then there is no $-\frac{1}{2}$ in the equation for u_j^B, as originally proposed by Bookstein (1986). We subtract the $\frac{1}{2}$ in u_j^B to also simplify the transformation to Kendall coordinates in Section 2.5, although precisely where we send the baseline is an arbitrary choice. These coordinates have been used widely in shape analysis for planar data. Bookstein coordinates are the most straightforward to use for a newcomer to shape analysis. However, because of the lack of symmetry in choosing a particular baseline and the fact that correlations are induced into the coordinates, many practitioners often prefer to use the Procrustes tangent coordinates (see Section 4.4).

Galton (1907) defined precisely the same coordinates [with baseline $(0,0)$, $(1,0)$] at the beginning of the last century, but the statistical details of using this approach needed to be worked out.

The construction of Bookstein coordinates is particularly simple if using complex arithmetic. Bookstein coordinates are obtained from the original complex coordinates $z_1^o, \ldots, z_k^o$, where $z_j^o \in \mathbb{C}, j = 1, \ldots, k$:

$$u_j^B + iv_j^B = \frac{z_j^o - z_1^o}{z_2^o - z_1^o} - \frac{1}{2} = \frac{2z_j^o - z_1^o - z_2^o}{2(z_2^o - z_1^o)}, \quad j = 3, \ldots, k, \tag{2.6}$$

where $i = \sqrt{-1}$. Consider how the formulae in Equation (2.5a) are obtained (Mardia 1991). To find (u_j^B, v_j^B) for a fixed $j = 3, \ldots, k$, we have to find the scale $c > 0$, the rotation A, and the translation $b = (b_1, b_2)^T$ such that

$$U = cA(X - b), \tag{2.7}$$

where $X = (x_j, y_j)^T$ and $U = (u_j^B, v_j^B)^T$ are the coordinates of the jth point before and after the transformation, $j = 3, \ldots, k$, and A is a 2×2 rotation matrix, namely

$$A = \begin{bmatrix} \cos\theta & \sin\theta \\ -\sin\theta & \cos\theta \end{bmatrix}, \quad |A| = 1, \quad A^T A = AA^T = I_2,$$

where we rotate clockwise by θ radians. Applying the transformation to landmarks 1 and 2 gives four equations in four unknowns (c, θ, b_1, b_2)

$$cA\left(\begin{bmatrix} x_1 \\ y_1 \end{bmatrix} - b\right) = \begin{bmatrix} -\frac{1}{2} \\ 0 \end{bmatrix}, \quad cA\left(\begin{bmatrix} x_2 \\ y_2 \end{bmatrix} - b\right) = \begin{bmatrix} \frac{1}{2} \\ 0 \end{bmatrix}.$$

Now, we can solve these equations and see that the translation is

$$b = (\tfrac{1}{2}(x_1 + x_2), \tfrac{1}{2}(y_1 + y_2))^T,$$

the rotation (in the appropriate quadrant) is

$$\theta = \arctan\{(y_2 - y_1)/(x_2 - x_1)\}$$

and the rescaling is $c = \{(x_2 - x_1)^2 + (y_2 - y_1)^2\}^{-1/2}$. So,

$$A = c \begin{bmatrix} x_2 - x_1 & y_2 - y_1 \\ -(y_2 - y_1) & x_2 - x_1 \end{bmatrix}. \tag{2.8}$$

Substituting $X = (x_j, y_j)^T (j = 3, \ldots, k)$ and c, A, b into Equation (2.7) we obtain the shape variables of Equation (2.5a). This solution can also be seen from Figure 2.6 using geometry.

Example 2.3   In Figure 2.7 we see scatter plots of the Bookstein coordinates for the T2 Small vertebrae of Section 1.4.1. We take landmarks 1 and 2 as the baseline.

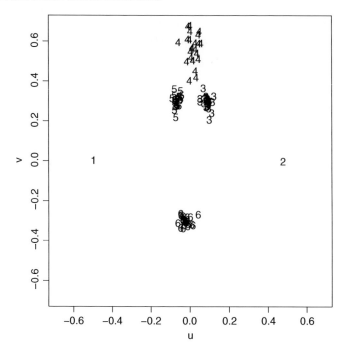

Figure 2.7   Scatter plots of the Bookstein coordinates of the T2 Small vertebrae registered on the baseline 1, 2.

Note that the marginal scatter plots at landmarks 4 and possibly 3 and 5 are elliptical in nature. The plot can be obtained using:

```
data(mice)
qset2<-mice$x[,,mice$group=="s"]
u<-bookstein2d(qset2)$bshpv
plotshapes(u,symbol=as.character(1:6))
```

For example, the first mouse has Bookstein shape variables given by the last four rows of:

```
> u[,,1]
, , 1
         [,1]   [,2]
[1,] -0.50000000 0.0000000
[2,]  0.50000000 0.0000000
[3,]  0.09121051 0.2826022
[4,]  0.04065577 0.5089238
[5,] -0.07367575 0.2735386
[6,] -0.02672182 -0.3054334
```

and this could also be obtained directly using the command:

```
bookstein.shpv(qset2[,,1])
```

For practical data analysis it is sensible to choose the baseline as landmarks that are not too close together, as in this example. As often happens with using Bookstein coordinates, the variability in the points away from the origin appears larger than the points nearer to the origin. This is an artifact of this coordinate system, and will be explored further in Sections 9.4 and 11.1.4. The centroid sizes of the bones are obtained using:

```
sz<-centroid.size(qset2)
```

A pairwise scatter plot of Bookstein's shape variance and centroid size is obtained using:

```
x<-rbind(u[3:6,1,],u[3:6,2,])
x<-rbind(matrix(sz,1,23),x)
pairs(t(x),labels=c("s","u3","u4","u5","u6","v3","v4","v5","v6"))
```

In Figure 2.8 we see pairwise scatter plots of Bookstein coordinates. There are strong positive correlations in the v_3^B, v_4^B and v_5^B coordinates. There are also positive correlations between v_3^B and v_6^B, between v_5^B and v_6^B, and between u_3^B and u_4^B. However, it is also an artifact of the coordinate system that correlations are induced into the shape variables, even when the landmark coordinates are uncorrelated (see Section 11.1.4), and so correlations can be difficult to interpret.

We can also examine the joint relationship between size and shape, which will be considered in more detail in Chapter 5. Scatter plots of the centroid size S versus each of the shape coordinates are also given in Figure 2.8. We see that there are quite strong positive correlations between S and each of v_3^B, v_4^B and v_5^B.   □

Throughout the text we shall often refer to the real $(2k - 4)$-vector of Bookstein coordinates $u^B = (u_3^B, \ldots, u_k^B, v_3^B, \ldots, v_k^B)^T$, stacking the coordinates in this particular order.

One approach to shape analysis is to use standard multivariate analysis of the Bookstein coordinates, ignoring the non-Euclidean nature of the space. Provided variations in the data are small, then the method is adequate for mean estimation and hypothesis testing. For example, we could obtain an estimate of the mean shape of the configuration by taking the arithmetic average of the Bookstein coordinates

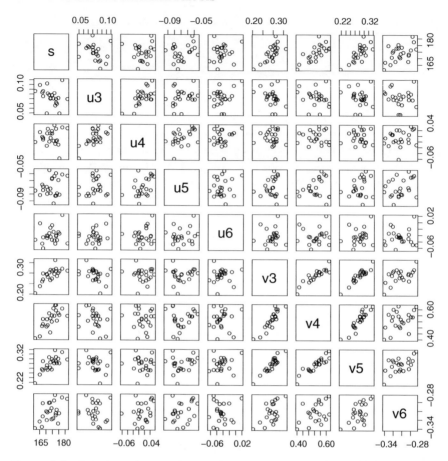

Figure 2.8   Pairwise scatter plots of centroid size S and Bookstein coordinates $u_3^B, \ldots, u_6^B, v_3^B, \ldots, v_6^B$ *for the T2 Small vertebrae.*

(the **Bookstein mean shape**). The Bookstein mean shape for the T2 Small mouse vertebrae of Example 2.7 is given by the last four rows of:

```
> bookstein2d(mice$x[,,mice$group=="s"])$mshape
        [,1] [,2]
[1,] -0.50000000 0.000000e+00
[2,]  0.50000000 -6.634366e-21
[3,]  0.08469746 2.933430e-01
[4,]  0.01215768 5.613175e-01
[5,] -0.06874750 2.991278e-01
[6,] -0.02502185 -3.041418e-01
```

which when rounded to three decimal places is:

$$\bar{u}^B = (0.085, 0.012, -0.069, -0.025, 0.293, 0.561, 0.299, -0.304)^{\mathrm{T}}.$$

The variability of the shape variables is less straightforward to interpret. Transforming the objects (or registering) to a given edge induces correlations into the shape variables in general and this can lead to spurious correlations (Kent 1994), see Sections 9.4 and 11.1.4. So the method should not be used to interpret the structure of shape variability unless there is a good reason to believe that two landmarks are essentially fixed (see Section 9.4).

2.4.2   Bookstein-type coordinates for 3D data

Registration on a base to filter out the similarity transformations can also be carried out for 3D data, in a similar way to edge registration in Bookstein coordinates for 2D data. Consider k landmarks $X_j = (x_{1j}, x_{2j}, x_{3j})^T \in \mathbb{R}^3, j = 1, \ldots, k$, in three dimensions. The number of shape variables is $3k - 7$ since we have $3k$ landmark coordinates but must remove 3 location, 1 scale and 3 rotation parameters. The Bookstein coordinates can be taken as

$$u_j = (u_{1j}, u_{2j}, u_{3j})^T = \frac{1}{\|X_2 - X_1\|} A \left(X_j - \frac{(X_1 + X_2)}{2} \right), \quad j = 3, \ldots, k, \qquad (2.9)$$

where A is a 3×3 rotation matrix (a function of (X_1, X_2, X_3)) and

$$X_1 \to (-1/2, 0, 0)^T, \quad X_2 \to (1/2, 0, 0)^T, \quad X_3 \to u_3 = (u_{13}, u_{23}, 0)^T,$$

where $u_{23} \geq 0, u_{33} = 0$ and $X_j \to u_j$ for $j = 4, \ldots, k$. We drop the superscript B denoting Bookstein coordinates in this section. Geometrically the figure has been translated, rotated and rescaled so that the baseline (landmarks 1 and 2) is of unit length, has midpoint at the origin and is rotated to lie along the x-axis. The figure is further rotated so that landmark 3 lies in the plane of the first two dimensions (X, Y), with $Y \geq 0$. The remaining non-fixed coordinates are the shape variables (cf. Mardia and Dryden 1997). The details are as follows: first translate the figure,

$$w_{li} = x_{li} - (x_{l1} + x_{l2})/2, \quad l = 1, 2, 3; \quad i = 2, \ldots, k.$$

We then rescale by dividing by the length of the baseline between points 1 and 2: $D_{12} = 2(w_{12}^2 + w_{22}^2 + w_{32}^2)^{1/2}$, and rotate through clockwise angles of θ about the z-axis, ω about the y-axis and ϕ about the x-axis where

$$\theta = \arctan(w_{22}/w_{12}), \quad \omega = \arctan\left(w_{32}/(w_{12}^2 + w_{22}^2)^{1/2} \right),$$

$$\phi = \arctan\left(\frac{(w_{12}^2 + w_{22}^2)w_{33} - (w_{12}w_{13} + w_{22}w_{23})w_{32}}{(w_{12}^2 + w_{22}^2 + w_{32}^2)^{1/2}(w_{12}w_{23} - w_{13}w_{22})} \right).$$

So, using matrix notation if

$$R_x = \begin{bmatrix} 1 & 0 & 0 \\ 0 & \cos\phi & \sin\phi \\ 0 & -\sin\phi & \cos\phi \end{bmatrix}, R_y = \begin{bmatrix} \cos\omega & 0 & \sin\omega \\ 0 & 1 & 0 \\ -\sin\omega & 0 & \cos\omega \end{bmatrix}, R_z = \begin{bmatrix} \cos\theta & \sin\theta & 0 \\ -\sin\theta & \cos\theta & 0 \\ 0 & 0 & 1 \end{bmatrix},$$

then the 3D Bookstein coordinates are:

$$u_j = R_x R_y R_z (w_{1j}, w_{2j}, w_{3j})^{\mathrm{T}} / D_{12}, \quad j = 3, \dots, k.$$

An explicit expression for the shape coordinates is:

$$u_{13} = \left(w_{12}w_{13} + w_{22}w_{23} + w_{32}w_{33} \right)/a,$$

$$u_{23} = \left((w_{12}w_{23} - w_{13}w_{22})^2 + (w_{12}w_{33} - w_{32}w_{13})^2 + (w_{22}w_{33} - w_{23}w_{32})^2 \right)^{1/2}/a,$$

$$u_{1j} = \left(w_{12}w_{1j} + w_{22}w_{2j} + w_{32}w_{3j} \right)/a,$$

$$u_{2j} = \frac{\left[\begin{array}{l} w_{12}^2 w_{23}w_{2j} + w_{12}^2 w_{33}w_{3j} - w_{12}w_{13}w_{22}w_{2j} - w_{12}w_{13}w_{32}w_{3j} \\ - w_{12}w_{22}w_{23}w_{1j} - w_{12}w_{32}w_{33}w_{1j} + w_{13}w_{22}^2 w_{1j} + w_{13}w_{32}^2 w_{1j} \\ + w_{22}^2 w_{33}w_{3j} - w_{22}w_{32}w_{23}w_{3j} - w_{22}w_{32}w_{33}w_{2j} + w_{32}^2 w_{23}w_{2j} \end{array} \right]}{(ab)^{1/2}},$$

$$u_{3j} = \frac{\left[\begin{array}{l} w_{12}w_{23}w_{3j} - w_{12}w_{33}w_{2j} - w_{13}w_{22}w_{3j} \\ + w_{13}w_{32}w_{2j} + w_{22}w_{33}w_{1j} - w_{32}w_{23}w_{1j} \end{array} \right]}{(2ab)^{1/2}},$$

where $a = 2(w_{12}^2 + w_{22}^2 + w_{32}^2)$, $j = 4, \dots, k$, and

$$b = w_{12}^2 w_{23}^2 + w_{12}^2 w_{33}^2 - 2w_{12}w_{13}w_{22}w_{23} - 2w_{12}w_{13}w_{32}w_{33}$$
$$+ w_{13}^2 w_{22}^2 + w_{13}^2 w_{32}^2 + w_{22}^2 w_{33}^2 - 2w_{22}w_{32}w_{23}w_{33} + w_{32}^2 w_{23}^2.$$

As in the $m = 2$ dimensional case, care must be taken when interpreting correlations between the shape variables, because spurious correlations can occur with edge registration methods such as this (see Sections 9.4 and 11.1.4).

Example 2.4   In Figure 2.9(a) we see a male macaque skull from the dataset described in Section 1.4.3, with 24 landmarks located in three dimensions and a wire box regular grid drawn over the skull. Let us consider the page to be the x–y plane and the z-axis is perpendicular to the page. The 3D Bookstein coordinates involve selecting two landmarks (e.g. landmarks 1 and 6 in Figure 1.11) to translate, rotate and rescale to lie horizontal in the plane of the page at points $(\frac{1}{2}, 0, 0)$ and $(-\frac{1}{2}, 0, 0)$, and a third landmark (e.g. landmark 10) is chosen to lie in the x–y plane, as displayed in Figure 2.9(b).   □

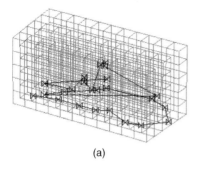
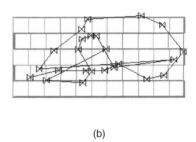

(a)   (b)

Figure 2.9   The 24 landmarks located on a macaque monkey skull where the page is regarded as the x–y plane and into the page is the z-axis. The skull is translated, rotated and rescaled so that landmarks 1 and 6 (see Section 1.2.8) lie in the x–y plane of the page, at unit length apart, and landmark 10 is also rotated to be in the x–y plane.

Important point: For **affine shape** (where invariance is under affine transformations rather than Euclidean similarity transformations – see Section 12.2), there are 12 rather than 7 constraints. Hence all the coordinates of the first four landmarks are sufficient to form a base in the affine case and the expressions are simpler than for similarity shape.

2.5   Kendall's shape coordinates

Kendall coordinates (Kendall 1984) are related to Bookstein coordinates but location is removed in a different manner. We first need to define the Helmert submatrix which is used to remove location.

The Helmert submatrix H is the $(k-1) \times k$ Helmert matrix without the first row. The full Helmert matrix H^F, which is commonly used in Statistics, is a square $k \times k$ orthogonal matrix with its first row of elements equal to $1/\sqrt{k}$, and the remaining rows are orthogonal to the first row. We drop the first row of H^F so that the transformed HX does not depend on the original location of the configuration. Note $H^T H = C$, where C is the centring matrix of (2.3).

Definition 2.5   *The jth row of the* **Helmert submatrix** H *is given by:*

$$(h_j, \ldots, h_j, -jh_j, 0, \ldots, 0), \quad h_j = -\{j(j+1)\}^{-1/2}, \tag{2.10}$$

and so the jth row consists of h_j repeated j times, followed by $-jh_j$ and then $k-j-1$ zeros, $j = 1, \ldots, k-1$.

For $k = 3$ the full Helmert matrix is explicitly

$$H^F = \begin{bmatrix} 1/\sqrt{3} & 1/\sqrt{3} & 1/\sqrt{3} \\ -1/\sqrt{2} & 1/\sqrt{2} & 0 \\ -1/\sqrt{6} & -1/\sqrt{6} & 2/\sqrt{6} \end{bmatrix}$$

and the Helmert submatrix is

$$H = \begin{bmatrix} -1/\sqrt{2} & 1/\sqrt{2} & 0 \\ -1/\sqrt{6} & -1/\sqrt{6} & 2/\sqrt{6} \end{bmatrix}.$$

For $k = 4$ points the full Helmert matrix is

$$H^F = \begin{bmatrix} 1/2 & 1/2 & 1/2 & 1/2 \\ -1/\sqrt{2} & 1/\sqrt{2} & 0 & 0 \\ -1/\sqrt{6} & -1/\sqrt{6} & 2/\sqrt{6} & 0 \\ -1/\sqrt{12} & -1/\sqrt{12} & -1/\sqrt{12} & 3/\sqrt{12} \end{bmatrix}$$

and the Helmert submatrix is

$$H = \begin{bmatrix} -1/\sqrt{2} & 1/\sqrt{2} & 0 & 0 \\ -1/\sqrt{6} & -1/\sqrt{6} & 2/\sqrt{6} & 0 \\ -1/\sqrt{12} & -1/\sqrt{12} & -1/\sqrt{12} & 3/\sqrt{12} \end{bmatrix}.$$

We can obtain the Helmert submatrix using defh(k-1) where k is the number of landmarks, for example:

```
> defh(2)
        [,1]      [,2]      [,3]
[1,] -0.7071068 0.7071068 0.0000000
[2,] -0.4082483 -0.4082483 0.8164966
> defh(3)
        [,1]      [,2]      [,3]     [,4]
[1,] -0.7071068 0.7071068 0.0000000 0.0000000
[2,] -0.4082483 -0.4082483 0.8164966 0.0000000
[3,] -0.2886751 -0.2886751 -0.2886751 0.8660254
> defh(4)
        [,1]      [,2]      [,3]     [,4]     [,5]
[1,] -0.7071068 0.7071068 0.0000000 0.0000000 0.0000000
[2,] -0.4082483 -0.4082483 0.8164966 0.0000000 0.0000000
[3,] -0.2886751 -0.2886751 -0.2886751 0.8660254 0.0000000
[4,] -0.2236068 -0.2236068 -0.2236068 -0.2236068 0.8944272
```

Consider the original complex landmarks $z^o = (z_1^o, \ldots, z_k^o)^\text{T}$ and remove location by pre-multiplying by the Helmert submatrix H to give $z_H = Hz^o = (z_1, \ldots, z_{k-1})^\text{T}$.

Definition 2.6   *The* **Kendall coordinates** *are given by:*

$$u_j^K + iv_j^K = \frac{z_{j-1}}{z_1}, \quad j = 3, \ldots, k. \tag{2.11}$$

There is a simple one to one correspondence between Kendall and Bookstein coordinates. If we write

$$w^B = (u_3^B + iv_3^B, \ldots, u_k^B + iv_k^B)^\text{T}$$

for Bookstein coordinates and

$$w^K = (u_3^K + iv_3^K, \ldots, u_k^K + iv_k^K)^\text{T}$$

for Kendall coordinates, then it follows that

$$w^K = \sqrt{2}H_1 w^B \tag{2.12}$$

where H_1 is the lower right $(k-2) \times (k-2)$ partition matrix of the Helmert submatrix H. Note that

$$H_1^\text{T}H_1 = I_{k-2} - \frac{1}{k}1_{k-2}1_{k-2}^\text{T}, \quad \|H_1\|^2 = 2/k, \quad (H_1^\text{T}H_1)^{-1} = I_{k-2} + \frac{1}{2}1_{k-2}1_{k-2}^\text{T}$$

so linear transformation from one coordinate system to the other is straightforward. The inverse transformation is:

$$w^B = (H_1^\text{T}H_1)^{-1}H_1^\text{T}w^K/\sqrt{2}.$$

For $k = 3$ we have the relationship:

$$u_3^B + iv_3^B = \frac{z_3^o - \frac{1}{2}(z_1^o + z_2^o)}{z_2^o\ z_1^o} = \frac{\sqrt{3}}{2}(u_3^K + iv_3^K)$$

and so Kendall coordinates in this case are the coordinates of the third landmark after transforming landmarks 1 and 2 to $(-1/\sqrt{3}, 0)$ and $(1/\sqrt{3}, 0)$ by the similarity transformations. The transformation from (z_1^o, z_2^o, z_3^o) to Kendall coordinates is:

$$z_1^o \to -\frac{1}{\sqrt{3}}, \quad z_2^o \to \frac{1}{\sqrt{3}}, \quad z_3^o \to u_3^K + iv_3^K.$$

Throughout the text we shall often refer to the real $(2k - 4)$-vector of Kendall coordinates $u^K = (u_3^K, \ldots, u_k^K, v_3^K, \ldots, v_k^K)^T$, stacking the coordinates in this particular order.

2.6   Triangle shape coordinates

2.6.1   Bookstein coordinates for triangles

For the case of a triangle of landmarks we have two shape coordinates.

Example 2.5   In Figure 2.3 we see the triangles with internal angles at points 1, 2, 3 given by:

A: $60°, 60°, 60°$ (equilateral),
B: $30°, 30°, 120°$ (isosceles) ,
C: $90°, 30°, 60°$ (right-angled),
D: $45°, 45°, 90°$ (right-angled and isosceles).

Bookstein coordinates for these three triangles with baseline points 1, 2 are:

A: $U_3^B = 0, V_3^B = \frac{\sqrt{3}}{2}$,
B: $U_3^B = 0, V_3^B = \frac{1}{2\sqrt{3}}$,
C: $U_3^B = -\frac{1}{2}, V_3^B = \frac{1}{\sqrt{3}}$,
D: $U_3^B = 0, V_3^B = \frac{1}{2}$. $\square$

A plot of the shape space of triangles for Bookstein coordinates is given in Figure 2.10. Each triangle shape is drawn with its centroid in the position of the Bookstein coordinates (U^B, V^B) corresponding to its shape in the shape space. For example, the equilateral triangles are at $(0, \sqrt{3}/2)$ and $(0, -\sqrt{3}/2)$ (marked E and F). The right-angled triangles are located on the lines $U^B = -0.5, U^B = 0.5$ and on the circle $(U^B)^2 + (V^B)^2 = 0.5$; the isosceles triangles are located on the line $U^B = 0$ and on the circles $(U^B + 0.5)^2 + (V^B)^2 = 1$ and $(U^B - 0.5)^2 + (V^B)^2 = 1$; and the flat triangles (3 points in a straight line) are located on $V^B = 0$.

The plane can be partitioned into two half-planes – all triangles below $V^B = 0$ can be reflected to lie above $V^B = 0$. In addition, each half-plane can be partitioned into six regions where the triangles have their labels permuted. If a triangle has baseline 1, 2 and apex 3 and side lengths d_{12}, d_{13} and d_{23}, then the six regions are:

$$d_{23} \geq d_{13} \geq d_{12}; \quad d_{23} \geq d_{12} \geq d_{13}; \quad d_{12} \geq d_{23} \geq d_{13};$$
$$d_{12} \geq d_{13} \geq d_{23}; \quad d_{13} \geq d_{12} \geq d_{23}; \quad d_{13} \geq d_{23} \geq d_{12}.$$

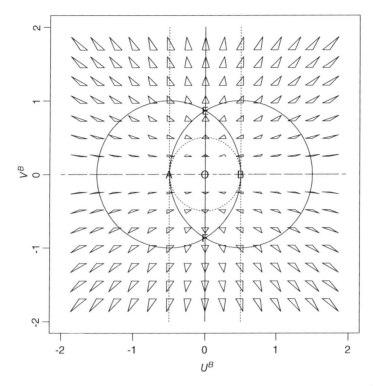

*Figure 2.10   The shape space of triangles, using the Bookstein coordinates (U^B, V^B).
Each triangle is plotted with its centre at the shape coordinates (U^B, V^B). The equilateral triangles are located at the points marked E and F. The isosceles triangles
are located on the unbroken lines and circles (——) and the right-angled triangles
are located on the broken lines and circles (- - - -). The flat (collinear) triangles
are located on the $V^B = 0$ line (the U^B-axis). All triangles could be relabelled and
reflected to lie in the region AOE, bounded by the arc of isosceles triangles AE, the
line of isosceles triangles OE and the line of flat triangles AO.*

Thus, if invariance under relabelling and reflection of the landmarks was required,
then we would be restricted to one of the 12 regions, for example the region AOE,
bounded by the arc of isosceles triangles AE, the line of isosceles triangles OE and
the line of flat triangles AO in Figure 2.10.

It is quite apparent from Figure 2.10 that a non-Euclidean distance in (U^B, V^B) is
most appropriate for the shape space. For example, two triangles near the origin that
are a Euclidean distance of 1 apart are very different in shape, but two triangles away
from the origin that are a Euclidean distance of 1 apart are quite similar in shape.
Non-Euclidean shape metrics are described in Section 4.1.1. Bookstein (1986) also
suggested using the hyperbolic Poincaré metric for triangle shapes, which is described
in Section 12.2.4.

2.6.2   Kendall's spherical coordinates for triangles

For $k = 3$ we will see in Section 4.3.4 that the shape space is a sphere with radius $\frac{1}{2}$. A mapping from Kendall coordinates to the sphere of radius $\frac{1}{2}$ is:

$$x = \frac{1 - r^2}{2(1 + r^2)}, \quad y = \frac{u_3^K}{1 + r^2}, \quad z = \frac{v_3^K}{1 + r^2}, \tag{2.13}$$

and $r^2 = (u_3^K)^2 + (v_3^K)^2$, so that $x^2 + y^2 + z^2 = \frac{1}{4}$, where u_3^K and v_3^K are Kendall coordinates of Section 2.5.

Definition 2.7   Kendall's spherical coordinates (θ, ϕ) *are given by the polar coordinates*

$$\frac{1}{2} \sin \theta \cos \phi = \frac{1 - r^2}{2(1 + r^2)}, \quad \frac{1}{2} \sin \theta \sin \phi = \frac{u_3^K}{1 + r^2}, \quad \frac{1}{2} \cos \theta = \frac{v_3^K}{1 + r^2}, \tag{2.14}$$

where $0 \leq \theta \leq \pi$ is the angle of latitude and $0 \leq \phi < 2\pi$ is the angle of longitude.

The relationship between (u_3^K, v_3^K) and the spherical shape variables (Mardia 1989b) is given by:

$$u_3^K = \frac{\sin \theta \sin \phi}{1 + \sin \theta \cos \phi},$$

$$v_3^K = \frac{\cos \theta}{1 + \sin \theta \cos \phi}. \tag{2.15}$$

The sphere can be partitioned into 6 lunes and 12 half-lunes. In order to make the terminology clear, let us note that one example full-lune is $0 \leq \phi \leq \pi/3, 0 \leq \theta \leq \pi$ and one example half-lune is $0 \leq \phi \leq \pi/3, 0 \leq \theta \leq \pi/2$.

In Figure 2.11 we see triangle shapes located on the spherical shape space. The equilateral triangle with anti-clockwise labelling corresponds to the 'North pole' ($\theta = 0$) and the reflected equilateral triangle (with clockwise labelling) is at the 'South pole' ($\theta = \pi$). The flat triangles (three collinear points) lie around the equator ($\theta = \pi/2$). The isosceles triangles lie on the meridians $\phi = 0, \pi/3, 2\pi/3, \pi, 4\pi/3, 5\pi/3$. The right-angled triangles lie on three small circles given by:

$$\sin \theta \cos \left(\phi - \frac{2k\pi}{3} \right) = \frac{1}{2}, \quad k = 0, 1, 2,$$

and we see the arc of unlabelled right-angled triangles on the front half-lune in Figure 2.11.

Reflections of triangles in the upper hemisphere at (θ, ϕ) are located in the lower hemisphere at $(\pi - \theta, \phi)$. In addition, permuting the triangle labels gives rise to points

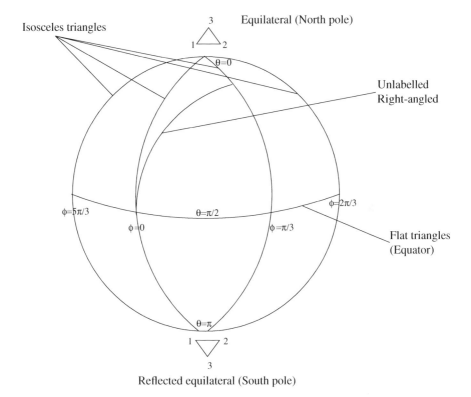

Figure 2.11   Kendall's spherical shape space for triangles in m = 2 dimensions. The shape coordinates are the latitude θ (with zero at the North pole) and the longitude φ.

in each of the six equal half-lunes in each hemisphere. Thus, if invariance under labelling and reflection was required, then we would be restricted to one of these half-lunes, for example the sphere surface defined by $0 \le \phi \le \pi/3, 0 \le \theta \le \pi/2$. Consider a triangle with labels A, B and C, and edge lengths AB, BC and AC. If the labelling and reflection of the points was unimportant, then we could relabel each triangle so that, for example, AB ≥ AC ≥ BC and point C is above the baseline AB.

If we have a triangle in three dimensions, then we see that we can translate, rotate and rescale so that

$$
\begin{bmatrix} x_1 \\ y_1 \\ z_1 \end{bmatrix} \rightarrow \begin{bmatrix} -\frac{1}{2} \\ 0 \\ 0 \end{bmatrix}, \quad \begin{bmatrix} x_2 \\ y_2 \\ z_2 \end{bmatrix} \rightarrow \begin{bmatrix} \frac{1}{2} \\ 0 \\ 0 \end{bmatrix}, \quad \begin{bmatrix} x_3 \\ y_3 \\ z_3 \end{bmatrix} \rightarrow \begin{bmatrix} u_{13} \\ u_{23} \\ 0 \end{bmatrix}
$$

where $u_{23} > 0$. Hence, the triangle has two shape coordinates $(u_{13}, u_{23})^{\mathrm{T}}$ which lie in a half-plane. We could transfer to Kendall's spherical shape coordinates of Equation (2.14), although the range of the latitude angle θ will be $[0, \frac{\pi}{2}]$, as $u_{23} > 0$. Hence, the shape space for triangles in three dimensions is the hemisphere with radius half, that is $S_+^2(\frac{1}{2})$, as we will see in Section 4.3.4. This is also the case for triangles in more than three dimensions.

2.6.3   Spherical projections

For practical analysis and the presentation of data it is often desirable to use a suitable projection of the sphere for triangle shapes. Kendall (1983) defined an equal area projection of one of the half-lunes of the shape sphere to display unlabelled triangle shapes. The projected lune is bell-shaped and this graphical tool is also known as 'Kendall's Bell' or the spherical blackboard (an example is given later in Figure 14.2).

An alternative equal-area projection is the Schmidt net (Mardia 1989b) otherwise known as the Lambert projection given by:

$$\xi = 2\sin\left(\frac{\theta}{2}\right), \quad \psi = \phi; \; 0 \le \xi \le \sqrt{2}, 0 \le \psi < 2\pi.$$

In Figure 2.12 we see a plot of one of the half-lunes on the upper hemisphere of shape space projected onto the Schmidt net. Example triangles are drawn with their centroids at polar coordinates (ξ, ψ) in the Schmidt net.

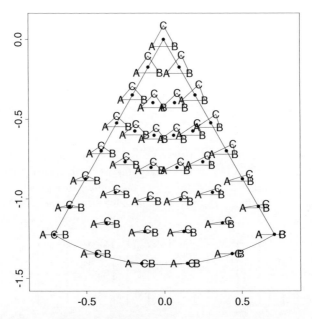

Figure 2.12   Part of the shape space of triangles projected onto the equal-area projection Schmidt net. If relabelling and reflection was not important, then all triangles could be projected into this sector.

2.6.4   Watson's triangle coordinates

Watson (1986) considers a coordinate system for triangle shape. Let $z = (z_1, z_2, z_3)^{\mathrm{T}} \in \mathbb{C}^3$ be the complex landmark coordinates. Let $\omega = \exp(i\pi/3)$ and $u = (1, \omega, \omega^2)^{\mathrm{T}}$, $\bar{u} = (1, \omega^2, \omega)^{\mathrm{T}}$. So, 1_3, u and $\bar{u}$ are orthogonal and have length $\sqrt{3}$. One can express the triplet of points z as:

$$z = \alpha 1_3 + \beta u + \gamma \bar{u}.$$

The vector 1_3 represents the degenerate triangle shape (all points coincident), u is an equilateral triangle and $\bar{u}$ is its reflection. Since z has the same shape as $cz + d1_3$, where $c, d \in \mathbb{C}$, the shape of the triangle can be obtained from $z = u + b\bar{u}$. The Watson shape coordinates are given by the complex number b. All unlabelled triangles can be represented in the sector of the unit disc, with $|b| \leq 1$ and $0 < \mathrm{Arg}(b) < \pi/3$. Watson (1986) constructs sequences of triangle shapes from products of circulant matrices.

3

Manifolds, shape and size-and-shape

In this chapter we introduce some differential geometrical aspects of shape and size-and-shape. After a brief review of Riemannian manifolds, we define what is meant by the pre-shape, shape and size-and-shape of a configuration.

3.1 Riemannian manifolds

Throughout this text the spaces of interest are primarily Riemannian manifolds, and we begin with some informal discussion about the topic. There are many treatments of differential geometry at various levels of formalism, and an excellent introduction is given by Bär (2010).

A **manifold** is a space which can be viewed locally as a Euclidean space. We first consider tangent spaces for a manifold M in general. Consider a differentiable curve in M given by $\gamma(t) \in M, t \in \mathbb{R}$ with $\gamma(0) = p$. The tangent vector at p is given by:

$$\gamma'(0) = \lim_{t \to 0} \frac{\mathrm{d}\gamma}{\mathrm{d}t},$$

and the unit tangent vector is $\xi = \gamma'(0)/\|\gamma'(0)\|$. The set of all tangent vectors $\gamma'(0)$ for all curves passing through p is called the **tangent space** of M at p, denoted by $T_p(M)$. If we consider a manifold M then if it has what is called an affine connection (a way of connecting nearby tangent spaces) then a geodesic can be defined.

A **Riemannian manifold** M is a connected manifold which has a positive-definite inner product defined on each tangent space $T_p(M)$, such that the choice varies smoothly from point to point. We write $g = \{g_{ij}\}$ to denote the positive definite tensor

Statistical Shape Analysis, with Applications in R, Second Edition. Ian L. Dryden and Kanti V. Mardia.
© 2016 John Wiley & Sons, Ltd. Published 2016 by John Wiley & Sons, Ltd.

which defines the inner-product on each tangent space for a given coordinate system. Specifically if we have coordinates $(x_1, \ldots, x_n)$ then the metric on the space is:

$$ds^2 = \sum_{i=1}^{n} \sum_{j=1}^{n} g_{ij} dx_i dx_j.$$

Consider a tangent vector $v \in T_p(M)$ in the tangent space at a point p on the manifold M, then there is a unique **geodesic** $\gamma(t)$ passing through p with initial tangent vector $\gamma'(0) = v$. The corresponding **exponential map** is defined as:

$$\exp_p(v) = \gamma(1) \tag{3.1}$$

and the **inverse exponential map**, or logarithmic map, is:

$$\exp_p^{-1}(\gamma(1)) = v. \tag{3.2}$$

The **Riemannian distance** between any two points in the Riemannian manifold is given by the arc length of the minimizing geodesic between two points, where the length of a parameterized curve $\{\gamma(t), \in [a, b]\}$ is defined as:

$$L = \int_a^b \|\gamma'(t)\|_g dt,$$

where $\|du\|_g = \{\sum_i \sum_j g_{ij} du_i du_j\}^{1/2}$ is the norm of a vector du induced by the inner product g. The length of the curve does not change under re-parameterizations. The Riemannian distance is an **intrinsic distance**, where the distance is obtained from the length of the geodesic path entirely within the manifold. An alternative is an **extrinsic** distance, where the distance is computed within a higher dimensional embedding space (usually Euclidean). In this case an embedding function is required, denoted by $j(X), X \in M$, together with a unique projection from the embedding space back to the manifold, denoted by $P()$ (see Chapter 6).

The Riemannian metric g defines a unique affine connection called the **Levi-Civita connection** ∇ (or covariant derivative), which enables us to say how vectors in tangent spaces change as we move along a curve from one point to another. Let $\gamma(t)$ be a curve on M and we want to move from $\gamma(t_0)$ to $\gamma(s)$ and see how the vector $\xi \in T_{\gamma(t_0)}(M)$ is transformed to in the new tangent space $T_{\gamma(s)}(M)$. The **parallel transport** of ξ along γ is a vector field $X(s)$ which satisfies

$$X(t_0) = \xi, \quad \nabla X(s) = 0, \tag{3.3}$$

and we write $P_{\gamma,\gamma(t_0),\gamma(s)} : T_{\gamma(t_0)}(M) \to T_{\gamma(s)}(M)$ for the parallel transport. In general, $X(s)$ is obtained by solving a system of ordinary differential equations, although for some manifolds the solution is analytic.

The **Riemannian curvature tensor** is used to describe the curvature of a manifold. The curvature tensor $R(U, V)$ is given in terms of the Levi-Civita connection

$$R(U, V)W = \nabla_U \nabla_V W - \nabla_V \nabla_U W - \nabla_{[U,V]} W \qquad (3.4)$$

where $[U, V] = UV - VU$ is the Lie bracket of vector fields (e.g. see Warner 1971).

For 2D landmark shapes the shape space is a Riemannian manifold (a complex projective space), as shown by Kendall (1984) and considered in Section 4.3. However, for the important case of three or more dimensional shapes the space is not a manifold, but rather it is a **stratified space**. For $m \geq 3$ dimensions the shape space has singularities (where configurations lie in a subspace of dimension $m - 2$ or less), and these singularities are strata of the space. We assume throughout that we are away from such degenerate shapes, which is often reasonable in practice, and so we restrict ourselves to the manifold part of the shape space (i.e. the non-degenerate configurations).

3.2 Shape

3.2.1 Ambient and quotient space

We have already noted that the shape of an object is given by the geometrical information that remains when we remove translation, rotation and scale information.

We shall remove the transformations in stages, which have different levels of difficulty. For example, removing location is very straightforward (e.g. by a specific linear transformation) as is the removal of scale (e.g. by rescaling to unit size). We say that a configuration which has been standardized by these two operations is called a 'pre-shape'. The pre-shape space is also sometimes called an 'ambient space', which contains the rotation information as well as the shape.

Removing rotation is more difficult and can be carried out by optimizing over rotations by minimizing some criterion. This final removal of rotation is often called 'quotienting out' in geometry and the resulting shape lies in the shape space, which is a type of 'quotient space'.

3.2.2 Rotation

A rotation of a configuration about the origin is given by post-multiplication of the $k \times m$ configuration matrix X by a $m \times m$ rotation matrix Γ.

Definition 3.1   *An $m \times m$ **rotation matrix** satisfies $\Gamma^T \Gamma = \Gamma \Gamma^T = I_m$ and $\det(\Gamma) = +1$. A rotation matrix is also known as a **special orthogonal matrix**, which is an orthogonal matrix with determinant +1. The set of all $m \times m$ rotation matrices is known as the special orthogonal group $SO(m)$.*

A rotation matrix has $\frac{1}{2}m(m - 1)$ degrees of freedom. For $m = 2$ dimensions the rotation matrix can be parameterized by a single angle θ, $-\pi \leq \theta < \pi$ in radians for rotating clockwise about the origin:

$$\Gamma = \begin{bmatrix} \cos\theta & \sin\theta \\ -\sin\theta & \cos\theta \end{bmatrix}.$$

In $m = 3$ dimensions one requires three Euler angles. For example one could consider a clockwise rotation of angle $\theta_1, -\pi \leq \theta_1 < \pi$ about the z-axis, followed by a rotation of angle $\theta_2, -\pi/2 \leq \theta_1 < \pi/2$ about the x-axis, then finally a rotation of angle $\theta_3, -\pi \leq \theta_3 < \pi$ about the z-axis. This parametrization is known as the x-convention and is unique apart from a singularity at $\theta_2 = -\pi/2$. The rotation matrix is:

$$\Gamma = \begin{bmatrix} \cos\theta_3 & \sin\theta_3 & 0 \\ -\sin\theta_3 & \cos\theta_3 & 0 \\ 0 & 0 & 1 \end{bmatrix} \begin{bmatrix} 1 & 0 & 0 \\ 0 & \cos\theta_2 & \sin\theta_2 \\ 0 & -\sin\theta_2 & \cos\theta_2 \end{bmatrix} \begin{bmatrix} \cos\theta_1 & \sin\theta_1 & 0 \\ -\sin\theta_1 & \cos\theta_1 & 0 \\ 0 & 0 & 1 \end{bmatrix}.$$

There are many choices of Euler angle representations, and all have singularities (Stuelpnagel 1964).

To complete the set of similarity transformations an isotropic scaling is obtained by multiplying X by a positive real number.

Definition 3.2   *The **Euclidean similarity transformations** of a configuration matrix X are the set of translated, rotated and isotropically rescaled X, that is*

$$\{\beta X \Gamma + 1_k \gamma^T : \beta \in \mathbb{R}^+, \Gamma \in SO(m), \gamma \in \mathbb{R}^m\}, \tag{3.5}$$

where $\beta \in \mathbb{R}^+$ is the scale, Γ is a rotation matrix and γ is a translation m-vector.

Definition 3.3   *The **rigid-body transformations** of a configuration matrix X are the set of translated and rotated X, that is*

$$\{X\Gamma + 1_k \gamma^T : \Gamma \in SO(m), \gamma \in \mathbb{R}^m\}, \tag{3.6}$$

where Γ is a rotation matrix and γ is a translation m-vector.

For $m = 2$ we can use complex notation. Consider $k \geq 3$ landmarks in $\mathbb{C}$, $z^o = (z_1^o, \dots, z_k^o)^T$ which are not all coincident. The Euclidean similarity transformations of z^o are:

$$\{\eta z^o + 1_k \xi : \eta = \beta e^{i\theta} \in \mathbb{C}, \xi \in \mathbb{C}\},$$

where $\beta \in \mathbb{R}^+$ is the scale, $0 \leq \theta < 2\pi$ is the rotation angle and $\xi \in \mathbb{C}$ is the translation. Here θ is an anti-clockwise rotation about the origin. The Euclidean similarity transformations of z^o are the set of the same complex linear transformations applied to each landmark z_j^o. Specifying the Euclidean similarity transformations as complex linear transformations leads to great simplifications in shape analysis for the 2D case, as we shall see in Chapter 8.

For $m = 3$ we can use unit quaternions to represent a 3D rotation, and this approach was used by Horn (1987), Theobald (2005) and Du *et al.* (2015) for example. A combined 3D rotation and isotropic scaling can be represented by a (non-unit) quaternion.

We can consider the shape of X as the equivalence class of the full set of similarity transformations of a configuration, and we remove the similarity transformations from the configuration in a systematic manner.

3.2.3   Coincident and collinear points

If all k points are coincident, then this has a special shape that must be considered as a separate case. We shall remove this case from the set of configurations. After removing translation the coincident points are represented by the origin, the m-vector $0_m = (0, \dots, 0)^{\mathrm{T}}$. The coincident case is not generally of interest except perhaps as a starting point in the study of the diffusion of shape (Kendall 1989).

The $m = 1$ case is also not of primary interest and it is simply seen (Kendall 1984) that the shape space is S^{k-2} (the sphere in $k - 1$ real dimensions) after translation and scale have been removed, and thus can be dealt with using directional data analysis techniques (e.g. Mardia 1972; Mardia and Jupp 2000). Our first detailed consideration is the shape space of Kendall (1984) for the case where $k > m$ and for $m \geq 2$.

3.2.4   Removing translation

A translation is obtained by adding a constant m-vector to the coordinates of each point. Translation is the easiest to remove from X and can be achieved by considering contrasts of the data, that is pre-multiplying by a suitable matrix. We can make a specific choice of contrast by pre-multiplying X with the Helmert submatrix of Equation (2.10). We write

$$X_H = HX \in \mathbb{R}^{(k-1)m} \setminus \{0\} \tag{3.7}$$

(the origin is removed because coincident landmarks are not allowed) and we refer to X_H as the **Helmertized landmark coordinates**.

The **centred landmark coordinates** are an alternative choice for removing location and are given by:

$$X_C = CX. \tag{3.8}$$

We can revert back to the centred landmark coordinates from the Helmertized landmark coordinates by pre-multiplying by H^{T}, as

$$H^{\mathrm{T}}H = I_k - \frac{1}{k}1_k 1_k^{\mathrm{T}} = C$$

and so

$$H^{\mathrm{T}}X_H = H^{\mathrm{T}}HX = CX.$$

Note that the Helmertized landmark coordinates X_H are a $(k - 1) \times m$ matrix, whereas the centred landmark coordinates X_C are a $k \times m$ matrix.

3.2.5   Pre-shape

We standardize for size by dividing through by our notion of size. We choose the centroid size [see Equation (2.2)] which is also given by:

$$\|X_H\| = \sqrt{\text{trace}(X^T H^T H X)} = \sqrt{\text{trace}(X^T C X)} = \|CX\| = S(X), \qquad (3.9)$$

since $H^T H = C$ is idempotent. Note that $S(X) > 0$ because we do not allow complete coincidence of landmarks. The pre-shape of a configuration matrix X has all information about location and scale removed.

Definition 3.4   *The **pre-shape** of a configuration matrix X is given by:*

$$Z = \frac{X_H}{\|X_H\|} = \frac{HX}{\|HX\|} \qquad (3.10)$$

which is invariant under the translation and scaling of the original configuration.

An alternative representation of pre-shape is to initially centre the configuration and then divide by size. The **centred pre-shape** is given by:

$$Z_C = CX/\|CX\| = H^T Z \qquad (3.11)$$

since $C = H^T H$. Note that Z is a $(k - 1) \times m$ matrix whereas Z_C is a $k \times m$ matrix.

Important point: Both pre-shape representations are equally suitable for the pre-shape space which has real dimension $(k - 1)m - 1$. The advantage in using Z is that the number of rows is less than that of Z_C (although of course they have the same rank). On the other hand, the advantage of working with the centred pre-shape Z_C is that a plot of the Cartesian coordinates gives a correct geometrical view of the shape of the original configuration.

Notation: We use the notation S_m^k to denote the pre-shape space of k points in m dimensions.

Definition 3.5   *The **pre-shape space** is the space of all pre-shapes. Formally, the pre-shape space S_m^k is the orbit space of the non-coincident k point set configurations in $\mathbb{R}^m$ under the action of translation and isotropic scaling.*

The pre-shape space $S_m^k \equiv S^{(k-1)m-1}$ is a hypersphere of unit radius in $(k - 1)m$ real dimensions, since $\|Z\| = 1$. The term 'pre-shape' signifies that we are one step away from shape: rotation still has to be removed. The term was coined by Kendall (1984), and it is a type of **ambient space**. The pre-shape space is of higher dimension and informally can be thought of as 'surrounding' the shape space, hence the use of the word 'ambient' here.

3.2.6   Shape

In order to also remove rotation information from the configuration we identify all rotated versions of the pre-shape with each other, and this set or equivalence class is the shape of X.

Definition 3.6   *The **shape** of a configuration matrix X is all the geometrical information about X that is invariant under location, rotation and isotropic scaling (Euclidean similarity transformations). The shape can be represented by the set [X] given by:*

$$[X] = \{Z\Gamma : \Gamma \in SO(m)\}, \tag{3.12}$$

where SO(m) is the special orthogonal group of rotations and Z is the pre-shape of X.

Notation: We use the notation Σ_m^k to denote the shape space of k points in m dimensions.

Definition 3.7   *The **shape space** is the space of all shapes. Formally, the shape space Σ_m^k is the orbit space of the non-coincident k point set configurations in $\mathbb{R}^m$ under the action of the Euclidean similarity transformations (translation, rotation and scale).*

In order to compare two different shapes we need to choose a particular relative rotation so that they are closest as possible in some sense (see Chapter 4). This optimization over rotation is also called 'quotienting out' the group of rotations, and so the shape space is a type of quotient space.

Important point: The dimension of the shape space is:

$$q = km - m - 1 - \frac{m(m-1)}{2},$$

and this can be simply seen as we initially have km coordinates and then must lose m dimensions for location, one dimension for isotropic scale and $\frac{1}{2}m(m-1)$ for rotation.

The shape of X is a set: an equivalence class under the action of the group of similarity transformations. In order to visualize shapes it is often convenient to choose a particular member of the shape set $[X]$.

Definition 3.8   *An **icon** is a particular member of the shape set [X] which is taken as being representative of the shape.*

The word icon can mean 'image or likeness' and it is appropriate as we use the icon to picture a representative figure from the shape equivalence class which has a resemblance to the other members, that is the objects of the class are all similar (Goodall 1995). The centred pre-shape Z_C is a suitable choice of icon.

3.3   Size-and-shape

We could change the order of removing the similarity transformations or only remove some of the transformations. For example, if location and rotation are removed but not scale, then we have the size-and-shape of X.

Definition 3.9   *The* **size-and-shape** *of a configuration matrix X is all the geometrical information about X that is invariant under location and rotation (rigid-body transformations), and this can be represented by the set $[X]_S$ given by:*

$$[X]_S = \{X_H \Gamma : \Gamma \in SO(m)\}, \tag{3.13}$$

where X_H are the Helmertized coordinates of Equation (3.7).

Notation: We use the notation $S\Sigma_m^k$ to denote the size-and-shape space of k points in m dimensions.

Definition 3.10   *The* **size-and-shape space** *is the space of all size-and-shapes. Formally, the size-and-shape space $S\Sigma_m^k$ is the orbit space of k point set configurations in $\mathbb{R}^m$ under the action of translation and rotation.*

Size-and-shape is also known as **form**, particularly in biology. We discuss size-and-shape in more detail in Chapter 5.

If size is removed from the size-and-shape (e.g. by rescaling to unit centroid size), then again we obtain the shape of X,

$$[X] = [X]_S / S(X) = \{Z\Gamma : \Gamma \in SO(m)\},$$

as in Equation (3.12).

3.4   Reflection invariance

We can also include invariances under reflections for shape or size-and-shape. A reflection can be obtained by multiplying one of the coordinate axes by -1. Joint rotations and reflections can be represented by an orthogonal matrix.

Definition 3.11   *An $m \times m$* **orthogonal matrix** *satisfies $\Gamma^T \Gamma = \Gamma \Gamma^T = I_m$ and $\det(\Gamma) = \pm 1$. The set of all $m \times m$ orthogonal matrices is known as the orthogonal group $O(m)$.*

The orthogonal group includes rotations (determinant $+1$) and rotations/reflections (determinant -1).

Definition 3.12   *The* **reflection shape** *of a configuration matrix X is all the geometrical information that is invariant under translation, rotation, scale and reflection. The reflection shape can be represented by the set*

$$[X]_R = \{ZR : R \in O(m)\}$$

where $O(m)$ is the set of $m \times m$ orthogonal matrices, satisfying $R^T R = I_m = RR^T$ and $\det(R) = \pm 1$, and Z is the pre-shape.

Definition 3.13   *The* **reflection size-and-shape** *of a configuration matrix X is all the geometrical information that is invariant under translation, rotation and reflection. The reflection size-and-shape can be represented by the set*

$$[X]_{RS} = \{X_H R : R \in O(m)\}$$

where $O(m)$ is the set of $m \times m$ orthogonal matrices and X_H are the Helmertized coordinates.

3.5   Discussion

3.5.1   Standardizations

In obtaining the shape, the removal of the location and scale could have been performed in a different manner. For example, Ziezold (1994) centres the configuration to remove location, CX, where C is given by Equation (2.3) and he uses the normalized size $\frac{1}{\sqrt{k}} \|CX\|$. We could alternatively have removed location by pre-multiplying by B where the jth row of B is:

$$
\begin{aligned}
(-1, 1, 0, \ldots, 0), \qquad & j = 1 \\
(-\tfrac{1}{2}, -\tfrac{1}{2}, 0, \ldots, 0, 1, 0, \ldots, 0), \qquad & 2 \le j \le k - 1,
\end{aligned}
\qquad (3.14)
$$

and the 1 is in the $(j + 1)$th column. The implication of pre-multiplication by B is that location is removed by sending the midpoint of the line between landmarks 1 and 2 to the origin, as is carried out when using Bookstein coordinates (see Section 2.4).

3.5.2   Over-dimensioned case

Consider a triangle of $k = 3$ points in $m = 3$ dimensions. It is clear in this case that the triangle can be rotated to lie in a particular 2D plane (say the x–y plane). A reflection of the triangle can be carried out by a 3D rotation, and hence shape is the same as reflection shape for a triangle in three dimensions.

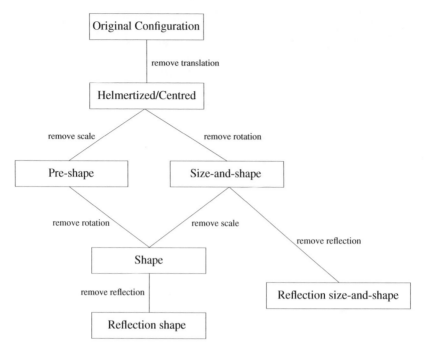

*Figure 3.1   The hierarchies of the various spaces. Source: Adapted from Goodall &
Mardia 1992.*

More generally, if $k \leq m$, then the pre-shape of X can be identified with another
pre-shape U in S^k_{k-1} by rotating the configuration to be in a fixed $k - 1$ dimensional
plane. Now U can be reflected in this plane without changing its shape and so the
shape of X is:

$$\{ UR : R \in O(k - 1) \} \tag{3.15}$$

where $O(k - 1)$ is the orthogonal group (rotation and reflection). Kendall (1984)
called the case $k \leq m$ 'over-dimensioned' and dealing with these shape spaces is
equivalent to dealing with Σ^k_{k-1}, and identifying reflections with each other. Throughout most of this book we shall deal with the case $k > m$.

3.5.3   Hierarchies

Important point: With quite a wide variety of terminology used for the different
spaces it may be helpful to refer to Figure 3.1 where we give a diagram indicating
the hierarchies of shape and size-and-shape spaces. In addition removing all affine
transformations leads to the affine shape space.

4

Shape space

In this chapter we consider the choice of shape distance, which is required to fully define the non-Euclidean shape metric space. We consider the partial Procrustes, full Procrustes and Riemannian shape distances. In the case of 2D point configurations the shape space is the complex projective space, and in the special case of triangles in two dimensions the shape space is a sphere. Finally, we consider different choices of coordinates in the tangent space to shape space, which are particularly useful for practical statistical inference.

4.1   Shape space distances

4.1.1   Procrustes distances

We shall initially describe the partial and full Procrustes distances. Consider two configuration matrices from k points in m dimensions X_1 and X_2 with pre-shapes Z_1 and Z_2. First we minimize over rotations to find the closest Euclidean distance between Z_1 and Z_2.

Definition 4.1   *The **partial Procrustes distance** d_P is obtained by matching the pre-shapes Z_1 and Z_2 of X_1 and X_2 as closely as possible over rotations. So,*

$$d_P(X_1, X_2) = \inf_{\Gamma \in SO(m)} \| Z_2 - Z_1 \Gamma \|,$$

where $Z_j = HX_j / \| HX_j \|$, $j = 1, 2$.

Here 'inf' denotes the infimum, and we will use 'sup' for the supremum.

Statistical Shape Analysis, with Applications in R, Second Edition. Ian L. Dryden and Kanti V. Mardia.
© 2016 John Wiley & Sons, Ltd. Published 2016 by John Wiley & Sons, Ltd.

Result 4.1   *The partial Procrustes distance is given by:*

$$d_P(X_1, X_2) = \sqrt{2} \left(1 - \sum_{i=1}^{m} \lambda_i \right)^{1/2}, \tag{4.1}$$

where $\lambda_1 \geq \lambda_2 \geq \dots \geq \lambda_{m-1} \geq |\lambda_m|$ are the square roots of the eigenvalues of $Z_1^T Z_2 Z_2^T Z_1$, and the smallest value λ_m is the negative square root if and only if $\det(Z_1^T Z_2) < 0$.

A proof of Result 4.1 is given after two useful lemmas.

Lemma 4.1

$$\sup_{\Gamma \in SO(m)} \text{trace}\left(Z_2^T Z_1 \Gamma\right) = \sum_{i=1}^{m} \lambda_i. \tag{4.2}$$

Proof: We follow the proof of Kendall (1984). First, consider a singular value decomposition of $Z_2^T Z_1$ given by:

$$Z_2^T Z_1 = V \Lambda U^T, \tag{4.3}$$

where $U, V \in SO(m)$, $\Lambda = \text{diag}(\lambda_1, \dots, \lambda_m)$ with $\lambda_1 \geq \lambda_2 \geq \dots \geq \lambda_{m-1} \geq |\lambda_m|$ and

$$\lambda_m < 0 \quad \text{iff} \quad \det\left(Z_2^T Z_1\right) < 0. \tag{4.4}$$

In this case the sequence of eigenvalues is called 'optimally signed' (Kent and Mardia 2001). Hence

$$\sup_{R \in SO(m)} \text{trace}\left(Z_2^T Z_1 R\right) = \sup_{R \in SO(m)} \text{trace}(R\Lambda) = \sup_{R \in SO(m)} \sum_{i=1}^{m} r_{ii} \lambda_i, \tag{4.5}$$

where $(r_{11}, \dots, r_{mm})$ are the diagonals of $R \in SO(m)$. Now the set of diagonals of R in $SO(m)$ is a compact convex set with extreme points

$$\{(\pm 1, \pm 1, \dots, \pm 1)\}$$

with an even number of minus signs (Horn 1954). Hence, it is clear in our case that the supremum is achieved when $r_{ii} = 1, i = 1, \dots, m$, and hence Equation (4.2) follows. $\qquad \square$

Lemma 4.2   *The optimal rotation is*

$$\hat{\Gamma} = \arg\sup \, \text{trace}\left(Z_2^T Z_1 \Gamma\right) = U V^T, \tag{4.6}$$

where U, V are given in Equation (4.3), *and the optimal rotation is unique if the singular values satisfy*

$$\lambda_{m-1} + \lambda_m > 0, \tag{4.7}$$

in which case Kent and Mardia (2001) *denote the sequence as 'non-degenerate'.*

Proof: The lemma follows because

$$\text{trace}\left(Z_2^\mathsf{T} Z_1 \hat{\Gamma}\right) = \text{trace}(V\Lambda U^\mathsf{T} U V^\mathsf{T}) = \text{trace}(\Lambda),$$

where $\hat{\Gamma}$ is the Procrustes rotation of Z_1 onto Z_2. For further details see Le (1991b) and Kent and Mardia (2001). ☐

Proof (of Result 4.1): Note that

$$
\begin{aligned}
d_P^2(X_1, X_2) &= \inf_{\Gamma \in SO(m)} \text{trace}\{(Z_2 - Z_1\Gamma)^\mathsf{T}(Z_2 - Z_1\Gamma)\} \\
&= \|Z_2\|^2 + \|Z_1\|^2 - 2 \sup_{\Gamma \in SO(m)} \text{trace}\left(Z_2^\mathsf{T} Z_1 \Gamma\right),
\end{aligned} \tag{4.8}
$$

and $\|Z_1\| = 1 = \|Z_2\|$. The supremum of $\text{trace}(Z_2^\mathsf{T} Z_1\Gamma)$ over $\Gamma \in SO(m)$ is given by Lemma 4.1, and the optimal rotation is given by Lemma 4.2. Hence

$$d_P^2(X_1, X_2) = \left(2 - 2\sum_{i=1}^m \lambda_i\right),$$

and the result follows. ☐

Definition 4.2 *The* **full Procrustes distance** *between X_1 and X_2 is*

$$d_F(X_1, X_2) = \inf_{\Gamma \in SO(m), \beta \in \mathbb{R}^+} \left\|Z_2 - \beta Z_1 \Gamma\right\|, \tag{4.9}$$

where $Z_r = HX_r / \|HX_r\|$, $r = 1, 2$.

Result 4.2 *The full Procrustes distance is*

$$d_F(X_1, X_2) = \left\{1 - \left(\sum_{i=1}^m \lambda_i\right)^2\right\}^{1/2}. \tag{4.10}$$

Proof: Note that

$$d_F^2(X_1, X_2) = \inf_{\Gamma \in SO(m), \beta} \text{trace}\{(Z_2 - \beta Z_1 \Gamma)^T(Z_2 - \beta Z_1 \Gamma)\}$$

$$= \inf_{\beta} \left(\|Z_2\|^2 + \beta^2 \|Z_1\|^2 - 2\beta \sup_{\Gamma \in SO(m)} \text{trace}(Z_2^T Z_1 \Gamma) \right),$$

and $\|Z_1\| = 1 = \|Z_2\|$. The optimization over rotation is given by Lemma 4.1, which is exactly the same as for the partial Procrustes distance. Hence

$$d_F^2(X_1, X_2) = \inf_{\beta} \left(1 + \beta^2 - 2\beta \sum_{i=1}^{m} \lambda_i \right). \tag{4.11}$$

By differentiation we have

$$\hat{\beta} = \sum_{i=1}^{m} \lambda_i,$$

which is clearly a minimum as $\partial^2(d_F^2)/\partial\beta^2 > 0$. Substituting $\hat{\beta}$ into Equation (4.11) leads to the expression for the full Procrustes distance, as required. □

We shall often use the full Procrustes distance in the shape space for practical statistical inference. Note that $0 \le \sum_{i=1}^{m} \lambda_i \le 1$ and so

$$0 \le d_F \le 1, \ 0 \le d_P \le \sqrt{2}.$$

The Procrustes distances are types of **extrinsic** distances as they are actually distances in an embedding of shape space. Another type of distance is an **intrinsic** distance defined in the space itself (see Section 3.1), and we introduce this distance in Section 4.1.4.

4.1.2   Procrustes

The term 'Procrustes' is used because the above matching operations are identical to those of Procrustes analysis, a commonly used technique for comparing matrices (up to transformations) in multivariate analysis (see Mardia *et al.* 1979, p. 416). In Procrustes analysis the optimal transformation parameters are estimated by minimizing a least squares criterion. The expression 'Procrustes analysis' was first used by Hurley and Cattell (1962) in factor analysis. In Greek mythology Procrustes was the nickname of a robber Damastes, who lived by the road from Eleusis to Athens. He would offer travellers a room for the night and fit them to the bed by stretching them if they were too short or chopping off their limbs if they were too tall. The analogy is rather tenuous but we can regard one configuration as the bed and the other as the person being 'translated', 'rotated' and possibly 'rescaled' so as to fit as close as possible to the bed. Procrustes was eventually captured and killed by Theseus, who fitted Procrustes to his own bed.

4.1.3   Differential geometry

The rotated Z on the pre-shape sphere is called a **fibre** of the pre-shape space S_m^k. Fibres on the pre-shape sphere correspond one to one with shapes in the shape space, and so we can think of a fibre as representing the shape of a configuration. The pre-shape sphere is partitioned into fibres by the rotation group $SO(m)$ and the fibre is the orbit of Z under the action of $SO(m)$. The fibres do not overlap.

In Figure 4.1 we see a diagrammatic view of the pre-shape sphere. Since the pre-shape sphere is a hypersphere embedded in $\mathbb{R}^{(k-1)m}$ we can consider the great circle distance between two points on a sphere. The great circle distance is an intrinsic distance, defined as the shortest geodesic distance between any two points on the pre-shape sphere.

Since the shapes of configurations are represented by fibres on the pre-shape sphere, we can define the distance between two shapes as the closest great circle distance between the fibres on the pre-shape sphere. This shape distance is called the **Riemannian distance** in shape space and is denoted by ρ. The Riemannian

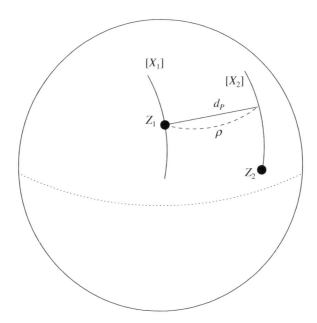

Figure 4.1   A diagrammatic schematic view of two fibres $[X_1]$ and $[X_2]$ on the pre-shape sphere, which correspond to the shapes of the original configuration matrices X_1 and X_2 which have pre-shapes Z_1 and Z_2. Also displayed are the smallest great circle ρ and chordal distance d_P between the fibres.

distance $\rho(X_1, X_2)$ is the distance inherited from the projection of the fibres on the pre-shape sphere to points in the shape space. The projection from pre-shape sphere to shape space is **isometric**, because distances are preserved, and is termed a Riemannian submersion. The Riemannian distance is an **intrinsic distance** in the shape space. The shape space is a type of **quotient space**, where the rotation has been quotiented out from the pre-shape sphere using optimization. We write $\Sigma_m^k = S_m^k / SO(m)$ and more formally we say that Σ_m^k is the quotient space of S_m^k under the action of $SO(m)$.

For $m > 2$ there are singularities in the shape space when the k points lie in an $m - 2$ dimensional subspace (see Kendall 1989). Here we assume that our shapes are away from any such singularities. If our data are modelled by a continuous probability distribution, then the Lebesgue measure of the singularity set is zero. For $m = 2$ there is no such problem with singularities.

Further discussion of geometrical properties of the shape space is given by Kendall (1984, 1989); Le and Kendall (1993); Small (1996); Kendall *et al.* (1999); and Kent and Mardia (2001).

4.1.4   Riemannian distance

In Figure 4.1 two minimum distances have been drawn diagrammatically on the pre-shape sphere between the fibres (shapes): ρ is the closest great circle distance; and d_P is the closest chordal distance. The Riemannian distance is measured in radians.

Definition 4.3   *The **Riemannian distance** $\rho(X_1, X_2)$ is the closest great circle distance between Z_1 and Z_2 on the pre-shape sphere, where $Z_j = HX_j / \|HX_j\|$, $j = 1, 2$. The minimization is carried out over rotations.*

Result 4.3   *The Riemannian distance ρ is*

$$\rho(X_1, X_2) = \arccos\left(\sum_{i=1}^{m} \lambda_i \right), \tag{4.12}$$

where the eigenvalues $\lambda_i, i = 1, \ldots, m$ are defined in Equation (4.3) and Equation (4.4).

Proof: From trigonometry on sections of the pre-shape sphere one can see that the Riemannian distance is related to the partial Procrustes distance as follows:

$$\rho(X_1, X_2) = 2\arcsin(d_P(X_1, X_2)/2) = \arccos\left(\sum_{i=1}^{m} \lambda_i \right),$$

as

$$d_P^2 = 4\sin^2(\rho/2) = 2 - 2\cos\rho = 2 - 2\sum_{i=1}^{m} \lambda_i. \qquad \square$$

The Riemannian distance is also known as an **intrinsic** distance, as it is the shortest geodesic distance within the shape space. The partial Procrustes and full Procrustes distances are both **extrinsic** distances, which are shortest distances in an embedding of the shape space. Hence the extrinsic distances are shortest paths which are typically outside the space for non-Euclidean spaces, whereas the intrinsic distances are shortest paths that remain wholly within the space.

4.1.5   Minimal geodesics in shape space

It of interest to study the shortest path in shape space between two shapes – the minimal geodesic path – the length of which is the Riemannian distance. In Euclidean space the geodesic path is just a straight line, but in a non-Euclidean space the geodesic path is curved. Working with geodesics directly on shape space is complicated (Le 1991a). However, a practical way to proceed is to work with the pre-shape sphere (which is just an ordinary sphere) and then find an isometric (i.e. distance preserving) representation on the pre-shape sphere of a geodesic on the shape space. This representation corresponds to horizontal lifts to the pre-shape sphere of minimal geodesics in the shape space.

A geodesic on the pre-shape sphere is a great circle. Any two orthogonal points p and w on the pre-shape sphere define a unit speed geodesic $\Gamma_{(p,w)}$ parameterized by:

$$\Gamma_{(p,w)}(s) = p \cos s + w \sin s, \quad -\frac{\pi}{2} < s \leqslant \frac{\pi}{2},$$

where $\text{trace}(pp^T) = \text{trace}(ww^T) = 1$ and $\text{trace}(pw^T) = 0$. If in addition $pw^T = wp^T$, then the geodesic is known as a 'horizontal' geodesic, which is invariant to rotations of p and w. See Chapter 6 of Kendall *et al.* (1999) for a full description of geodesics in shape space. The quantity $|s|$ is the great circular arc distance from $\Gamma_{(p,w)}(0) = p$ to $\Gamma_{(p,w)}(s)$, and for a horizontal geodesic the length of the great circular arc is equal to the Riemannian distance $\rho([p], [w])$.

4.1.6   Planar shape

In the planar shape case with $m = 2$ then the use of complex arithmetic helps to simplify the expressions for distances and geodesics.

Result 4.4   *If p and w are two complex $(k-1)$-vectors on the pre-shape sphere, then the Riemannian distance is given by:*

$$\rho(p, w) = \arccos|p^*w|, \quad 0 \leq \rho < \pi/2, \tag{4.13}$$

where p^* is the transpose of the complex conjugate of p, and $|z| = \sqrt{z^*z}$ is the modulus of z.

Proof: We have

$$d_P^2 = \inf_{\theta} \|p - we^{i\theta}\|^2$$
$$= \inf_{\theta} \left\{ p^*p + w^*w - p^*we^{i\theta} - w^*p^{-i\theta} \right\},$$
$$= 2 - \sup_{\theta} \left(p^*we^{i\theta} + w^*p^{-i\theta} \right).$$

The minimum is given when the last two terms are real, that is when $\theta = -\mathrm{Arg}(p^*w)$, and hence

$$d_P^2 = 2(1 - |p^*w|),$$

which in turn leads to

$$\rho = \arccos\left(1 - d_P^2/2\right) = \arccos(|p^*w|),$$

as required.   □

The horizontal geodesic on the pre-shape sphere is:

$$\Gamma_{(p,w)}(s) = p\cos s + w\sin s, \quad -\frac{\pi}{2} < s \leqslant \frac{\pi}{2}, \tag{4.14}$$

with conditions $p^*p = w^*w = 1$ and $p^*w = 0$, with the final condition ensuring that the geodesic is horizontal.

The exponential map is a bijection if $\|v\| \in [0, \pi/2)$ and is given by:

$$\exp_p(v) = p\cos(s) + \frac{v}{\|v\|}\sin(s),$$

where $s = \|v\|$, with conditions $p^*p = 1$ and $p^*v = 0$. The inverse exponential map is:

$$\exp_p^{-1}(w) = \frac{\rho}{\sin(\rho)}(w - p\cos\rho),$$

with $w^*p > 0$.

If we parallel transport a vector $v \in T_{z_1}(M)$ along the shortest geodesic γ from z_1 to z_2 the result in $T_{z_2}(M)$ is:

$$P_{\gamma,z_1,z_2}(v) = v - \frac{z_2^*v(z_1 + z_2)}{1 + \cos\rho},$$

where $z_2^*z_1 > 0$. Note that here the parallel transport is a linear isometry (i.e. a linear transformation which preserves distance).

The Riemannian curvature tensor (or Ricci curvature tensor) (Su *et al.* 2012) for $X, Y, Z \in T_z(M)$ is given by:

$$R(X, Y)Z = - < Y, Z > X + < X, Z > Y + < Y, iZ > iX$$
$$- < X, iZ > iY - 2 < X, iY > iZ, \tag{4.15}$$

where $i = \sqrt{-1}$ and $< Y, Z > = Z^* Y$.

4.1.7   Curvature

Le and Kendall (1993) studied the curvature of the shape space and gave explicit formulae for the sectional curvatures, scalar curvature and an average curvature. The curvatures depend on the singular values of the preshape $\lambda_1, \ldots, \lambda_m$, assuming $k > m$. Although in general the formulae are complicated, an expression for an average section curvature when $k = 4$ and $m = 3$ has a concise form, and is given by:

$$1 + \frac{3}{10} \frac{1 + 2\left(\lambda_2^2 \lambda_3^2 + \lambda_3^2 \lambda_1^2 + \lambda_1^2 \lambda_2^2\right)}{\left(\lambda_2^2 + \lambda_3^2\right)\left(\lambda_3^2 + \lambda_1^2\right)\left(\lambda_1^2 + \lambda_2^2\right)}.$$

It is clear that the average curvature is greater than 1. Also, if $\lambda_2 \to 0$ and hence $\lambda_3 \to 0$ then the average curvature blows up to infinity, which occurs at each singularity of the shape space.

4.2   Comparing shape distances

4.2.1   Relationships

In Figure 4.2 we see a cross-section of the pre-shape sphere illustrating the relationships between d_F, d_P and ρ, where

$$d_F(X_1, X_2) = \sin \rho,$$
$$d_P(X_1, X_2) = 2 \sin(\rho/2).$$

Note the Riemannian distance ρ can be considered as the smallest angle (with respect to rotations of the pre-shapes) between the vectors corresponding to Z_1 and Z_2 on the pre-shape sphere.

Important point: For shapes which are close together there is very little difference between the shape distances, since

$$d_P = d_F + O(d_F^3), \quad \rho = d_F + O(d_F^3).$$

Consequently for many practical datasets with small variability there is very little difference in the analyses when using different Procrustes distances. However, the distinction between the distances is worth making and the terminology is summarized

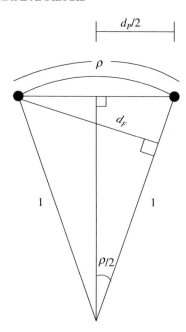

Figure 4.2   Section of the pre-shape sphere, illustrating schematically the relationship between the distances d_F, d_P and ρ.

in Table 4.1. We discuss some practical implications of different choices of distance in estimation in Chapter 6.

4.2.2   Shape distances in R

The shape distances can be calculated in R using the command `riemdist` from the `shapes` library. We consider three triangles, which are two equilateral triangles (which are reflections of each other) and a collinear triangle.

```
> a1<-matrix( c(-0.5,0.5,0,0,0,sqrt(3)/2),3,2)
> a2<-matrix( c(-0.5,0.5,0,0,0,-sqrt(3)/2),3,2)
> a3<-matrix( c(-0.5,0.5,0,0,0,0),3,2)
> a1
```

Table 4.1   Distances in the shape space.

Distance	Notation	Formula	Range
Full Procrustes distance	d_F	$\{1-(\sum_{i=1}^{m}\lambda_i)^2\}^{1/2}$	$0 \le d_F \le 1$
Partial Procrustes distance	d_P	$\sqrt{2(1-\sum_{i=1}^{m}\lambda_i)^{1/2}}$	$0 \le d_P \le \sqrt{2}$
Riemannian distance	ρ	$\arccos(\sum_{i=1}^{m}\lambda_i)$	$0 \le \rho \le \pi/2$

```
        [,1] [,2]
[1,]  -0.5 0.0000000
[2,]   0.5 0.0000000
[3,]   0.0 0.8660254
> a2
        [,1] [,2]
[1,]  -0.5 0.0000000
[2,]   0.5 0.0000000
[3,]   0.0 -0.8660254
> a3
        [,1] [,2]
[1,]  -0.5 0
[2,]   0.5 0
[3,]   0.0 0
```

The Riemannian distances between the three pairs are:

```
> riemdist(a1,a2)
[1] 1.570796
> riemdist(a1,a3)
[1] 0.7853982
> riemdist(a2,a3)
[1] 0.7853982
```

and so here the first two triangles are maximally far apart at $\rho = \pi/2$. The partial and full Procrustes distances between the last two triangles are:

```
# partial Procrustes
> 2*sin(riemdist(a2,a3)/2)
[1] 0.7653669
# full Procrustes
> sin(riemdist(a2,a3))
[1] 0.7071068
```

A useful method of displaying data on Riemannian manifolds is to use **classical multidimensional scaling (MDS)** with a suitable choice of distance (e.g. Mardia *et al.* 1979, p. 397). The technique, also known as **principal coordinate analysis** (Gower 1966), involves representing the data in a low dimensional Euclidean subspace, such that the distances between pairs of observations in the original space are well approximated by Euclidean distances in the lower dimensional subspace. We display the first two principal coordinates of the schizophrenia dataset of Section 1.4.5 using the Riemannian distance and the full Procrustes distance in Figure 4.3. Clearly the plots are extremely similar as the distances are so similar.

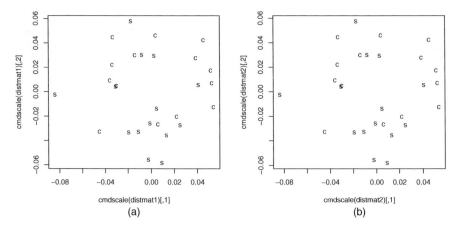

Figure 4.3   Multidimensional scaling plots using (a) Riemannian distance and (b) full Procrustes distance.

```
data(schizophrenia)
distmat1 <- matrix( 0, 28, 28)
distmat2 <- matrix( 0, 28, 28)
for (i in 1:28){
for (j in 1:28){
rho <- riemdist( schizophrenia$x[,,i],schizophrenia$x[,,j] )
distmat1[i,j] <- rho
distmat2[i,j] <- sin(rho)
}
}
par(mfrow=c(1,2))
plot(cmdscale(distmat1),pch=as.character(schizophrenia$group))
plot(cmdscale(distmat2),pch=as.character(schizophrenia$group))
```

Also, there are other functions in the shapes library which compute shape distances, for example the routine procGPA includes the Riemannian distances of each observation to an estimated mean shape.

If it is of interest to compare configurations with reflection invariance then the option reflect=TRUE is used. For example, comparing the two equilateral triangles with reflection invariance:

```
> riemdist(a1,a2,reflect=TRUE)
[1] 1.490116e-08
```

which is effectively zero, and so the two equilateral triangles have the same reflection shape.

4.2.3   Further discussion

The minimization over rotations (and reflections) in the Procrustes distance calculations could also be obtained using calculus (e.g. see Mardia *et al.* 1979, p. 416) and we consider this calculation for minimizing over rotations in Section 7.2.1.

We illustrate the geometrical steps involved in obtaining the full and partial Procrustes distance in Figure 4.4. The first row shows the original figure X_1 (left), the centred X_1 (middle) and the rescaled X_1, which we call Z_1 (right). The second row shows the original figure X_2 (left), the centred X_2 (middle) and the rescaled X_2, which we call Z_2 (right). In the third row (left) Z_2 is rotated to Z_1 to minimize the sum of

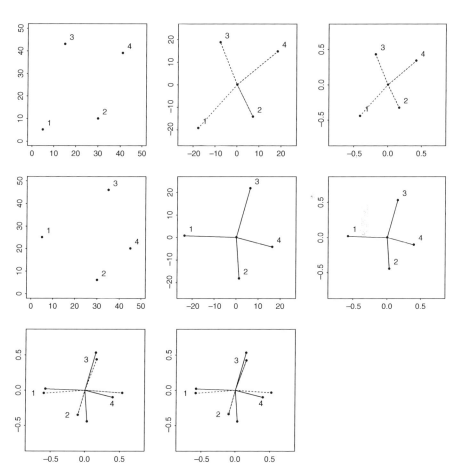

Figure 4.4   The geometry of Procrustes fits in calculating Procrustes distances. The first two rows show each configuration (left), centred (middle) and scaled (right). The last row shows the partial Procrustes fit (left) and full Procrustes fit (middle) of the centred pre-shape from configuration 1 (- - - -) onto the centred pre-shape of configuration 2 (———). Source: Adapted from Bookstein 1996b.

squared distances between pairs of landmarks – the partial Procrustes distance is the Euclidean distance between these fitted configurations. In the third row (middle) Z_2 is rotated and rescaled to Z_1 to minimize the sum of squared distances between pairs of landmarks – the full Procrustes distance is the Euclidean distance between these fitted configurations.

4.3   Planar case

4.3.1   Complex arithmetic

We now consider the case where the $k \geq 3$ landmarks are in $m = 2$ dimensions. In this case complex arithmetic enables us to deal with shape analysis very effectively. Consider $k \geq 3$ landmarks in $\mathbb{C}$, $z^o = (z_1^o, \dots, z_k^o)^{\mathrm{T}}$ which are not all coincident, where $z_j^o = x_j^o + i y_j^o, j = 1, \dots, k, i = \sqrt{-1}$. Location is removed by pre-multiplying by the Helmert submatrix H giving the complex Helmertized landmarks

$$z_H = H z^o,$$

where H is defined in (2.10). The centroid size is:

$$S(z^o) = \{(z^o)^* C z^o\}^{1/2} = \|z_H\| = \sqrt{(z_H)^* z_H}, \qquad (4.16)$$

where $(z^o)^*$ denotes the complex conjugate of the transpose of z^o and C is the centring matrix of (2.3). Hence the complex pre-shape z is obtained by dividing the Helmertized landmarks by the centroid size,

$$z = z_H / S(z^o), \quad z \in S_2^k.$$

We see that the pre-shape space S_2^k is the complex sphere in $k - 1$ complex dimensions:

$$\mathbb{C}S^{k-2} = \{z : z^* z = 1, z \in \mathbb{C}^{k-1}\}, \qquad (4.17)$$

which is the same as the real sphere of unit radius in $2k - 2$ real dimensions, S^{2k-3}. In order to remove rotation we identify all rotated versions of z with each other, that is the shape of z^o is:

$$[z^o] = \{z e^{i\theta} : 0 \leq \theta < 2\pi\}.$$

The complex sphere $\mathbb{C}S^{k-2}$ which has points z identified with $z e^{i\theta}$ ($0 \leq \theta < 2\pi$) is the **complex projective space** $\mathbb{C}P^{k-2}$, as we see later in Result 4.5.

Example 4.1   Consider one particular T2 mouse vertebra in Figure 4.5, which has $k = 6$ landmarks in $m = 2$ dimensions, and is viewed at an observed location, scale

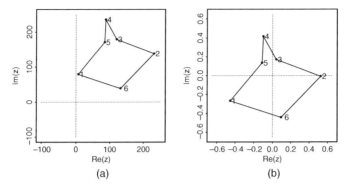

Figure 4.5 (a) *Six landmarks from a T2 mouse vertebra viewed at an observed location, scale and rotation.* (b) *An icon for the T2 vertebra, which is the centred pre-shape [with the same rotation as in (a)].*

and rotation. The full dataset has already been described in Section 1.4.1. The original complex coordinates of this vertebra are:

$$z^o = (7 + 79i, 231 + 138i, 120 + 179i, 88 + 235i, 85 + 171i, 132 + 39i)^{\mathrm{T}}.$$

The Helmertized landmarks are:

$$z_H = Hz^o = (158.39 + 41.72i, 0.82 + 57.56i, -27.14 + 89.2i,$$
$$-23.70 + 11.85i, 23.55 - 110.82i)^{\mathrm{T}}.$$

The centroid size is $S(z^o) = 228.85$ and so the pre-shape is given by $z = z_H/228.85$. The shape of z^o is the set $\{ze^{i\theta} : 0 \le \theta < 2\pi\}$. We can plot the figure by considering a suitable icon. A choice of icon is the centred pre-shape given by:

$$z_C = H^{\mathrm{T}}z = Cz^o/\|Cz^o\|$$
$$= (-0.452 - 0.267i, 0.527 - 0.009i, 0.042 + 0.170i,$$
$$-0.098 + 0.414i, -0.111 + 0.134i, 0.093 - 0.442i)^{\mathrm{T}},$$

where C is the centring matrix of Equation (2.3). In Figure 4.5 we see a plot of the centred pre-shape icon, with its centroid at the origin and with unit size – the rotation is unchanged from the original configuration z^o. □

4.3.2 Complex projective space

Kendall (1984) has shown that the shape space Σ_2^k with Riemannian metric ρ can be regarded as $\mathbb{C}P^{k-2}(4)$, the complex projective space with maximal sectional curvature 4, and $\rho(X_1, X_2)$ is the Fubini–Study metric on $\mathbb{C}P^{k-2}(4)$ (see Kobayashi and Nomizu, 1969, Chapter IX).

Result 4.5   *In the 2D case the shape space is:*

$$\Sigma_2^k = S_2^k / SO(2) = \mathbb{C}P^{k-2}(4),$$

the complex projective space with maximal sectional curvature 4.

Proof: Let the original landmarks be $z^o = (z_1^o, \ldots, z_k^o)^{\mathrm{T}}$. Consider the Helmertized complex landmarks,

$$z_H = H z^o = (z_1, \ldots, z_{k-1})^{\mathrm{T}} \in \mathbb{C}^{k-1} \setminus \{0\} \tag{4.18}$$

(not allowing complete coincidence of landmarks). Multiplying z_H by some non-zero complex number $\lambda = \beta e^{i\theta}$, $\beta \in \mathbb{R}^+$, $\theta \in [0, 2\pi)$, is the same as rescaling and rotating z_H, and so λz_H has the same shape as z_H. Thus, the shape of z^o is represented by the set:

$$\{\lambda z_H : \lambda \in \mathbb{C}, z_H \in \mathbb{C}^{k-1} \setminus \{0\}\}, \tag{4.19}$$

a complex line through the origin (but not including it) in $k - 1$ dimensions. The union of all such sets is the complex projective space $\mathbb{C}P^{k-2}$ and the maximal sectional curvature is 4 to make sure the Riemannian distance $\rho(X_1, X_2)$ is the same in both pre-shape and shape space. Full details are given by Kendall (1984) and Stoyan *et al.* (1995). $\square$

From Equation (4.13) the Riemannian distance (and hence the partial and full Procrustes distance) is particularly simple to calculate for $m = 2$ dimensional data when we can use complex notation.

Consider the centred, unit size configurations (centred pre-shapes) $y = (y_1, \ldots, y_k)^{\mathrm{T}}$ and $w = (w_1, \ldots, w_k)^{\mathrm{T}}$, where $\|y\| = 1 = \|w\|$ and $y^* 1_k = 0 = w^* 1_k$. It follows from Equation (4.13) that

$$d_F^2 = 1 - \left| \sum_{j=1}^{k} y_j \bar{w}_j \right|^2, \tag{4.20}$$

where d_F is the full Procrustes distance between the two shapes. Hence, we can regard

$$\cos \rho = \left(1 - d_F^2\right)^{1/2}$$

as the modulus of the complex correlation between the y_j and w_j.

For planar data the pre-shape space is a complex sphere $\mathbb{C}S^{k-2}$ of unit radius in $k - 1$ complex dimensions, defined in Equation (4.17). The angle between the complex pre-shapes y and w is:

$$\rho = \arccos(|y^* w|). \tag{4.21}$$

This quantity is unaffected by rotations of y and w. Hence, we can explicitly see that the Riemannian distance ρ is the angle between complex pre-shapes y and w. Also, since the radius of the pre-shape sphere is 1 we can consider ρ to be the great circle distance on the pre-shape sphere, using simple geometry.

The squared chordal distance between pre-shapes y and w is:

$$d_P^2 = 2(1 - |y^*w|) = 2(1 - \cos\rho), \qquad (4.22)$$

and d_P is also invariant under rotations of y and w. The partial Procrustes distance d_P can be regarded as the chordal distance between the complex pre-shapes y and w.

Important point: We have a clear analogy with directional data analysis (Mardia and Jupp, 2000). The angle between the directions l and v on the real sphere is given by $\theta = \arccos(l^T v)$, and the chordal distance between l and v is $2(1 - l^T v) = 2(1 - \cos\theta)$. Hence, from Equation (4.21) and Equation (4.22) the rôle of the cosine of Riemannian distance $\cos\rho = |y^*w|$ is similar to that of

$$|\cos\theta| = |l^T v|$$

in directional data analysis, where $|\cos\theta|$ is useful when axial or bipolar data are available. Hence, one can see that the complex pre-shapes y and w are closely analogous to the axes $\pm l$ and $\pm v$ on the real sphere. As we shall see, techniques in shape analysis using the pre-shape sphere often have similarities with techniques in directional data analysis.

4.3.3   Kent's polar pre-shape coordinates

Kent (1994) proposed some non-standard polar coordinates on the pre-shape sphere for 2D data. Given a point $(z_1, \ldots, z_{k-1})^T$ on $\mathbb{C}S^{k-2}$ we transform to $(s_1, \ldots, s_{k-2}, \theta_1, \ldots, \theta_{k-1})$, where

$$\mathrm{Re}(z_j) = s_j^{1/2}\cos\theta_j, \quad \mathrm{Im}(z_j) = s_j^{1/2}\sin\theta_j, \qquad (4.23)$$

for $j = 1, \ldots, k-1$, $s_j \geq 0$, $0 \leq \theta_j < 2\pi$ and $s_{k-1} = 1 - s_1 - \ldots - s_{k-2}$. A description of this and other polar coordinate systems was given by Shelupsky (1962). The coordinates $s_1, \ldots, s_{k-2}$ are on the $k-2$ dimensional unit simplex, S_{k-2}. By identifying the complex pre-shape sphere with $S_{k-2} \times [0, 2\pi)^{k-1}$ we have the volume measure of $\mathbb{C}S^{k-2}$ as:

$$2^{2-k}\mathrm{d}s_1 \ldots \mathrm{d}s_{k-2}\mathrm{d}\theta_1 \ldots \mathrm{d}\theta_{k-1}. \qquad (4.24)$$

The total volume is

$$\frac{2\pi^{k-1}}{(k-2)!}$$

since the volume of the j-dimensional simplex is $1/j!$, $j = 1, 2, 3, \ldots$.

Important point: This set of coordinates has the advantage that the uniform density on the pre-shape sphere is uniform in these coordinates (see Section 10.1). Hence, the coordinates are particularly useful for distributional results.

Shape coordinates can be obtained by rotating z to a fixed axis. For example, considering the rotation information of the original figure to be in θ_{k-1}, then the $2k - 4$ shape coordinates are:

$$(s_1, \ldots, s_{k-2}, \phi_1, \ldots, \phi_{k-2}), \tag{4.25}$$

where $\phi_j = \theta_j - \theta_{k-1}$, $j = 1, \ldots, k - 2$. So the volume measure on the shape space $\Sigma_2^k = \mathbb{C}P^{k-2}(4)$ is:

$$2^{2-k}ds_1 \ldots ds_{k-2}d\phi_1 \ldots d\phi_{k-2}, \tag{4.26}$$

and the total volume is:

$$\frac{\pi^{k-2}}{(k-2)!}.$$

4.3.4   Triangle case

Working with shape spaces for triangles in two dimensions is particularly simple.

Result (Kendall 1983)   *The shape space for triangles in a plane is:*

$$\Sigma_2^3 = \mathbb{C}P^1(4) = S^2\left(\tfrac{1}{2}\right),$$

the sphere in three dimensions with radius $\tfrac{1}{2}$.

The proof follows from the result in differential geometry that the complex projective space $\mathbb{C}P^1(4)$ can be identified with $S^2(\tfrac{1}{2})$ (Kendall 1983).

The radius is required to have length $\tfrac{1}{2}$ in order to ensure that the Procrustes (smallest great circle) distance between two fibres on the pre-shape sphere is equal to the great circle distance between corresponding shape points on the shape sphere. This important result enables us to work with the shapes of triangles as points on $S^2(\tfrac{1}{2})$. Hence, statistical shape analysis for triangles in a plane is equivalent to directional data analysis on a sphere in $\mathbb{R}^3$ with radius $\tfrac{1}{2}$. Mardia (1989b) explored this connection statistically after identification of the spherical shape space by Kendall (1983). Mardia (1989b) gave an explicit expression for the transformation from Bookstein or Kendall coordinates to the sphere [see Equation (2.15)]. For an alternative simple proof that the triangle shape space is $S^2(\tfrac{1}{2})$ see Green (1995) and Mardia (1996b).

Result 4.7   *The shape space for triangles in $m \geq 3$ dimensions is:*

$$\Sigma_m^3 = S_+^2\left(\tfrac{1}{2}\right),$$

the hemisphere in three dimensions with radius $\tfrac{1}{2}$.

Proof: If $k = 3$ and $m \geq 3$, then, since in this case the triangle shapes are invariant under reflection (see Section 3.5.2), there is a one to one correspondence between shapes in the upper and lower hemispheres of the shape space (they are reflections and hence 3D rotations of each other), and so $\Sigma_m^3 = S_+^2(\tfrac{1}{2})$, the hemisphere in three dimensions of radius $\tfrac{1}{2}$.   □

We now illustrate the various distances for the triangle case. In Figure 4.6 we have a section of the shape space sphere (with radius half) for triangles in $m = 2$ dimensions. In this case, we see that the chordal distance is given by:

$$d_F = \sin \rho, \quad 0 \leq \rho \leq \pi/2,$$

which is the full Procrustes distance and note that ρ is measured in radians. The partial Procrustes distance d_P between two shapes $[z^o]$ and $[w^o]$ on $S^2(\tfrac{1}{2})$ is the sum of the Euclidean chord lengths from each shape point to half way along the great circle route from $[z^o]$ to $[w^o]$, see Figure 4.6.

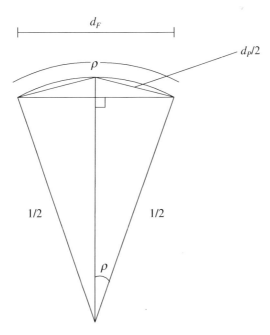

Figure 4.6   Section of the shape sphere for triangles, illustrating schematically the relationship between the Procrustes distances d_P, ρ and d_F.

Important point: Note that ρ is the great circle distance, but because the radius of the sphere is $\frac{1}{2}$ the angle between $[z^o]$ and $[w^o]$ is 2ρ.

Further discussion of coordinates for the triangle case is given in Section 2.6.2.

A measure of shape distance, such as the full Procrustes distance, gives us a numerical measure of overall shape comparison. However, a global shape measure such as shape distance does not indicate locally where the objects differ and the manner of the difference. The deformation methods of Chapter 12 are particularly useful for describing such differences.

4.4   Tangent space coordinates

4.4.1   Tangent spaces

The tangent space to shape space is the linearized version of the shape space in the vicinity of a particular point of shape space (the pole of the tangent projection). The pole is usually chosen to be an average shape obtained from the dataset of interest, and hence this choice of coordinates depends on the dataset under study. It turns out that working with the tangent space to shape space directly is complicated (Le 1991a) so instead we consider a tangent space to the pre-shape sphere, and identify the 'horizontal part' which is invariant to rotation. The Euclidean distance in the tangent space is a good approximation to the Procrustes distances d_F, d_P and ρ in shape space in the vicinity of the pole. In fact we will show that the Euclidean distance in the tangent space to the pole of the projection is exactly the same as the full Procrustes distance to the pole. So, if most of the objects in a dataset are quite close in shape, then using the Euclidean distance in the tangent space will be a good approximation to the shape distances in the shape space. Hence, for practical shape analysis we shall see that the tangent space can be extremely important and useful.

For illustration purposes we first discuss the tangent space to a real sphere, and then concentrate on tangent spaces to the pre-shape sphere which has an added layer of complication.

Example 4.2   Consider a real sphere $S^{q-1} = \{x : x^T x = 1, x \in \mathbb{R}^q\}$. Let $p \in S^{q-1}$ and $v \in T_p(S^{q-1})$. Let $\delta = \|v\| \geq 0$. The geodesic γ with $\gamma(0) = p$ and $\dot{\gamma}(0) = v$ is given by the great circle:

$$\gamma(t) = \cos(\delta t)p + \sin(\delta t)\frac{v}{\|v\|}.$$

and the exponential map is:

$$\exp_p(v) = \begin{cases} \cos(\|v\|)p + \sin(\|v\|)\frac{v}{\|v\|}, & \text{if } v \neq 0 \\ p & v = 0. \end{cases}$$

The inverse exponential map $\exp_p^{-1}$ is the mapping from the sphere to the tangent space, which satisfies $\exp_p^{-1}(\exp_p(v)) = v$. The Riemannian distance between p and $\exp_p(v)$ is the arc length of the minimal geodesic, and is given by δ.

If $q = 3$ and we use spherical coordinates (θ, ϕ) where θ is angle with the north pole and ϕ is the latitude then the Riemannian metric is:

$$ds^2 = d\phi^2 + \sin \theta^2 d\theta^2,$$

and the metric tensor is:

$$g = \begin{bmatrix} 1 & 0 \\ 0 & \sin \theta \end{bmatrix}.$$

There are other choices of tangent coordinates for the sphere, for example using an orthogonal projection. The tangent normal decomposition of x at μ is given by:

$$x = t\xi + (1 - t^2)^{1/2}\mu,$$

where $\xi^T \mu = 0$. It is clear that $(1 - t^2)^{1/2} = x^T \mu$ and so the orthogonal projection tangent coordinates for x are:

$$v_\perp = t\xi = (I_q - \mu\mu^T)x. \tag{4.27}$$

The matrix $I_q - \mu\mu^T$ is the projection matrix onto the space orthogonal to μ. See Figure 4.7 for a geometrical explanation. Note that the inverse exponential map tangent coordinates v and the orthogonal projection tangent coordinates $v_\perp$ are related by:

$$v = v_\perp \delta / \sin \delta \ \ 0 \le \delta \le \pi.$$

A disadvantage of the orthogonal projection is that points at distance further than $\pi/2$ are closer in the projection than a point at $\pi/2$.   □

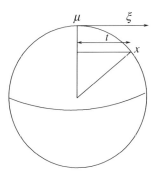

Figure 4.7   A geometrical view of the tangent coordinates on the real sphere.

4.4.2   Procrustes tangent coordinates

The tangent space to the shape space itself is complicated (Le 1991a), but it is convenient to work with tangent coordinates to the pre-shape sphere. The tangent space to the pre-shape sphere is exactly as given in Equation (4.27) with $q = (k - 1)m$. In order to use the space for shape analysis the tangent space can be decomposed into two complementary subspaces: the **horizontal** tangent space of dimension $(k - 1)m - m(m - 1)/2 - 1$ which does not depend on rotation, and the **vertical** tangent space of dimension $m(m - 1)/2$ which contains the rotation information. For shape analysis we work with coordinates in the horizontal tangent space to the pre-shape sphere.

Constructing the Procrustes tangent space involves first choosing a pole p, a $k \times m$ matrix, which is assumed to be non-degenerate (Kent and Mardia 2001) which means that the eigenvalues of $p^T p$ satisfy $\lambda_{m-1} > 0$ (i.e. the rank of p is at least $m - 1$). At degenerate p the shape space has a singularity (Kendall 1984), and so we assume that p is away from such singularities. For $m = 2$ all pre-shapes are non-degenerate (Kent and Mardia 2001). After choosing a non-degenerate pole p we carry out Procrustes rotation of X onto p (where p and X have been centred or Helmertized) to give $X^P = X\hat{\Gamma}$, where $\hat{\Gamma}$ is given in Equation (4.6). The Procrustes tangent matrix is then (Kent and Mardia 2001):

$$V = X\hat{\Gamma} - \alpha p,$$

where $\alpha = \cos \rho(X, p) > 0$, with $\rho(X, p)$ the Riemannian distance between X and p, and X and p are non-degenerate [see Equation (4.7)]. Note that we do need to restrict ourselves to $\alpha > 0$, and so the projection cannot be used for two shapes which are maximally remote at distance $\rho = \pi/2$ apart. We can write:

$$X = (\cos \rho \, p + \sin \rho V_0)\hat{\Gamma}^T.$$

where $V_0 = V/\sin \rho$, and $0 \leq \rho < \pi/2$. From Kent and Mardia (2001) the Procrustes tangent matrix satisfies the following properties:

1. V is centred $V^T 1 = 0$.

2. V is orthogonal to p: trace$(p^T V) = 0$.

3. $p^T V$ is symmetric: $p^T V = V^T p$.

4. V is bounded: $\|V\| < 1$.

5. The eigenvalues of the symmetric matrix $p^T [(1 - \|V\|^2)^{1/2} p + V]$ are optimally signed and non-degenerate, see Equation (4.4) and Equation (4.7).

Note that the number of linear constraints on V from (i)–(iii) is $m + 1 + m(m - 1)/2$. Property (iii) ensures horizontality. Property (iv) ensures that $\alpha > 0$. The Procrustes tangent coordinates are also known as Kent's partial Procrustes tangent coordinates (Kent 1995), as the matching is only done over the 'partial' transformation group of rotations and not scale.

4.4.3   Planar Procrustes tangent coordinates

We now provide explicit expressions for the Procrustes tangent coordinates for planar data. Consider complex landmarks $z^o = (z_1^o, \ldots, z_k^o)^T$ with pre-shape

$$z = (z_1, \ldots, z_{k-1})^T = Hz^o / \|Hz^o\|.$$

Let γ be a complex pole on the complex pre-shape sphere usually chosen to correspond to an average shape [e.g. corresponding to the full Procrustes mean shape from Equation (6.11)].

Let us rotate the configuration by an angle θ to be as close as possible to the pole and then project onto the **tangent plane** at γ, denoted by T_γ. Note that $\hat{\theta} = -\mathrm{Arg}(\gamma^* z)$ minimizes $\|\gamma - ze^{i\theta}\|^2$ (see Result 4.4).

Definition 4.4   *The* **Procrustes tangent coordinates** *for a planar shape are given by:*

$$v_P = e^{i\hat{\theta}}[I_{k-1} - \gamma\gamma^*]z, \ v \in T_\gamma, \quad (4.28)$$

where $\hat{\theta} = -\mathrm{Arg}(\gamma^* z)$. Procrustes tangent coordinates involve only rotation (and not scaling) to match the pre-shapes.

Note that $v_P^* \gamma = 0$ and so the complex constraint means we can regard the tangent space as a real subspace of $\mathbb{R}^{2k-2}$ of dimension $2k - 4$. The matrix $I_{k-1} - \gamma\gamma^*$ is the matrix for complex projection into the space orthogonal to γ. In Figure 4.8 we see a section of the pre-shape sphere showing the tangent plane coordinates. Note that the inverse projection from v_P to $ze^{i\hat{\theta}}$ is given by:

$$ze^{i\hat{\theta}} = [(1 - v_P^* v_P)^{1/2}\gamma + v_P], \ z \in \mathbb{C}S^{k-2}. \quad (4.29)$$

Hence an icon for Procrustes tangent coordinates is given by $X_I = H^T z$.

Example 4.3   For the T2 mouse vertebral data of Section 1.4.1 we consider in Figure 4.9 a plot of the icons for the Procrustes tangent coordinates when using a pole corresponding to the full Procrustes mean shape defined later in Equation (6.11). One must remember that the dimension is $2k - 4$, as each configuration is centred, is of unit centroid size and is rotated to be as close as possible to the pole. There is clearly non-circular variability in landmarks 1, 2 and 4. Pairwise plots of the icon coordinates and centroid size are given in Figure 4.10 and we see strong structure in the plot – there are strong correlations between some of the icon coordinates, for example x_1 and x_2 with y_4; and centroid size S with x_1, x_2, y_1, y_2 and y_4.   □

Result 4.8   *The Euclidean norm of a point v in the partial Procrustes tangent space is equal to the full Procrustes distance from the original configuration z^o corresponding to v to an icon of the pole $H^T\gamma$, that is*

$$\|v_P\| = d_F(z^o, H^T\gamma).$$

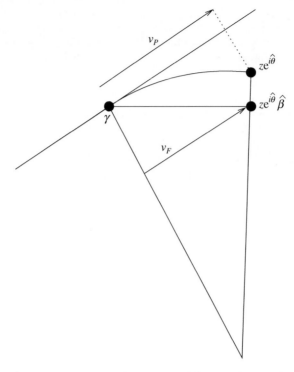

Figure 4.8   A diagrammatic view of a section of the pre-shape sphere, showing the partial tangent plane coordinates v_P and the full Procrustes tangent plane coordinates v_F discussed in Section 25.

Proof: It can be seen that

$$\|v_P\|^2 = 1 - z^* \gamma \gamma^* z = \sin^2 \rho(z^o, H^T \gamma) = d_F^2(z^o, H^T \gamma). \qquad (4.30)$$

$\square$

Important point: This result means that multivariate methods in tangent space which involve calculating distances to the pole γ will be equivalent to non-Euclidean shape methods which require the full Procrustes distance to the icon $H^T \gamma$. Also, if X_1 and X_2 are close in shape, and v_1 and v_2 are the tangent plane coordinates, then

$$\|v_1 - v_2\| \approx d_F(X_1, X_2) \approx \rho(X_1, X_2) \approx d_P(X_1, X_2). \qquad (4.31)$$

For practical purposes this means that standard multivariate statistical techniques in tangent space can be good approximations to non-Euclidean shape methods, provided the data are not too highly dispersed.

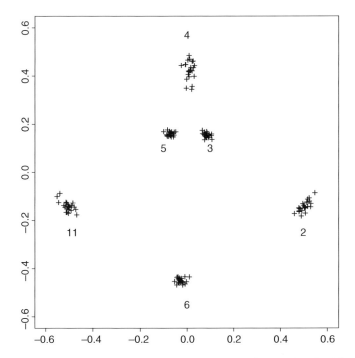

Figure 4.9   Icons for partial Procrustes tangent coordinates for the T2 vertebral data (Small group).

4.4.4   Higher dimensional Procrustes tangent coordinates

The Procrustes tangent coordinates can be extended simply into $m \geq 3$ dimensions, as proposed by Dryden and Mardia (1993). We first need to define the vectorize operator and its inverse.

Definition 4.5   *The* **vectorize** *operator* $\text{vec}(X)$ *of an* $l \times m$ *matrix* X *stacks the columns of* X *to give an lm-vector, i.e. if* X *has columns* $x_1, x_2, \ldots, x_m$, *then*

$$\text{vec}(X) = \left(x_1^{\text{T}}, x_2^{\text{T}}, \ldots, x_m^{\text{T}} \right)^{\text{T}}. \tag{4.32}$$

The inverse vectorize operator $\text{vec}_m^{-1}(\cdot)$ *is the inverse operation of* $\text{vec}(\cdot)$, *forming a matrix of m columns, that is if* $\text{vec}(Y) = X$, *then* $\text{vec}_m^{-1}(X) = Y$, *where* Y *is an* $l \times m$ *matrix.*

To obtain the tangent coordinates, a pole γ [a $(k-1) \times m$ matrix] is chosen on the pre-shape sphere; for example, this could correspond to the full Procrustes mean shape of Equation (6.11). Given a pre-shape Z this is rotated to the pole to be as close

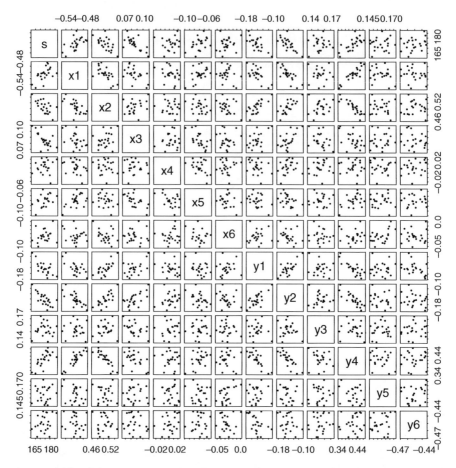

Figure 4.10   Pairwise scatter plots for centroid size (S) and the (x, y) coordinates of icons for the partial Procrustes tangent coordinates for the T2 vertebral data (Small group).

as possible, and we write the rotated pre-shape as $Z\hat{\Gamma}$, where $\hat{\Gamma} \in SO(m)$. We choose $\hat{\Gamma}$ to minimize

$$\|\gamma - Z\Gamma\|^2$$

which is obtained using Equation (4.6). Then we project onto the tangent plane at γ,

$$v_P = [I_{km-m} - \text{vec}(\gamma)\text{vec}(\gamma)^T]\text{vec}(Z\hat{\Gamma}). \tag{4.33}$$

where $\text{vec}(X)$ is the vectorize operator. The inverse transformation is given by

$$Z = \text{vec}_m^{-1}\left[\left(1 - v_P^T v_P\right)^{1/2}\text{vec}(\gamma) + v_P\right]\hat{\Gamma}^T. \tag{4.34}$$

The tangent space is an $q = (k - 1)m - m(m - 1)/2 - 1$ dimensional subspace of $\mathbb{R}^{(k-1)m}$ and the dimensionality of the tangent space can be seen from the fact there are m location constraints, one size constraint and $m(m - 1)/2$ rotational constraints, after choosing $\hat{\Gamma}$. For $m = 2$ the complex Procrustes tangent coordinates of Equation (4.28) are identical to those in Equation (4.33).

On a historical note, Goodall (1991) provided one of the earliest discussions of tangent coordinates, although he concentrated on the Procrustes residuals which are approximate tangent coordinates (see Section 4.4.6). Kent (1992) gave the first treatment of Procrustes tangent coordinates to the pre-shape sphere, which was extended to higher dimensions by Dryden and Mardia (1993).

4.4.5   Inverse exponential map tangent coordinates

In Riemannian geometry a natural projection from the metric space to a tangent space is via the inverse of the exponential map (or logarithmic map) as discussed in Section 3.1. The exponential map provides a projection from the tangent space to the Riemannian manifold which preserves the Riemannian distance from each point to the pole [see Equation (3.1)]. Let $\gamma(t)$ be the unique geodesic starting at p with initial tangent vector $v_E = \gamma'(0)$ passing through the shape corresponding to X at $\gamma(1)$. If v_P are the partial tangent space coordinates corresponding to pre-shape X and pole p then the inverse exponential map tangent coordinates of X with pole p are given by:

$$v_E = \exp_p^{-1}[\gamma(1)] = \frac{\rho}{\sin \rho} v_P. \tag{4.35}$$

These coordinates are also known as the logarithmic map coordinates. Note

$$\rho(M, X) = \|v_E\|,$$

and so the Riemannian distance ρ to the pole is preserved, whereas for partial Procrustes tangent coordinates the full Procrustes distance $d_F = \sin \rho$ to the pole is preserved. The main practical difference between the partial Procrustes and inverse exponential map tangent coordinates are that the more extreme observations are pulled in more towards the pole with the partial Procrustes tangent coordinates.

4.4.6   Procrustes residuals

We now give an approximation to the Procrustes tangent space (Goodall 1991; Cootes *et al.* 1992), which can be formulated in a straightforward manner for practitioners when carrying out Procrustes analysis (see Chapter 7). The Procrustes residuals are:

$$r = \hat{\beta} z \hat{\Gamma} - \hat{\mu},$$

where $\hat{\Gamma}$ is the optimal rotation, and $\hat{\beta}$ the optimal scale for matching z to $\hat{\mu}$, and $\hat{\mu}$ is the full Procrustes mean shape of Equation (6.11). The Procrustes residuals are only approximate tangent coordinates, as they are not orthogonal to the pole, that is trace$(r^T \hat{\mu}) \neq 0$. For $m = 2$ dimensions for pre-shape z and pole $\hat{\mu}$ the Procrustes residual is:

$$r = z(z^* \hat{\mu}) - \hat{\mu},$$

which do not satisfy $r^* \hat{\mu} = 0$ in general, but approximately hold for z close to the pole.

If a shape is close to the pole (which we will usually have in practice), then the differences between the choices of tangent coordinates will be very small.

4.4.7 Other tangent coordinates

Other choices of tangent projection can be used. For example, the full Procrustes tangent coordinates allow additional scaling to the pole and are given by:

$$v_F = (\cos \rho) v_P, \qquad\qquad (4.36)$$

where v_P are the partial tangent coordinates. However, these coordinates are only sensible to use for $0 \leq \rho \leq \pi/4$ as the distance to the pole becomes monotonic decreasing for larger ρ. Another possibility are the coordinates

$$v_S = \frac{\sin 2\rho}{\sin \rho (1 - \cos 2\rho)} v_P,$$

which for triangle shapes reduces to coordinates from a stereographic projection on $S^2(1/2)$, which is a conformal mapping which preserves angles.

Tangent spaces are particularly useful for practical data analysis, provided the data are fairly concentrated. This approach to data analysis is considered in Chapter 9 where standard multivariate methods can be carried out on the tangent plane coordinates in this linear space. As a summary the different types of tangent space coordinates are given in Table 4.2.

Table 4.2 Different types of tangent space coordinates/approximate tangent coordinates.

Notation	Name	Length	Section	R option
v_P	Partial Procrustes	$\sin \rho$	4.4.2	partial
v_E	Inverse exponential map	ρ	4.4.5	expomap
r	Procrustes residuals	$\sin \rho$	4.4.6	residual
v_F	Full Procrustes	$\cos \rho \sin \rho$	4.4.7	

4.4.8   Tangent space coordinates in R

In the shapes package in R we can choose which tangent coordinates to work with. In the function procGPA which carries out Procrustes analysis there is an option tangentcoords in which the user specifies the type of tangent coordinates and the pole is a mean shape. If the option scale=TRUE (which is used for shape analysis as opposed to size-and-shape analysis of Chapter 5) then the options are for tangent-coords are partial (Kent's partial tangent coordinates), expomap (tangent coordinates from the inverse of the exponential map) and residual (Procrustes residuals). For example the partial tangent coordinates for the small group of mouse vertebrae that are used in the example in Section 4.4.2 are obtained from:

```
ans<-procGPA(qset2.dat,tangentcoords="partial",eigen2d=TRUE)}
```

and ans$tan is a 10×23 matrix of tangent coordinates, for the $2k - 2$ dimensional tangent coordinates of the $n = 23$ mouse vertebrae. To plot the icons of the tangent coordinates:

```
x<-array(0,c(6,2,23))
H<-defh(5)
pole<-ans$mshape/centroid.size(ans$mshape)
x[,1,]<-t(H)%*%ans$tan[1:5,]+pole[,1]
x[,2,]<-t(H)%*%ans$tan[6:10,]+pole[,2]
plotshapes(x,symbol=3)
```

and the resulting plot is identical to Figure 4.9.

5

Size-and-shape space

5.1 Introduction

In the previous chapters we have primarily considered the geometric invariance of objects under translation, rotation and scale. In some applications the joint study of size and shape is appropriate, and we use the term **size-and-shape** in this case. Another common term used for size-and-shape is **form**, and this is particularly common in biological applications.

In molecule comparisons (see Sections 1.4.4 and 1.4.9) the relative scales of the molecules are known, and so we do not need scale invariance. Hence, molecules are usually compared geometrically in size-and-shape space, rather than shape space.

Another example where size-and-shape is appropriate is the analysis of the microfossil data of Section 1.4.16. It is of interest to examine whether the size is related to shape. The form of the microfossils is represented in the size-and-shape space, which is a product of the positive real line (for size) and shape space.

The definition of **size-and-shape** of an object was given in Definition 3.9. When carrying out size-and-shape analysis all objects must be recorded on the same scale, which we denote as being commensurate in scale. In many applications we do have scale information so a choice needs to be made as to whether to work in size-and-shape space, or to work in shape space and consider the size variable separately.

5.2 Root mean square deviation measures

One way of measuring size-and-shape differences between sets of points is to consider the root mean square deviation (RMSD) between configurations after Procrustes matching. Let X_1^o and X_2^o be two configurations and write $X_1 = CX_1^o$, $X_2 = CX_2^o$ for the centred configurations. If we match X_1 to X_2 by Procrustes analysis using rotation and translation only, then denote the fitted configuration as X_1^P. In particular,

---

Statistical Shape Analysis, with Applications in R, Second Edition. Ian L. Dryden and Kanti V. Mardia.
© 2016 John Wiley & Sons, Ltd. Published 2016 by John Wiley & Sons, Ltd.

$X_1^P = X_1 \hat{\Gamma}$ where $\hat{\Gamma} = UV^T$ and $X_2^T X_1 = VDU^T$ with $U, V \in SO(m)$ and D diagonal with positive elements except possibly the smallest singular value (see Lemma 4.2).

The RMSD is:

$$RMSD = \left\{ \frac{1}{k} \|X_2 - X_1^P\|^2 \right\}^{1/2} = \left\{ \frac{1}{k} \text{trace}\{(X_2 - X_1^P)^T (X_2 - X_1^P)\} \right\}^{1/2}. \quad (5.1)$$

The RMSD has been used in structural bioinformatics to measure size-and-shape difference (Levitt 1976, p. 77), and it is related to the Riemannian size-and-shape distance d_S that we define below in Equation (5.5), where

$$RMSD = \frac{1}{\sqrt{k}} d_S.$$

Also,

$$RMSD^2 = \sum_{i=1}^{k} \|x_i - \hat{\tau} - \hat{A} y_i\|^2 / k, \quad (5.2)$$

for the two configurations (m-vectors) (x_i) and (y_i) $i = 1, \ldots, k$ in m dimensions, where $\hat{A}$ is the minimizing rotation and $\hat{\tau}$ is the minimizing translation from Procrustes analysis.

There are in fact two RMSD measures used in the early bioinformatics literature for comparing size-and-shape (Mardia 2010). As well as the coordinate-based $RMSD$ another measure, denoted by $RMSD_D$, is obtained using the distances between pairs of atoms $\delta_{ij}^{(1)} = \|x_i - x_j\|$, $\delta_{ij}^{(2)} = \|y_i - y_j\|$, and the measure is given by:

$$RMSD_D^2 = \sum_{i=1}^{k} \sum_{j=1}^{k} (\delta_{ij}^{(1)} - \delta_{ij}^{(2)})^2 / k^2.$$

It has been found from empirical studies by Cohen and Sternberg (1980) that

$$RMSD_D \simeq 0.75 RMSD + c, \quad (5.3)$$

where c ranges from $(0, 0.2)$. A similar relation can be obtained under an independent isotropic Gaussian model for the points (Mardia 2010) where

$$E(RMSD_D^2) / E(RMSD^2) = 2/3,$$

$$E(RMSD_D) / E(RMSD) \simeq \sqrt{\frac{2}{3}} \approx 0.82. \quad (5.4)$$

The result was suggested in Levitt (1976, p. 77) and it is similar to Equation (5.3). Note that $RMSD$ is now used exclusively in bioinformatics rather than $RMSD_D$, partly for its computational efficiency.

5.3   Geometry

Consider two k-point configurations in m dimensions, $X_1^o, X_2^o \in \mathbb{R}^{km}(k \times m)$ matrices.

Result 5.1   *The Riemannian distance in size-and-shape space is given by:*

$$d_S(X_1^o, X_2^o) = \sqrt{S_1^2 + S_2^2 - 2S_1 S_2 \cos \rho(X_1^o, X_2^o)}, \qquad (5.5)$$

where S_1, S_2 are the centroid sizes of X_1^o, X_2^o, and ρ is the Riemannian shape distance of Equation (4.12).

Proof: We initially remove location by pre-multiplying by the Helmert submatrix to give the Helmertized coordinates, $X_1 = HX_1^o$, $X_2 = HX_2^o$ as in Equation (3.7). We have sizes $S_1 = \|X_1\| = \|CX_1^o\|$ and $S_2 = \|X_2\| = \|CX_2^o\|$. The Riemannian distance between the size-and-shape of the configurations is found by minimizing the Euclidean distance over rotations (Le 1988, 1995; Ziezold 1994), that is

$$
\begin{aligned}
d_S^2(X_1^o, X_2^o) &= \inf_{\Gamma \in SO(m)} \|X_2 - X_1 \Gamma\|^2 \\
&= \text{trace}(X_1^T X_1) + \text{trace}(X_2^T X_2) - 2 \sup_{\Gamma \in SO(m)} \text{trace}\left(X_2^T X_1 \Gamma\right) \\
&= S_1^2 + S_2^2 - 2S_1 S_2 \sup_{\Gamma \in SO(m)} \text{trace}\left(\frac{X_2^T}{S_2} \frac{X_1 \Gamma}{S_1}\right).
\end{aligned}
\qquad (5.6)
$$

We have already seen in Equation (4.2) and Equation (4.12) that

$$\sup_{\Gamma \in SO(m)} \text{trace}\left(\frac{X_2^T}{S_2} \frac{X_1 \Gamma}{S_1}\right) = \cos \rho(X_1, X_2),$$

where $\rho(X_1, X_2)$ is Riemannian distance in shape space, hence the result follows. $\square$

The Riemannian distance in size-and-shape space $S\Sigma_m^k$ was derived by Le (1988, 1995), and it is also known as the Procrustes distance or intrinsic size-and-shape distance. The Riemannian size-and-shape distance can be computed in R using the function `ssriemdist`. For example, from

```
ssriemdist(macf.dat[,,1],macf.dat[,,2])
[1] 13.4061
ssriemdist(macf.dat[,,1],macm.dat[,,1])
[1] 18.29003
```

we see that the first and second Macaque female skulls are closer in size-and-shape distance than the first female and first male skulls.

We can also consider classical MDS of the schizophrenia data of Section 1.4.5 using size-and-shape distance:

```
data(schizophrenia)
distmat <- matrix( 0, 28, 28)
for (i in 1:28){
for (j in 1:28){
rho <- ssriemdist( schizophrenia$x[,,i],schizophrenia$x[,,j] )
distmat[i,j] <- rho
}
}
par(mfrow=c(1,2))
plot(cmdscale(distmat),pch=as.character(schizophrenia$group),
                        xlab="MDS1",ylab="MDS2")
plot(schizophrenia$group,centroid.size(schizophrenia$x))
```

From Figure 5.1 we see that there is a clear difference between the two groups, with the control group being larger in centroid size. Note the difference in the multi-dimensional scaling (MDS) plot compared with Figure 4.3 where shape distances were used (with scale invariance).

Instead of the Helmertized coordinates we could work with the centred configurations $X_1 = CX_1^o = H^T H X_1^o$, $X_2 = CX_2^o = H^T H X_2^o$. Note the similarity between Equation (5.5) and the law of cosines. In the $m = 2$ case complex notation can be

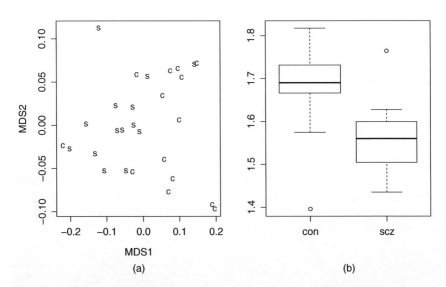

Figure 5.1   (a) Multidimensional scaling plot using Riemannian size-and-shape distance and (b) boxplots of centroid size for controls and schizophrenia patients.

used, and if Z_1 and Z_2 are the pre-shapes, then we see that ρ can be regarded as the angle between the complex vectors $S_1 Z_1$ and $S_2 Z_2$. Equation (5.6) is the sum of squares from ordinary partial Procrustes matching, as seen later in Section 7.2.3 and Equation (7.8).

The differential geometric properties of the space are as in the following (Le 1988). The size-and-shape space is a cone with a warped-product metric. The vertex of the cone corresponds to complete coincidence of the points and the unit section (corresponding to unit centroid size) is the shape space Σ_m^k. The singularities in the space are the vertex of the cone and all the points of the rays which meet the unit section at a singularity of Σ_m^k. As in the pure shape case the singularity set has Lebesgue measure zero if dealing with continuous distributions, and so can easily be dealt with in practical applications.

On a historical note Herbert Ziezold presented work in 1974 at the European Meeting of Statisticians which also contained metrics for comparing the size-and-shape of point configurations based on Fréchet (1948) means, and the proceedings were published in 1977. The size-and-shape distance of Equation (5.5) was studied by Ziezold (1977), who considered the m-dimensional case and also studied the 2D case in detail.

5.4   Tangent coordinates for size-and-shape space

Let X and p be Helmertized or centred versions of the configurations X^o, p^o. The Procrustes tangent coordinates of X in the tangent space to the size-and-shape space at pole p are given by:

$$V = X\hat{\Gamma} - p, \tag{5.7}$$

where $\hat{\Gamma}$ is the Procrustes rotation of X onto p of Equation (4.6). Note that $\|V\| = d_S(X^o, p^o)$, and these tangent coordinates are also the inverse exponential map coordinates and also the Procrustes residual coordinates. So, for size-and-shape space this is a very natural choice of tangent coordinates.

Note that V satisfies the following conditions:

1. V is centred $V^T 1 = 0$.

2. $p^T V$ is symmetric: $p^T V = V^T p$.

3. The eigenvalues of the symmetric matrix $p^T((1 - \|V\|^2)^{1/2} p + V)$ are optimally signed and non-degenerate, see Equation (4.4) and Equation (4.7).

The number of linear constraints on V is $m + m(m - 1)/2$. The Procrustes size-and-shape tangent coordinates are available in the procGPA routine in the R shapes library, by using the option scale=FALSE. For example,

```
V <- procGPA( digit3.dat , scale=FALSE )$tan
```

gives the Procrustes size-and-shape tangent coordinates for the digit 3 data of Section 1.4.2 with the pole as the sample Procrustes size-and-shape mean (see Section 6.7).

5.5   Geodesics

Minimal geodesics in size-and-shape space have a simple form. In particular, consider the minimal geodesic which passes through the size-and-shapes of Helmertized/centred configurations X and p corresponding to original configurations X^o and p^o. This geodesic passes through p in the direction of $W = V/\|V\|$ and is given by:

$$p + sW, \quad 0 \le s \le \|V\|,$$

where $V = X\hat{\Gamma} - p$ are size-and-shape tangent coordinates of X with pole p given by Equation (5.7). The length of the minimal geodesic between X and p is $\|V\| = d_S(X^o, p^o)$. Geodesics can be useful, although in practice more complicated models are often required to described changes on manifolds.

5.6   Size-and-shape coordinates

In order to represent size-and-shape we can use any choice of shape coordinates together with any size variable, for example centroid size or its logarithm. A common approach in morphometrics is to use shape tangent coordinates jointly with log centroid size (Mitteroecker *et al.*, 2004).

5.6.1   Bookstein-type coordinates for size-and-shape analysis

Following Mardia (2009) we indicate how to construct Bookstein-type coordinates for size-and-shape (form) analysis which have been used in registering backbones in structural bioinformatics (e.g. see Killian *et al.* 2007). Let $X(k \times 3)$ be the configuration matrix with rows x_i. First note that there are six unknowns in the rotation matrix A (3 Eulerian angles) and a translation vector, but x_1 and x_2 are not enough to determine the form coordinates, since the distance between x_1 and x_2 is fixed, so we have to use x_3 for the one remaining constraint. We first use the point x_1 as the origin so that

$$y_i = x_i - x_1, \quad i = 1, \ldots, k.$$

Then use y_2 (i.e. x_2) to fix the co-latitude and longitude of the points, that is let (θ_2, ϕ_2) be the polar coordinates of y_2 (z-axis is the north pole), then

$$u_i = R(\theta_2, \phi_2)y_i, \quad i = 1, \ldots, k$$

where

$$R(\theta, \phi) = \begin{bmatrix} \cos\theta \cos\phi & \cos\theta \sin\phi & -\sin\theta \\ -\sin\phi & \cos\phi & 0 \\ \sin\theta \cos\phi & \sin\theta \sin\phi & \cos\theta \end{bmatrix}.$$

Now rotate the new coordinates around the coordinates u_3 (or x_3), that is if

$$u_3 = \begin{bmatrix} a \\ b \\ c \end{bmatrix} = \begin{bmatrix} \sin \theta_3 \sin \phi_3 \\ \sin \theta_3 \cos \phi_3 \\ \cos \theta_3 \end{bmatrix}$$

then the Bookstein type coordinates for the form in 3D (angle ϕ_3 with new x-axis) are:

$$v_i = S(\phi_3)u_i, \quad i = 1, \dots, k,$$

where

$$S(\phi) = \begin{bmatrix} 1 & 0 & 0 \\ 0 & \cos \phi & \sin \phi \\ 0 & -\sin \phi & \cos \phi \end{bmatrix}.$$

In bioinformatics, these are termed bond–angle–torsion (BAT) coordinates (e.g. see Killian *et al.* 2007) and can be given a tree structure.

For 2D, the first two new coordinates are $(0, 0)$ and $(d, 0)$ using x_1 and x_2. That is

$$u_i = S(\phi)y_i, \quad y_i = x_i - x_1, \quad i = 1, \dots, n,$$

where now $S(\phi)$ is a 2×2 rotation matrix

$$S(\phi) = \begin{bmatrix} \cos \phi & \sin \phi \\ -\sin \phi & \cos \phi \end{bmatrix}$$

such that

$$u_2^T = (|x_2 - x_1|, 0) = (|y_2|, 0).$$

5.6.2 Goodall–Mardia QR size-and-shape coordinates

Let us consider now general $m \geq 2$ dimensions for k point configurations where $k > m$. Given a $k \times m$ matrix of landmark coordinates the location information can be removed by pre-multiplying by a suitable matrix, say $X_H = HX$, where H is the Helmert submatrix, defined in Equation (2.10). A **QR decomposition** of X_H is given by:

$$X_H = T\Gamma, \quad \Gamma \in SO(m),$$

where the $(k - 1) \times m$ lower triangular matrix T contains size-and-shape coordinates. Note that T is invariant under the original location and rotation of the configuration.

The matrix T has zero entries above the leading diagonal and therefore a total of $(k-1)m - m(m-1)/2$ size-and-shape coordinates.

To obtain the shape coordinates we divide by the centroid size to give $W = T/\|T\|$ and so there are $q = (k-1)m - m(m-1)/2 - 1$ shape coordinates. Note that $\|T\| = \|X_H\|$ is the centroid size. This coordinate system was developed by Goodall and Mardia (1992, 1993).

Slight variants of this coordinate system include using a different translation matrix from H [such as B from Equation (3.14)] or dividing by a different scale such as the $(1, 1)$th element of T (a baseline size). In particular, if $k = 3$ and $m = 2$ and B is used to remove location, then the QR decomposition leads to the size-and-shape variables given by:

$$Y = BX = \begin{bmatrix} t_{11} & 0 \\ t_{21} & t_{22} \end{bmatrix} A,$$

where A is given by Equation (2.8). Removing size by dividing by t_{11} we have:

$$\begin{bmatrix} 1 & 0 \\ t_{21}/t_{11} & t_{22}/t_{11} \end{bmatrix},$$

where $u_3^B = t_{21}/t_{11}$ and $v_3^B = t_{22}/t_{11}$ are Bookstein coordinates of Equation (2.5). Hence, for $m = 2$ (and $m = 3$) the QR decomposition is closely related to Bookstein coordinates. As is the case for Bookstein coordinates, one should be careful in interpreting correlations between QR shape variables.

For the general number of $m \geq 2$ dimensions the Goodall–Mardia polar shape coordinates are obtained from the QR decomposition as in the following. If $X_H = HX = TT$ is the QR decomposition and

$$W = T/\|T\|,$$

then write:

$$\text{vec}(W) = (\cos \rho, \sin \rho \mathbf{v}^{\mathrm{T}})^{\mathrm{T}}, \quad \mathbf{v} \in S^{(k-1)m-m(m-1)/2-1}, \tag{5.8}$$

and let ϕ be the generalized spherical polar coordinates of $\mathbf{v}$. The size-and-shape of X is therefore represented by $\{\|T\|, \rho, \phi\}$ and the shape of X is represented by $\{\rho, \phi\}$.

Goodall–Mardia polar coordinates were used by Ball et al. (2008) to construct Ornstein–Uhlenbeck processes for modelling cell shape motion, where a drift was specified to a specific reference shape.

Note that this spherical representation is different from the coordinates for Kendall's shape sphere for $k = 3$, from Equation (2.14). Also, ρ is the Riemannian distance from the shape of X to the matrix μ which contains zeros apart from the $(1, 1)$th entry, which is 1.

5.7 Allometry

Definition 5.1   Allometry *involves the study of the relationships between shape and size, and in particular the manner in which shape depends on size.*

Traditional methods in allometry involve the fitting of linear or non-linear regression equations between size and/or shape measures, see, for example, Sprent (1972). The notion of allometry was introduced by Huxley (1924, 1932).

We can also investigate allometry using our geometrical framework, for example using linear regression of shape coordinates on size. A shape coordinate is chosen as a response variable and a choice of size variable is used as the explanatory variable, and we could use standard normal-based inference. Simple exploratory plots of the data are useful and the variables might need to be transformed to model a linear relationship.

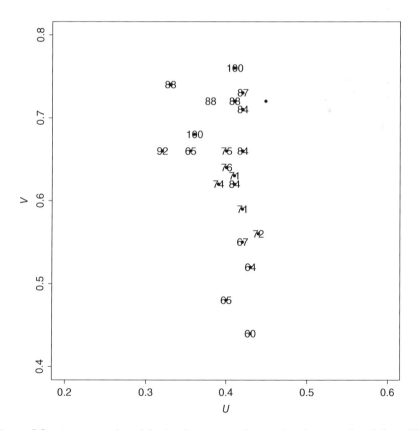

Figure 5.2   A scatter plot of the Bookstein coordinates for the microfossil data. The square root of the sample centroid size has been plotted at each point, and it is clear that size increases with V.

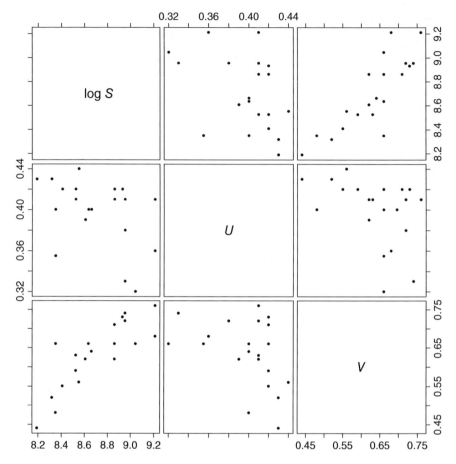

Figure 5.3   Pairwise scatter plots of log S versus the Bookstein coordinates U and V for the microfossil data.

Example 5.1   Consider the microfossil data of Section 1.4.16. We have actually added 0.5 to the variable U^B to tie in with Bookstein's original analysis of the data with the baseline sent to $(0,0)$ and $(1,0)$, rather than $(-\frac{1}{2}, 0)$ and $(\frac{1}{2}, 0)$ (Bookstein 1986). There are $n = 21$ triangles in the dataset. Scatter plots are show in Figure 5.2 and Figure 5.3. We regress $U = U^B + \frac{1}{2}$ and $V = V^B$ on $\sqrt{S}$ and then on log S. There is a slight suggestion of non-linearity in the residual plot from V on $\sqrt{S}$ and perhaps slightly less of a parabolic structure from V regressed on log S, see Figure 5.4. We fit the regression lines

$$U = \alpha_1 + \beta_1 \log S, \quad V = \alpha_2 + \beta_2 \log S$$

and the fitted values (with standard errors) are $\hat{\alpha}_1 = 0.77(0.20)$, $\hat{\beta}_1 = -0.04(0.02)$, $\hat{\alpha}_2 = -1.47(0.32)$ and $\hat{\beta}_2 = 0.24(0.04)$. Hence there is strong evidence for allometry between V and log S, as a standard test of $H_0 : \beta_2 = 0$ would be rejected, with a t

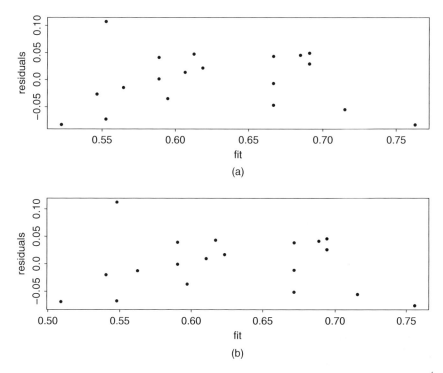

Figure 5.4   Plot of residuals versus fitted values for the regression of (a) V on $\sqrt{S}$ and (b) V on log S.

statistic of $t = 6.6$. Examination of a plot of the residuals (Figure 5.4) versus fitted values does not show much structure, although there is possibly a suggestion that one of the observations is an outlier (the observation with size 65 and $U = 0.35$, $V = 0.66$).   □

From Mitteroecker *et al.* (2004) allometry can also be investigated by appending the logarithm of centroid size to the shape tangent coordinates, and principal components analysis then provides an estimate of the allometric equation from the coefficients of the first principal component. This approach is commonly used in geometric morphometrics. Another example investigating allometry is given in Section 9.5, and further discussion of allometry is given by Klingenberg (1996); Bookstein (2013a); and Cardini *et al.* (2015) among many others.

6

Manifold means

6.1 Intrinsic and extrinsic means

The structure of shape and size variability is the primary concern in most applications (see Bookstein, 2016). However, the concept of a mean shape or mean size-and-shape also has an underpinning role to such analysis, for example when choosing a tangent space projection. We consider a situation where a population of shapes or size-and-shapes can be modelled by a probability distribution. Early definitions of a mean in a general manifold were given by Fréchet (1948) and Grove and Karcher (1973).

Definition 6.1 *If d is a choice of distance in a manifold M then a* **population mean** *is given by:*

$$\mu_M = \underset{\mu}{\text{arginf}} \ E_X[d^2(X, \mu)], \tag{6.1}$$

where $E_X[\cdot]$ is the expectation operator with respect to the random quantity $X \in M$.

If μ_M is a global minimum then it has been called a 'Fréchet mean' in the literature, and the term 'Karcher mean' has been used when it is a local minimum, after the work of Grove and Karcher (1973) and Karcher (1977) on Riemannian centres of mass. Although the terminology is not universally applied, we shall use these terms in order to help distinguish global and local minimizers. A detailed history is given by Afsari (2011).

Definition 6.2 *A* **population variance** *is:*

$$\sigma_M^2 = \underset{\mu}{\inf} \ E_X[d^2(X, \mu)], \tag{6.2}$$

and if it is a global minimum then σ_M^2 is called the Fréchet variance.

Statistical Shape Analysis, with Applications in R, Second Edition. Ian L. Dryden and Kanti V. Mardia.
© 2016 John Wiley & Sons, Ltd. Published 2016 by John Wiley & Sons, Ltd.

In the case of a Euclidean space the population mean is simply

$$\mu_M = E_X[X]$$

and the population variance is:

$$\sigma_M^2 = E_X \left(\|X - \mu_M\|^2 \right).$$

However, for more general manifolds we have many choices. The first distinction to make is that of an **intrinsic mean** versus an **extrinsic mean**, which are defined using either an intrinsic or extrinsic distance, respectively. An intrinsic distance is the length of the shortest geodesic path in the manifold; an extrinsic distance is induced by a Euclidean distance in an embedding of the manifold; and another type of distance is a residual distance in a tangent space (Huckemann 2012).

6.2 Population mean shapes

Definition 6.3 *If $d = \rho$ is an intrinsic distance in the shape space (e.g. the Riemannian metric) then the population mean is called the* **intrinsic mean shape** *and is denoted by:*

$$\mu_I = \operatorname*{arginf}_{\mu} E_X[\rho^2(X, \mu)]. \tag{6.3}$$

Intrinsic means are unique provided the distribution is sufficiently concentrated, (Karcher 1977; Kendall 1990b; Le 2001; Afsari 2011). Further discussion is given in Section 13.2.

Definition 6.4 *The* **extrinsic mean shape** *is given by $P(\mu_E)$ where*

$$\mu_E = \operatorname*{arginf}_{\mu} E_X[\rho_J(X, \mu)^2] = \operatorname*{arginf}_{\mu} E_X[\|j(X) - \mu\|^2] \tag{6.4}$$

and $j(X)$ is an embedding into a Euclidean space, $P(\mu)$ is a unique projection from the embedding space to M, and ρ_J is an extrinsic distance. By 'unique' we mean that the projection maps to a unique closest point in M.

There are many choices of Euclidean embedding for an extrinsic mean and Bhattacharya and Patrangenaru (2003, 2005) focus particularly on the Veronese–Whitney embedding for shape in two dimensions and reflection shape in three dimensions, although any embedding can be used to find an extrinsic distance. For a complex pre-shape $(k - 1)$-vector z, the Veronese–Whitney embedding is $j(z) = zz^*$, and the Euclidean distance using this embedding is the full Procrustes distance for 2D shapes. For a real pre-shape $(k - 1) \times m$ matrix Z the Veronese–Whitney embedding is $j(Z) = ZZ^T$, which gives rise to an embedding used for computing a MDS mean for reflection shape (see Section 15.3). Kent (1992) provides a general family of equivariant embeddings for 2D shapes which give rise to the extrinsic distances:

$$\rho_J = (1 - \cos^{2h} \rho)^{1/2} / \sqrt{h}, \tag{6.5}$$

where h is a positive integer.

Definition 6.5 *The* **population partial Procrustes mean shape** *is given by:*

$$\mu_P = \operatorname*{arginf}_{\mu} E_X[d_P^2(X, \mu)]. \tag{6.6}$$

Huckemann (2012) uses the term 'Ziezold mean' for the mean which involves minimizing the expected square Euclidean distance on a pre-shape sphere, after Ziezold (1977). The partial Procrustes mean is a Ziezold mean.

Definition 6.6 *The* **population full Procrustes mean shape** *is given by:*

$$\mu_F = \operatorname*{arginf}_{\mu} E_X[d_F^2(X, \mu)]. \tag{6.7}$$

The full Procrustes mean is an extrinsic mean for planar landmark data using the complex Veronese–Whitney embedding on pre-shape space. Huckemann (2012) uses the term 'residual mean' for the mean which minimizes the expected square Euclidean distance in the tangent space, and the full Procrustes mean is also a residual mean.

We can also define the population mean in the size-and-shape space, as seen in Section 6.7.

6.3 Sample mean shape

Consider a random sample of data available $X_1, \ldots, X_n \in M$, where M is a manifold.

Definition 6.7 *A* **sample mean shape** *is:*

$$\hat{\mu}_M = \arg\inf_{\mu} \frac{1}{n} \sum_{i=1}^{n} d^2(X_i, \mu). \tag{6.8}$$

If the the minimizer is a global minimum then the term **sample Fréchet mean shape** *is used, and if a local minimizer is available then we use the term* **sample Karcher mean shape***.*

The minimum of the objective function is a sample variance, as we include the divisor n in (6.8).

Also, we can have intrinsic sample means or extrinsic sample means depending on the choice of distance, in a similar manner as for population mean shapes.

Definition 6.8   *The* **sample intrinsic mean shape** *is given by:*

$$\hat{\mu}_I = \arg\inf_{\mu} \frac{1}{n} \sum_{i=1}^{n} \rho^2(X_i, \mu), \tag{6.9}$$

Definition 6.9   *The* **sample partial Procrustes mean shape** *is given by:*

$$\hat{\mu}_P = \arg\inf_{\mu} \frac{1}{n} \sum_{i=1}^{n} d_P^2(X_i, \mu), \tag{6.10}$$

Definition 6.10   *The* **sample full Procrustes mean shape** *is given by:*

$$\hat{\mu}_F = \arg\inf_{\mu} \frac{1}{n} \sum_{i=1}^{n} d_F^2(X_i, \mu), \tag{6.11}$$

Different types of distance give rise to different means, as in Table 6.1. In some situations it is numerically much more difficult to work with the intrinsic means compared with the extrinsic means, but there are also many situations where the computational differences are not so large. We will demonstrate in some examples that the means are often very similar, although they can differ for example when outliers are present.

Example 6.1   We begin with an illustrative example which highlights the intrinsic and extrinsic sample means in the very simple case of a circle S^1 (the shape space of $k = 3$ points in $m = 1$ dimension). Consider angles $\theta_1, \dots, \theta_n$ on a circle. The intrinsic distance between θ_1, θ_2 is:

$$d_C(\theta_1, \theta_2) = \min\{|\theta_1 - \theta_2|, |\theta_1 - \theta_2 + 2\pi|, |\theta_1 - \theta_2 - 2\pi|\}.$$

Table 6.1   Some shape space distances. Different types of means are obtained with different choices of distance.

Name	Distance	Notation for mean	Type of mean
Riemannian	ρ	$\hat{\mu}_I$	Intrinsic
Partial Procrustes	$2\sin(\rho/2)$	$\hat{\mu}_P$	Extrinsic/Ziezold
Full Procrustes	$\sin\rho$	$\hat{\mu}_F$	Extrinsic/Residual
Kent's family	$(1 - \cos^{2h}\rho)^{1/2}/\sqrt{h}$	$\hat{\mu}_J$	Extrinsic

The intrinsic sample mean is:

$$\hat{\theta}_I = \arg\inf_{\theta} \sum_{j=1}^{n} d_C(\theta_j, \theta)^2,$$

which must be found numerically. For an extrinsic mean we embed the angles into the complex plane $z_j = \exp(i\theta_j) \in \mathbb{C}, j = 1, \ldots, n$. The extrinsic mean is given by minimizing

$$\sum_{j=1}^{n} |z_j - \mu|^2,$$

with respect to μ and hence $\hat{\mu} = \frac{1}{n} \sum_{j=1}^{n} z_j$. Finally we project back onto the circle to give the extrinsic mean as the argument of the complex mean $\hat{\mu}$:

$$\hat{\theta}_E = \text{Arg}(\hat{\mu}). \qquad \qquad \square$$

Hotz and Huckemann (2015) discuss the intrinsic sample mean on the circle in particular detail, including discussion of uniqueness and the development of asymptotic results.

6.4   Comparing mean shapes

Dryden *et al.* (2014) consider a generalization of family of distances given by Kent (1992) in Equation (6.5) to the functions

$$d_h = (1 - \cos^{2h} \rho)^{1/2} / \sqrt{h}, \qquad (6.12)$$

on the shape space Σ_m^k of k points in m dimensions, where $h \in \mathbb{R}^+$ and ρ is the Riemannian shape distance. The resulting mean shape obtained by minimizing $E_X[d_h(X, \mu)^2]$ has different robustness properties depending on the value of h. For larger h the mean is more resistant to outlier shapes and for smaller h it is less resistant to outlier shapes. The partial Procrustes mean shape is equivalent to using $h = \frac{1}{2}$ and the full Procrustes mean shape is equivalent to using $h = 1$. The intrinsic mean is very similar to using $h - \frac{1}{3}$, as we shall see in Result 6.1. Note that the intrinsic mean shape is less resistant than the partial Procrustes mean shape. If outliers are present then larger h values are preferred, e.g. $h = 1$ rather than $h = 1/2$ or $h = 1/3$.

Example 6.2   We consider an example which helps to illustrate how similar the types of mean shapes can be in practice, but it also highlights some differences between them. We calculate a variety of sample means for the male gorilla data of Section 1.4.8 which has small shape variability. There are $k = 8$ landmarks in $m = 2$ dimensions on $n = 29$ male gorillas.

We calculate 13 means with different choices of distances, including nine means indicated by values of h using the distances in Equation (6.12). In particular, we label the means as follows: 1, $h = 0.001$; 2, $h = 0.1$; 3, $h = \frac{1}{3}$; 4, intrinsic; 5, $h = \frac{1}{2}$ (partial Procrustes); 6, $h = 0.75$; 7, $h = 1$ (full Procrustes); 8, $h = 1.5$; 9, $h = 2$; 10, $h = 5$; 11, MDS; 12, MDS($\alpha = 1$); and 13, isotropic maximum likelihood estimate (MLE) (shape of the means – see Section 6.6). Here MDS refers to the mean shape from Equation (14.3) and MDS($\alpha = 1$) refers to the mean shape from Equation (14.4). Both MDS means have been reflected if necessary to minimize the Riemannian distance to the intrinsic mean shape.

The Riemannian shape distance matrix is computed between all pairs of 29 skulls and the 13 mean shapes, and classical MDS is carried out (Mardia *et al.* 1979). The first two principal coordinates are displayed in Figure 6.1a. All the mean shapes are

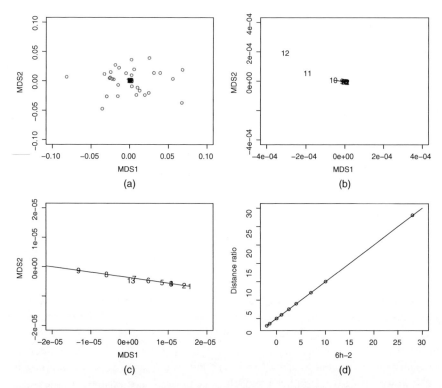

Figure 6.1   Multidimensional scaling plot of the male gorilla data and 13 means using Riemannian distance to form the dissimilarity matrix. (a) The principal coordinates of the 29 skulls (circles) and the 13 means (numbers). (b) Zoomed in version of (a) near the origin. (c) Another zoomed in version of (a) near the origin. In each of (a)–(c) the mean shapes are indicated by their number label, and the first ten means are joined by straight lines. (d) A plot of the observed ratios and expected approximate ratios of Riemannian distances for different h. For a colour version of this figure, see the colour plate section.

actually very close, as seen in Figure 6.1a. In Figure 6.1a–c the observations (circles) and the means (using their number label) are plotted and the means 1–10 are connected with piecewise straight lines. In Figure 6.1b we zoom in to see that means 10, 11 and 12 are further away from the remainder. Zooming in further in Figure 6.1c we see that means 1–10 are all close to being on a common straight line. Also, mean 13 is very close to mean 7 (but not on the line). Also, means 3 and 4 are indistinguishable even at this fine scale. Huckemann (2012) derived an intriguing result for concentrated distributions that the intrinsic, partial Procrustes (Ziezold mean) and full Procrustes estimators approximately lie on a geodesic, with the intrinsic and partial Procrustes mean closer together than the partial and full Procrustes (ratio of 1:3 approximately). In this example the ratio is 1:3.00171. □

In the following result we assume that the shape distribution is concentrated on Σ_m^k and sufficiently regular such that the objective functions for obtaining the means are approximately quadratic in the vicinity of the unique minimum.

Result 6.1 *The family of mean shapes using d_h in Equation (6.12) indexed by $h \in \mathbb{R}^+$ are approximately located on a minimal geodesic passing through the intrinsic mean μ_I. In addition we have*

$$\frac{\rho(\mu_I, \mu_h)}{\rho(\mu_I, \mu_{1/2})} \approx |6h - 2|. \tag{6.13}$$

A proof of this result can be obtained from Taylor series expansions of the objective function, and details are given in Dryden *et al.* (2014).

The above result will hold for either the population or sample mean shapes provided the shape distribution or sample data are concentrated and sufficiently regular. For $h = 1$ it follows that $R = \rho(\mu_I, \mu_1)/\rho(\mu_I, \mu_{1/2}) \approx 4$, which corresponds to Huckemann's (2012) ratio of 1:3. Also, for $h = \frac{1}{3}$ we have $R \approx 0$, and hence the intrinsic mean shape and the $h = \frac{1}{3}$ extrinsic mean are expected to be extremely close, as seen in our example. As $h \to 0$ the furthest that μ_h can be from the intrinsic mean is approximately twice the distance as that of the intrinsic mean to the partial Procrustes mean in the opposite direction along the minimal geodesic. Finally note that if a single outlier is present in the data then the less resistant means are pulled more towards the outlier, and hence in this case the outlier determines the particular geodesic on which the means approximately lie.

In Figure 6.1d the value of the empirical ratio $\hat{R}$ obtained from the sample means in the male gorilla data is plotted versus the theoretical approximate ratio $R = 6h - 2$ for each of the 9 values of h. Note that if $h < \frac{1}{3}$ we have also negated the empirical ratio. The agreement is extremely close in this example, and indeed it is very close in many other examples, although for less concentrated data there are more differences for larger values of h.

6.5   Calculation of mean shapes in R

To illustrate some commands in R for calculating mean shapes, we calculate a variety of sample means for the digit 3 data of Section 1.4.2 which has moderate variability, and the first observation might be considered an outlier as it is over 0.7 in terms of Riemannian distance from the mean. We calculate 10 means (1, $h = 0.001$; 2, intrinsic; 3, partial Procrustes; 4, isotropic MLE (see Section 11.3); 5, full Procrustes; 6, $h = 2$; 7, $h = 3$; 8, $h = 5$; 9, $h = 10$; 10, $h = 20$), which are successively more resistant to outliers. The full Procrustes mean shape is given in Figure 7.15.

A command used in R to calculate each mean is `frechet` which involves numerical optimization with the non-linear minimization function `nlm`.

```
frechet(digit3.dat,mean="intrinsic")
frechet(digit3.dat,mean="partial.procrustes")
frechet(digit3.dat,mean="mle")
frechet(digit3.dat,mean="full.procrustes")
frechet(digit3.dat,mean=2)
```

This calculates the mean shapes labelled 2, 3, 4, 5, 6 respectively. Some of the options only work for $m = 2$ dimensions, for example `mle`. A more reliable way of computing the Procrustes mean shapes is with the generalized Procrustes algorithm of Chapter 7. So, for example, the full Procrustes mean shape of the digit 3 data is given by:

```
procGPA(digit3.dat)$mshape
```

For the digit 3 example a distance matrix is computed between all 30 observations and the 10 mean shapes, and classical MDS is carried out (using `cmdscale` in R) and displayed in Figure 6.2. All the mean shapes are actually very close. However, on a fine scale we see that the less resistant estimators are pulled towards the outlier, and again the intrinsic, partial Procrustes and full Procrustes estimators lie approximately on a geodesic, with the intrinsic and partial Procrustes mean closer together than the partial and full Procrustes (ratio of 1:3 approximately). Again this provides another demonstration of Result 6.1.

As another example of how similar the mean shapes can be we estimate the intrinsic, partial and full Procrustes mean shapes for the Small group of mouse vertebrae in Section 1.4.1. Given that the estimates are invariant to rotation and scale we standardize them to Bookstein shape variables:

```
> f1<-frechet(qset2.dat,mean="intrinsic")
> f2<-frechet(qset2.dat,mean="partial.procrustes")
> f3<-frechet(qset2.dat,mean="full.procrustes")
> bookstein.shpv(f1$mshape)
          [,1]   [,2]
[1,] -0.50000000 0.0000000
[2,]  0.50000000 0.0000000
```

```
[3,]  0.08490820  0.2924684
[4,]  0.01245608  0.5589496
[5,] -0.06869796  0.2982314
[6,] -0.02512807 -0.3044915
> bookstein.shpv(f2$mshape)
       [,1]      [,2]
[1,] -0.50000000  0.0000000
[2,]  0.50000000  0.0000000
[3,]  0.08490589  0.2924798
[4,]  0.01245726  0.5589728
[5,] -0.06869559  0.2982405
[6,] -0.02513298 -0.3044881
> bookstein.shpv(f3$mshape)
       [,1]      [,2]
[1,] -0.50000000  0.0000000
[2,]  0.50000000  0.0000000
[3,]  0.08489897  0.2925142
[4,]  0.01246093  0.5590424
[5,] -0.06868845  0.2982678
[6,] -0.02514771 -0.3044776
```

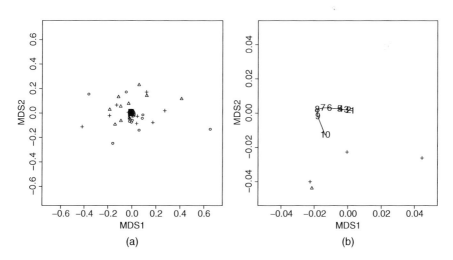

Figure 6.2   (a) Multidimensional scaling plot of the digit 3 data and ten means using Riemannian distance to form the dissimilarity matrix. The ten sample mean shapes are: 1, h = 0.001; 2, intrinsic; 3, partial Procrustes; 4, isotropic MLE; 5, full Procrustes; 6, h − 2, 7, h = 3; 8, h = 5; 9, h = 10; and 10, h = 20. (b) A zoomed in view showing the ten means and lines joining them in order of resistance to outlier shapes, with 10 being the most resistant.

These mean shape estimates are all extremely similar; the same to four decimal places using Bookstein shape variables. The Fréchet variances are all very similar:

```
> f1$var
[1] 0.09448654
> f2$var
[1] 0.09442313
> f3$var
[1] 0.09423311
```

Also, the Riemannian distances between the mean shape estimates are very close:

```
> riemdist(f1$mshape,f2$mshape)
[1] 1.737509e-05
> riemdist(f1$mshape,f3$mshape)
[1] 6.962873e-05
> riemdist(f2$mshape,f3$mshape)
[1] 5.225367e-05
```

and the ratio of Riemannian distances from the intrinsic to partial Procrustes to full Procrustes means is again approximately 1:3 here (Huckemann 2012).

Although for many practical examples there is very little difference between the estimates, in certain pathological examples they can be very different indeed (Huckemann 2012), and in particular the intrinsic and partial Procrustes means are manifold stable, whereas the full Procrustes mean may not be. Manifold stable means the following: given a distribution on the manifold part of the space, the mean also lies in the manifold part of the space with probability 1, that is there is zero probability of the mean being in the non-manifold strata (Huckemann 2012).

There are many other mathematical statistical aspects of mean shapes and estimation on manifolds that have been explored in the last couple of decades. Two papers by Bhattacharya and Patrangenaru (2003, 2005) lay out the framework for consistency for intrinsic and extrinsic mean estimators, and provide central limit theorems for extrinsic means in particular detail. Huckemann and Hotz (2014) discuss means and asymptotic results for circles and shape spaces. In particular they consider strata of shape spaces and demonstrate that means at singularities must be controlled. More discussion of these theoretical issues will be given in Chapter 13.

6.6   Shape of the means

A different notion of mean shape is where a probability distribution for the original configuration is given with, say, mean $E[X] = \mu$. The population mean shape is then given by the shape of μ, written as $[\mu]$. This type of shape has been called the **shape of the means** (Kendall *et al.* 1999, Chapter 9; Le and Kume 2000b). This type of mean shape was considered by Mardia and Dryden (1989a,b), where independent isotropic

Gaussian perturbations of planar landmarks about a mean configuration were considered. Further discussion is given in Section 11.1.2. Also, the extension to non-isotropic errors was considered by Dryden (1989) and Dryden and Mardia (1991b) (see Section 11.2), and to higher dimensions by Goodall and Mardia (1992, 1993) (see Section 11.6).

In general the population Fréchet mean and the shape of the means are different, although there are some special cases where they coincide. For example, for independent isotropic Gaussian errors for planar landmark data Kent and Mardia (1997) show that the shape of the means is the same as the population full Procrustes mean.

6.7   Means in size-and-shape space

6.7.1   Fréchet and Karcher means

Definitions of Fréchet and Karcher means for size-and-shape follow in exactly the same manner as in the shape space, except that the size-and-shape distance d_S is used in place of a shape distance. From Chapter 5 for the size-and-shape space the Riemannian size-and-shape distance coincides with the Procrustes distance and the residual distance, where optimization is carried out over location and rotation only in calculating the distance. So, for size-and-shape the population intrinsic Fréchet mean is identical to the population extrinsic Procrustes/Ziezold mean which is the same as the extrinsic residual mean.

Definition 6.11   *The* **population Fréchet mean size-and-shape** *is:*

$$\mu_S = \underset{\mu}{\arg\inf}\ E[d_S^2(Y, \mu)], \tag{6.14}$$

which is also known as the population Procrustes mean size-and-shape. Here the optimum is global.

Again, the term Karcher mean size-and-shape is used for a local minimum.

Definition 6.12   *The* **sample Fréchet mean size-and-shape** *from a random sample* $X_1, \ldots, X_n$ *is:*

$$\hat{\mu}_S = \arg\inf_{\mu}\ \frac{1}{n} \sum_{i=1}^{n} d_S^2(X_i, \mu), \tag{6.15}$$

which is also known as the sample Procrustes mean size-and-shape if it is a global optimum.

The term Karcher mean size-and-shape is used for a local minimum. The sample Fréchet or Karcher mean size-and-shape is an intrinsic, extrinsic and residual mean size-and-shape.

6.7.2   Size-and-shape of the means

The **size-and-shape of the means** is another type of mean and this is different from the Procrustes mean in general. In this situation a model for the configuration X is proposed with mean $E[X] = \mu$, and then the size-and-shape of the means is the size-and-shape of μ written as $[\mu]_S$, using the notation of Section 3.3. See Section 11.5 for further details.

6.8   Principal geodesic mean

An alternative measure of mean shape is where a geodesic or curve is first fitted to a dataset, and then a mean is estimated to minimize the sum of square distances constrained to lie within the geodesic path. This approach was considered by Huckemann and Ziezold (2006), Huckemann *et al.* (2010) and Kenobi *et al.* (2010). The type of mean can be very different from the Fréchet and other means, for example for data widely dispersed in a 'girdle' distribution, where the data are distributed around an approximate geodesic.

A further generalization was given by Jung *et al.* (2012) who considered a 'backwards' approach in fitting successively lower dimensional shape spaces, called **principal nested shape spaces**. The method starts with fitting the highest dimensional subspace, and then fits a sequence of lower dimensional subspaces, with the final iteration involving the fitting of a **principal nested shape space mean**. This method contrasts with the more usual 'forwards' methods of starting with a mean and then fitting successively more complicated, higher dimensional structures. The principal nested shape space mean can again be very different from the usual means, including for girdle distributed datasets. See Section 13.4.5 for more details, and also see Marron *et al.* (2010) for further discussion about the advantages of the backwards approach to PCA.

6.9   Riemannian barycentres

Kendall (2015) described the use of Riemannian barycentres, which provide a straightforward calculation for computing a mean shape on Riemannian manifolds. For a sphere we use:

$$E[X]/\|E[X]\|,$$

where the expectation is taken in the real embedding space ($\mathbb{R}^3$ for a sphere in three dimensions). Hence the use of the Riemannian barycentre is useful for analysing triangle shape, where the shape space is a sphere (see Section 4.3.4). Sample estimates are extremely straightforward to compute as one simply uses arithmetic averages of the coordinates in the real embedding space. The method enables sophisticated inference to be implemented, for example Kendall (2015) investigated clustering of hurricane tracks and whether tracks were correlated in time. For the complex pre-shape sphere we could use:

$$E[Z]/\|E[Z]\|,$$

where $Z \in \mathbb{C}S^{(k-1)}$, although it is important to remove orientation (e.g. by Procrustes registration).

7

Procrustes analysis

7.1 Introduction

This chapter outlines various methods based on Procrustes methods, which are very practical tools for analysing landmark data. Procrustes methods have earlier been seen to be useful for assessing distances between shapes in Chapter 3. In this chapter we provide a more comprehensive treatment of Procrustes methods suitable for two and higher dimensional shape analysis.

Procrustes methods are useful for estimating an average shape and for exploring the structure of shape variability in a dataset. The techniques described in this chapter are generally of a descriptive nature and more explicit emphasis on shape models and inference will be considered in Chapters 9 and 10.

Procrustes analysis involves matching configurations with similarity transformations to be as close as possible according to Euclidean distance, using least squares techniques. Procrustes analysis using orthogonal (rotation/reflection) matrices was developed initially for applications in psychology, and early papers on the topic appeared in the journal *Psychometrika*. The technique can be traced back to Boas (1905) and Mosier (1939) and later principal references include Green (1952); Cliff (1966); Schönemann (1966, 1968); Gruvaeus (1970); Schönemann and Carroll (1970); Gower (1971, 1975); Ten Berge (1977); Sibson (1978, 1979); Langron and Collins (1985); and Goodall (1991). In addition, Sneath (1967) considered a similar least squares matching procedure, with applications to biological shape comparison in mind. Gower (1975); Kendall (1984); Goodall (1991); Ziezold (1994); and Le (1995) also discuss the pure rotation case that is of interest in shape analysis. McLachlan (1972) derived the rigid body transformation case for comparing proteins. Ziezold (1977) considered a similar procedure for matching configurations using translation and rotation, in a mathematically rigorous manner. Some other books that include introductions to Procrustes analysis for comparing matrices (e.g. in MDS) are

Statistical Shape Analysis, with Applications in R, Second Edition. Ian L. Dryden and Kanti V. Mardia.
© 2016 John Wiley & Sons, Ltd. Published 2016 by John Wiley & Sons, Ltd.

Mardia *et al.* (1979, p. 416); Cox and Cox (1994); Krzanowski and Marriott (1994); Borg and Groenen (1997); Everitt and Rabe-Hesketh (1997); and Koch (2014).

We begin by describing ordinary Procrustes analysis (OPA) which is used for matching two configurations in Section 7.2. When at least two configurations are available we can use the technique of generalized Procrustes analysis (GPA) to obtain an average shape, as in Section 7.3. The implementation of ordinary and generalized Procrustes analyses is particularly simple in $m = 2$ dimensions, when complex arithmetic allows Euclidean similarity matching to be expressed as a complex linear regression problem, as seen later in Chapter 8. Generalized Procrustes analysis leads to an explicit eigenvector solution for planar data (see Result 8.2), but a numerical algorithm is required for higher dimensions. We also describe some variants of Procrustes analysis.

After an estimate of mean shape has been obtained we often wish to explore the structure of shape variability in a dataset. Principal component analysis of the tangent shape coordinates using the Procrustes mean as the pole sometimes provides a suitable method. Various graphical procedures for displaying the principal components are presented. If groups are available then canonical variate analysis and linear or quadratic discriminant analysis could be used in the tangent space.

7.2   Ordinary Procrustes analysis

7.2.1   Full OPA

Let us first consider the case where two configuration matrices X_1 and X_2 are available (both $k \times m$ matrices of coordinates from k points in m dimensions) and we wish to match the configurations as closely as possible, up to similarity transformations. **In this chapter we assume without loss of generality that the configuration matrices X_1 and X_2 have been centred** using Equation (3.8).

Definition 7.1   *The method of **full OPA** involves the least squares matching of two configurations using the similarity transformations. Estimation of the similarity parameters γ, Γ and β is carried out by minimizing the squared Euclidean distance*

$$D^2_{OPA}(X_1, X_2) = \|X_2 - \beta X_1 \Gamma - 1_k \gamma^T\|^2, \tag{7.1}$$

*where $\|X\| = \{\text{trace}(X^T X)\}^{1/2}$ is the Euclidean norm, Γ is an $(m \times m)$ rotation matrix ($\Gamma \in SO(m)$), $\beta > 0$ is a scale parameter and γ is an $(m \times 1)$ location vector. The minimum of Equation (7.1) is written as $OSS(X_1, X_2)$, which stands for **ordinary (Procrustes) sum of squares**.*

In Section 4.1.1, when calculating distances, we were interested in the minimum value of an expression similar to Equation (7.1) except X_1 and X_2 were of unit size.

Result 7.1   *The full ordinary Procrustes solution to the minimization of (7.1) is given by $(\hat{\gamma}, \hat{\beta}, \hat{\Gamma})$ where*

$$\hat{\gamma} = 0 \tag{7.2}$$

$$\hat{\Gamma} = UV^{\mathrm{T}} \tag{7.3}$$

where

$$X_2^{\mathrm{T}} X_1 = \|X_1\| \|X_2\| V \Lambda U^{\mathrm{T}}, \quad U, V \in SO(m) \tag{7.4}$$

with Λ a diagonal $m \times m$ matrix of positive elements except possibly the last element, and so the singular values are optimally signed, see Equation (4.4). The solution is unique if X_2 is non-degenerate in the sense of condition (4.7). Also, recall that X_1 and X_2 are centred already, and so the centring provides the optimal translation. Furthermore,

$$\hat{\beta} = \frac{\mathrm{trace}(X_2^{\mathrm{T}} X_1 \hat{\Gamma})}{\mathrm{trace}(X_1^{\mathrm{T}} X_1)}, \tag{7.5}$$

and

$$OSS(X_1, X_2) = \|X_2\|^2 \sin^2 \rho(X_1, X_2), \tag{7.6}$$

where $\rho(X_1, X_2)$ is the Riemannian shape distance of Equation (4.12).

Proof: We wish to minimize

$$\begin{aligned} D_{OPA}^2 &= \|X_2 - \beta X_1 \Gamma - 1_k \gamma^{\mathrm{T}}\|^2 \\ &= \|X_2\|^2 + \beta^2 \|X_1\|^2 - 2\beta \mathrm{trace}(X_2^{\mathrm{T}} X_1 \Gamma) + k\gamma^{\mathrm{T}}\gamma, \end{aligned}$$

where X_1 and X_2 are centred. It simple to see that we must take $\hat{\gamma} = 0$. If

$$Z_i = HX_i / \|X_i\|, \quad i = 1, 2,$$

are the pre-shapes of X_i, then we need to minimize

$$\|X_2\|^2 + \beta^2 \|X_1\|^2 - 2\beta \|X_1\| \|X_2\| \mathrm{trace}(Z_2^{\mathrm{T}} Z_1 \Gamma)$$

and we find the minimizing Γ from Lemma 4.2, Equation (4.6). Differentiating with respect to β we obtain:

$$\frac{\partial D_{OPA}^2}{\partial \beta} = 2\beta \mathrm{trace}(X_1^{\mathrm{T}} X_1) - 2\mathrm{trace}(\|X_1\| \|X_2\| Z_2^{\mathrm{T}} Z_1 \hat{\Gamma}).$$

Hence,

$$\hat{\beta} = \frac{\|X_2\|\text{trace}(Z_2^T Z_1 \hat{\Gamma})}{\|X_1\|} = \frac{\|X_2\|}{\|X_1\|}\text{trace}(\Lambda) = \frac{\|X_2\|}{\|X_1\|}\cos\rho(X_1, X_2). \qquad (7.7)$$

Substituting $\hat{\gamma}, \hat{\Gamma}$ and $\hat{\beta}$ into Equation (7.1) leads to:

$$OSS(X_1, X_2) = \|X_2\|^2 + \hat{\beta}^2\|X_1\|^2 - 2\hat{\beta}\|X_1\| \, \|X_2\|\cos\rho$$

and so the result of Equation (7.6) follows. $\qquad\qquad\qquad\qquad\qquad\qquad$ □

Note that λ_m will be negative in the cases where an orthogonal transformation (reflection and rotation) would produce a smaller sum of squares than just a rotation. In practice, for fairly close shapes λ_m will usually be positive – in which case the solution is the same as minimizing over the orthogonal matrices $O(m)$ instead of $SO(m)$.

Definition 7.2   *The **full Procrustes fit** (or full Procrustes coordinates) of X_1 onto X_2 is:*

$$X_1^P = \hat{\beta}X_1\hat{\Gamma} + 1_k\hat{\gamma}^T,$$

*where we use the superscript 'P' to denote the Procrustes registration. The **residual matrix** after Procrustes matching is defined as:*

$$R = X_2 - X_1^P.$$

Sometimes examining the residual matrix can tell us directly about the difference in shape, for example if one residual is larger than others or if the large residuals are limited to one region of the object or other patterns are observed. In other situations it is helpful to use further diagnostics for shape difference such as the partial warps from thin-plate spline transformations, discussed later in Section 12.3.4.

In order to find the decomposition of Equation (7.4) in practice, one carries out the usual singular value decomposition $V\Lambda U^T$ and if either one of U or V has determinant -1, then its mth column is multiplied by -1 and λ_m is negated.

In general if the rôles of X_1 and X_2 are reversed, then the ordinary Procrustes registration will be different. Writing the estimates for the reverse order case as $(\hat{\gamma}^R, \hat{\beta}^R, \hat{\Gamma}^R)$ we see that $\hat{\gamma}^R = -\hat{\gamma}$, $\hat{\Gamma}^R = (\hat{\Gamma})^T$ but $\hat{\beta}^R \neq 1/\hat{\beta}$ in general. In particular

$$OSS(X_2, X_1) \neq OSS(X_1, X_2)$$

unless the figures are both of the same size, and so one cannot use $\sqrt{OSS(X_1, X_2)}$ as a distance. If the figures are normalized to unit size, then we see that

$$OSS(X_1/\|X_1\|, X_2/\|X_2\|) = 1 - \left\{ \sum_{i=1}^{m} \lambda_i \right\}^2 = \sin^2 \rho(X_1, X_2) = d_F^2(X_1, X_2)$$

and in this case $\sqrt{OSS(X_1/\|X_1\|, X_2/\|X_2\|)} = \sin \rho(X_1, X_2)$ is a suitable choice of shape distance, and was denoted as $d_F(X_1, X_2)$ in Equation (4.10) – the full Procrustes distance. In Example 7.1 we see the ordinary Procrustes registration of a juvenile and an adult sooty mangabey in two dimensions.

Since each of the figures can be rescaled, translated and rotated (the full set of similarity transformations) we call the method **full Procrustes analysis**. There are many other variants of Procrustes matching and these are discussed in Section 7.6.

The term **ordinary Procrustes** refers to Procrustes matching of one observation onto another. Where at least two observations are to be matched to a common unknown mean the term **GPA** is used, which is discussed in Section 7.3.

Full Procrustes shape analysis for 2D data is particularly straightforward using complex arithmetic and details are given in Chapter 8.

Example 7.1 Consider a juvenile and an adult from the sooty mangabey data of Section 1.4.12. The unregistered outlines are shown in Figure 7.1. In Figure 7.2 we see the full Procrustes fit of the adult onto the juvenile (Figure 7.2a) and the full Procrustes

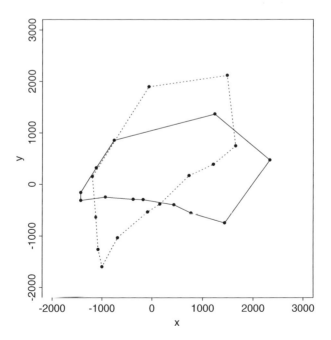

Figure 7.1   Unregistered sooty mangabeys: juvenile (——); and adult (- - -).

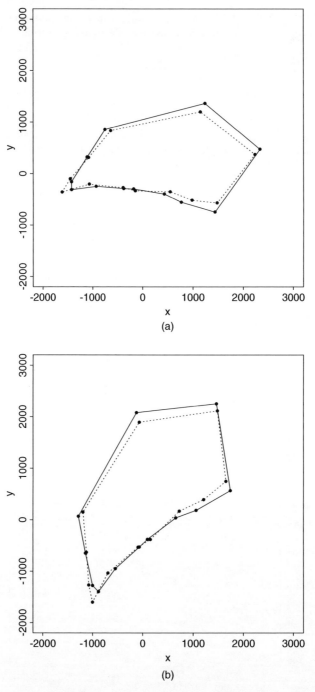

Figure 7.2   The Procrustes fit of (a) the adult sooty mangabey (- - -) onto the juvenile (——) and (b) the juvenile onto the adult.

fit of the juvenile onto the adult (Figure 7.2b). In matching the juvenile onto the adult $\hat{\theta} = 45.53°$ and $\hat{\beta} = 1.1309$. We see that the estimate of scale in matching the adult onto the juvenile is $\hat{\beta}^R = 0.8745$ and the rotation is $\hat{\theta}^R = -45.53°$. Note that $\hat{\beta}^R \neq 1/\hat{\beta}$ because the adult and juvenile are not the same size (the matching is not symmetric). Computing the measure of full Procrustes shape distance we see that $d_F = 0.1049$.   □

7.2.2   OPA in R

In the shapes library in R the function procOPA is used for OPA. The command for ordinary Procrustes registration of B onto A is procOPA(A,B). To match the juvenile Sooty Mangabey onto the adult male Sooty Mangabey we have:

```
data(sooty)
juvenile <- sooty[,,1]
adult<- sooty[,,2]
ans <- procOPA(adult, juvenile )
ans$Bhat
           [,1] [,2]
 [1,] -879.30998 -1396.74936
 [2,] -998.77789 -1276.29016
 [3,] -1143.71649 -647.43917
 [4,] -1287.85042 67.78971
 [5,] -119.56578 2079.45536
 [6,] 1465.48601 2253.98504
 [7,] 1740.93692 565.09765
 [8,] 1060.70877 180.56805
 [9,] 662.17427 36.14918
[10,] 102.18926 -379.26311
[11,] -63.53411 -533.53575
[12,] -538.74055 -949.76743

ans$OSS
[1] 298926.7

ans$rmsd
[1] 157.8308

ans$R
         [,1] [,2]
[1,] 0.7005709 0.7135828
[2,] -0.7135828 0.7005709

print(atan2(ans$R[1,2],ans$R[1,1])*180/pi) #rotation angle in degrees
[1] 45.52717

ans$s
[1] 1.130936
```

Hence, we see that the value of the *OSS* is 298926.7 and the RMSD is $\sqrt{OSS/k} = 157.8308$. The rotation angle is 45.53^o and the scaling is 1.1309, as seen in Example 7.1.

7.2.3   Ordinary partial Procrustes

One may be interested in size-and-shape (form), in which case it is not of interest to consider scaling. The objects must be measured on the same scale.

Definition 7.3   *Partial OPA involves registration over translation and rotation only to match two configurations, and scaling is not required. Minimization is required of the expression (Boas, 1905)*

$$\|X_2 - X_1\Gamma - 1_k\gamma^T\|^2.$$

The same solution for the location vector and the rotation matrix as in the full Procrustes case [Equation (7.2) and Equation (7.3)] gives the minimum

$$OSS_p(X_1, X_2) = \|X_1\|^2 + \|X_2\|^2 - 2\|X_1\| \|X_2\| \cos \rho(X_1, X_2). \qquad (7.8)$$

which is the square of the Riemannian size-and-shape distance of Equation (5.5).

Partial Procrustes analysis on the original centred configurations is particularly appropriate when studying joint size-and-shape, as considered in Chapter 5. If the two configurations are of unit size, then Equation (7.8) is equal to $d_p^2(X_1, X_2)$, the square of the partial Procrustes distance.

Ordinary partial Procrustes matching can be carried out in R using `procOPA` but with option `scale=FALSE`. For the partial Procrustes matching of the juvenile sooty mangebey to the adult using partial Procrustes analysis we have $OSS_p = 659375$ and RMSD = 234.4097, and the rotation is again 45.52717^o of course. The commands are:

```
ans <- procOPA(adult, juvenile , scale=FALSE)

ans$OSS
[1] 659375

ans$rmsd
[1] 234.4097

ans$R
[,1] [,2]
[1,]  0.7005709 0.7135828
[2,] -0.7135828 0.7005709

ans$s
[1] 1
```

7.2.4   Reflection Procrustes

A further type of ordinary Procrustes matching is where reflections are also allowed. The use of Procrustes methods in shape analysis is generally different from their more traditional use in multivariate analysis, where rotation and reflection are used instead of pure rotation (e.g. see Mardia *et al.* 1979, p. 416).

If reflections are not important than we can also include reflection invariance by using orthogonal matrices $\Gamma \in O(m)$, where $\det(\Gamma) = \pm 1$ in the matching, instead of special orthogonal matrices $\Gamma \in SO(m)$, where $\det(\Gamma) = +1$. We call this **reflection Procrustes analysis**. For datasets with small variability (and full rank configurations) there will usually be no difference between the two approaches.

Details of the minimization with $\Gamma \in O(m)$ (an orthogonal matrix) follow in a similar manner to that of $\Gamma \in SO(m)$ (Gower 1975; Sibson 1978, 1979; Goodall 1991; Cox and Cox 1994; Krzanowski and Marriott 1994). The method when allowing for reflections is almost identical to the rotation only case, except that a singular value decomposition with all positive diagonal elements is used for the reflection/rotation case, whereas the smallest singular value can be negative in the rotation case.

We can use calculus to carry out the minimization over the orthogonal matrices. We need to minimize

$$\text{trace}(X_2^T X_2) + \beta^2 \text{trace}(X_1^T X_1) - 2\beta\text{trace}(X_2^T X_1 \Gamma) \tag{7.9}$$

with respect to $\Gamma \in O(m)$ (cf. Mardia *et al.* 1979, p. 416). Consider using the decomposition of $X_2^T X_1$ in Equation (7.4). Hence, we must maximize $\text{trace}(V\Lambda U^T\Gamma)$ with respect to Γ, subject to the constraints that $\Gamma\Gamma^T = I_m$ [which gives $m(m + 1)/2$ constraints, because Γ is symmetric]. Let $\frac{1}{2}L$ be a symmetric $m \times m$ matrix of Lagrange multipliers for these constraints. The aim is to maximize

$$\text{trace}\left[V\Lambda U^T\Gamma - \tfrac{1}{2}L(\Gamma\Gamma^T - I)\right]. \tag{7.10}$$

By direct differentiation it can be shown that

$$\frac{\partial\text{trace}(V\Lambda U^T\Gamma)}{\partial\Gamma} = U\Lambda V^T , \quad \frac{\partial\text{trace}\left[\tfrac{1}{2}L(\Gamma\Gamma^T - I)\right]}{\partial\Gamma} = L\Gamma.$$

Hence on differentiating Equation (7.10) and setting derivatives equal to zero we find that

$$U\Lambda V^T = L\Gamma. \tag{7.11}$$

Note that

$$L^2 = (U\Lambda V^T)(V\Lambda U^T) = (U\Lambda U^T)(U\Lambda U^T)$$

and so we can take $L = U\Lambda U^T$. Substituting L into Equation (7.11) we see that

$$\hat{\Gamma} = UV^T.$$

Note that $\hat{\Gamma} \in O(m)$, and so we have found the minimizing rotation/reflection. In practice, for fairly close shapes the solutions for minimizing over the orthogonal matrices $O(m)$ instead of $SO(m)$ are the same.

In R we can carry out reflection shape analysis using the option `reflect=TRUE`, and for the sooty mangabey data there is no difference between the matching whether using reflections or not.

```
ans <- procOPA(adult, juvenile , reflect=TRUE)

ans$OSS
[1] 298926.7
```

7.3   Generalized Procrustes analysis

7.3.1   Introduction

Consider now the general case where $n \geq 2$ configuration matrices are available $X_1, \ldots, X_n$. For example, the configurations could be a random sample from a population with population $[\mu]$, and we wish to estimate the shape of the population mean with an 'average' shape from the sample.

For shape analysis our objects need not be commensurate in scale. However, for size-and-shape analysis our objects do need to be commensurate in scale.

A least squares approach to finding an estimate of $[\mu]$ is that of generalized Procrustes analysis (GPA), a direct generalization of OPA. We shall see that GPA provides a practical method of computing the sample full Procrustes mean, defined in Section 6.2.

Definition 7.4   *The method of full GPA involves translating, rescaling and rotating the configurations relative to each other so as to minimize a total sum of squares*

$$G(X_1, \ldots, X_n) = \sum_{i=1}^{n} \|(\beta_i X_i \Gamma_i + 1_k \gamma_i^T) - \mu\|^2 \qquad (7.12)$$

with respect to $\beta_i, \Gamma_i, \gamma_i, i = 1, \ldots, n$ and μ, subject to an overall size constraint. The constraint on the sizes can be chosen in a variety of ways. For example, we could choose

$$\sum_{i=1}^{n} S^2(\beta_i X_i \Gamma_i + 1_k \gamma_i^T) = \sum_{i=1}^{n} S^2(X_i). \qquad (7.13)$$

Full generalized Procrustes matching involves the registration of all configurations to optimal positions by translating, rotating and rescaling each figure so as to minimize the sum of squared Euclidean distances between each transformed figure and the estimated mean. The constraint of Equation (7.13) prevents the $\hat{\beta}_i$ from all becoming close to 0.

Definition 7.5   *The **full Procrustes fit** (or full Procrustes coordinates) of each of the X_i is given by:*

$$X_i^P = \hat{\beta}_i X_i \hat{\Gamma}_i + 1_k \hat{\gamma}_i^T, \quad i = 1, \dots, n, \tag{7.14}$$

where $\hat{\Gamma}_i \in SO(m)$ (rotation matrix), $\hat{\beta}_i > 0$ (scale parameter), $\hat{\gamma}_i^T$ (location parameters), $i = 1, \dots, n$, are the minimizing parameters.

An algorithm to estimate the transformation parameters $(\gamma_i, \beta_i, \Gamma_i)$ is described in Section 7.4. The parameters $(\Gamma_i, \beta_i, \gamma_i)$ have been termed 'nuisance parameters' by Goodall (1991) because they are not the parameters of primary interest in shape analysis.

Result 7.2   *The point in shape space corresponding to the arithmetic mean of the Procrustes fits,*

$$\bar{X} = \frac{1}{n} \sum_{i=1}^{n} X_i^P, \tag{7.15}$$

has the same shape as the full Procrustes mean, which was defined in Equation (6.11).

Proof: The result follows because we are minimizing sums of Euclidean square distances in GPA. The minimum of

$$\sum \| X_i^P - \mu \|^2$$

over μ is given by $\hat{\mu} = n^{-1} \sum_i X_i^P$.   □

Hence, estimating the transformation parameters by GPA is equivalent to minimizing

$$G(X_1, \dots, X_n) = \inf_{\beta_i, \Gamma_i, \gamma_i} \sum_{i=1}^{n} \left\| (\beta_i X_i \Gamma_i + 1_k \gamma_i^T) - \frac{1}{n} \sum_{j=1}^{n} (\beta_j X_j \Gamma_j + 1_k \gamma_j^T) \right\|^2,$$

$$= \inf_{\beta_i, \Gamma_i, \gamma_i} \frac{1}{n} \sum_{i=1}^{n} \sum_{j=i+1}^{n} \| (\beta_i X_i \Gamma_i + 1_k \gamma_i^T) - (\beta_j X_j \Gamma_j + 1_k \gamma_j^T) \|^2, \tag{7.16}$$

subject to the constraint (7.13).

After a collection of objects has been matched into optimal full Procrustes position with respect to each other, calculation of the full Procrustes mean shape is simple; it is computed by taking the arithmetic means of each coordinate. Full Procrustes matching also been called 'Procrustes-with-scaling' by Dryden (1991) and Mardia and Dryden (1994). We see that full GPA is analogous to minimizing sums of squared distances in the shape space d_F^2 defined in Equation (4.10).

The full Procrustes mean shape has to be found iteratively for $m = 3$ and higher dimensional data, and an explicit eigenvector solution is available for 2D data, which will be seen in Result 8.2.

Example 7.2   In Figure 7.3 we see the registered male and female macaques from the dataset described in Section 1.4.3, using full Procrustes registration. There are $k = 7$ landmarks in $m = 3$ dimensions. The R commands are:

```
> outm<-procGPA(macm.dat)
> outf<-procGPA(macf.dat)
> wire<-c((1:7),1,6,4,1,4,3,7,3,5)
> shapes3d(outm$rotated,joinline=wire)
> shapes3d(outf$rotated,joinline=wire)
> Bhat<-procOPA(outm$mshape,outf$mshape)$Bhat
> shapes3d(abind(outm$mshape,Bhat),joinline=wire,
            col=c(rep(2,times=7),rep(4,times=7)))
> sin(riemdist(outm$mshape,Bhat))
[1] 0.05351208
> outm$rmsd
[1] 0.07872565
> outf$rmsd
[1] 0.05811237
```

In two of the males the highest landmark in the 'y' direction (*bregma*) is somewhat further away (in the 'x' direction) than in the rest of the specimens. This landmark is highly variable in primates. The full Procrustes means (normalized to unit size) are displayed in Figure 7.4 and the female mean has been registered onto the male mean by OPA. The full Procrustes estimated mean shapes for the males [$\hat{\mu}_1$] and females [$\hat{\mu}_2$] are full Procrustes distance $d_F = 0.0535$ apart, and the root mean square of d_F to the estimated mean shape within each group is 0.0787 for the males and 0.0581 for the females. The males are a little more variable in shape than the females, although the non-isotropic nature of the variation also needs to be considered. Formal tests for mean shape difference are considered in Chapter 9.   □

7.4   Generalized Procrustes algorithms for shape analysis

For practical implementation one can use the GPA algorithm of Gower (1975), modified by Ten Berge (1977). The idea of GPA was originally proposed by Kristof and

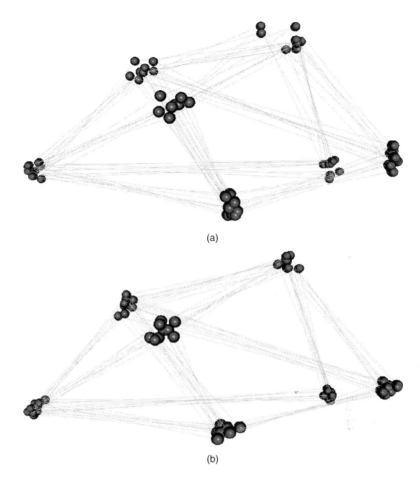

Figure 7.3   The male (a) and female (b) macaque skulls registered by full GPA. For a colour version of this figure, see the colour plate section.

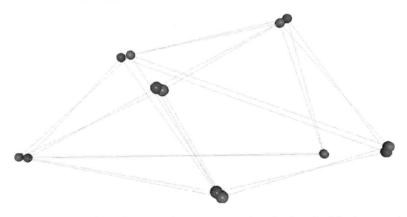

Figure 7.4   The male (red) mean shape registered to the female (blue) mean shape of the macaque skulls registered by OPA. For a colour version of this figure, see the colour plate section.

Wingersky (1971) and adapted to this situation by Gower (1975). Langron and Collins (1985) gave some useful distributional results based on perturbation theory. Goodall and Bose (1987) and Goodall (1991) adapted the method explicitly for shape analysis.

7.4.1   Algorithm: GPA-Shape-1

1. **Translations.** Centre the configurations to remove location. Initially let

$$X_i^P = CX_i \ , \quad i = 1, \dots, n,$$

where C is the centring matrix of Equation (2.3).

2. **Rotations.** For the ith configuration let

$$\overline{X}_{(i)} = \frac{1}{n-1} \sum_{j \neq i} X_j^P,$$

then the new X_i^P is taken to be the ordinary Procrustes registration, involving only rotation, of the old X_i^P onto $\overline{X}_{(i)}$. The n figures are rotated in turn. This process is repeated until the Procrustes sum of squares of Equation (7.12) cannot be reduced further (i.e. the difference is less than a tolerance parameter $tol1$). Hence, the matrices $(X_i^P)^T \overline{X}_{(i)}$ are symmetric positive semi-definite.

3. **Scaling.** Let Φ be the $n \times n$ correlation matrix of the $\mathrm{vec}(X_i^P)$ (with the usual rôles of variable and observation labels reversed) with eigenvector $\phi = (\phi_1, \dots, \phi_n)^T$ corresponding to the largest eigenvalue. Then from Ten Berge (1977) take

$$\hat{\beta}_i = \left(\frac{\sum_{k=1}^{n} \|X_k^P\|^2}{\|X_i^P\|^2} \right)^{1/2} \phi_i,$$

which is repeated for all i.

4. Repeat steps 2 and 3 until the Procrustes sum of squares of Equation (7.12) cannot be reduced further (i.e. the difference in sum of squares is less than another tolerance parameter $tol2$).

The algorithm usually converges very quickly. The two tolerance parameters for deciding when a step has converged are clearly data dependent, and specific choices are discussed in Section 7.4.3. Note that the resulting registration satisfies the constraint (7.13). From Goodall (1991) typical implementations will include a total of between $3 \times n \times 2$ and $5 \times n \times 3$ OPA steps. Groisser (2005) discussed several variants of GPA, proved that GPA converges, and provided error estimates and convergence times.

This algorithm, as described by Gower (1975) and Ten Berge (1977), included reflection invariances, although mention was made of the adaptation to the rotation only case in the papers. For shape analysis there must be modification to ensure $\Gamma_i \in SO(m)$ rather than $O(m)$. For many datasets with small variability in shape, the algorithms will give the same solution, whether modified to exclude reflections or not.

Note that the rotation step of this algorithm is based on minimizing the sum of squares with respect to rotations given that $\hat{\mu}$ is given by Equation (7.15). In particular,

$$\sum_{i=1}^{n} \|X_i\hat{\Gamma}_i - \bar{X}\|^2 = \left(\frac{n-1}{n}\right)^2 \left\|X_i\hat{\Gamma}_i - \frac{1}{n-1}\sum_{j\neq i} X_j\hat{\Gamma}_j\right\|^2,$$

and hence as each rotation matrix is updated the sum of squares is reduced until a minimim is reached.

A numerical procedure based on a Newton algorithm for Procrustes matching with translation and rotation was given by Fright and Linney (1993). Other simple implementations also often work well, such as using the first configuration as the initial estimated mean or combining steps 2 and 3 to carry out a full Procrustes match to the current estimated mean. This latter approach leads to a second algorithm.

7.4.2   Algorithm: GPA-Shape-2

1. **Translations.** Centre the configurations to remove location. Initially let

$$X_i^P = CX_i \ , \ \ i = 1, \ldots, n,$$

where C is the centring matrix of (2.3).

2. Initialize $\hat{\mu}$ for example at $\frac{1}{n}\sum_{i=1}^{n} X_i^P$.

3. **Rotations and scale** For the ith configuration ($i = 1, \ldots, n$) carry out an ordinary Procrustes match by rotating and scaling to $\hat{\mu}$,

$$X_i^P = \hat{\beta}_i CX_i\hat{\Gamma}_i.$$

4. Update $\hat{\mu} = \frac{1}{n}\sum_{i=1}^{n} X_i^P$.

5. Repeat steps 3 and 4 until the Procrustes sum of squares of Equation (7.12) cannot be reduced further (i.e. the difference in sum of squares is less than tolerance parameter *tol2*).

Both GPA algorithms should converge to the same mean shape, although the final value $\hat{\mu}$ will usually differ up to a rotation and scale (which is irrelevant).

Note that in practice the two algorithms behave very similarly. There are some practical advantages in the second algorithm in some cases as it is easily parallelizable and there is some flexibility in the choice of start point for the algorithm. If n is very large, as for example in molecular dynamics analysis (Sherer *et al.* 1999; Harris *et al.* 2001), there is a significant advantage in being able use parallel rotation updates on multiple CPUs, and only requiring the updated mean to be kept in memory after each iteration (rather than all the current fitted values).

When $n = 2$ objects are available we can consider OPA or GPA to match them. The advantage of using GPA is that the matching procedure is symmetrical in the ordering of the objects, that is GPA of X_1 and X_2 is the same as GPA of X_2 and X_1. As we have seen in Section 7.2.1, OPA is not symmetrical in general, unless the objects have the same size.

7.4.3   GPA in R

To carry out GPA in the shapes package in R one can use the command procGPA. This function is perhaps the most useful of all the commands in the shapes library, and carries out Procrustes registration using the GPA-Shape-1 algorithm of Section 7.4.1, computes the Procrustes mean, and various summary statistics. The default setting of the procGPA function carries out rotation and scaling. In order to remove the scaling step one uses the option scale=FALSE and examples are given in Section 7.5.3.

To carry out GPA on the macaque data, with scaling included:

```
ansm <- procGPA(macm.dat)
ansf <- procGPA(macf.dat)
```

To compute the Riemannian distance between the full Procrustes mean shape estimates, and the r.m.s.d. of the full Procrustes shape distances to the mean we have:

```
riemdist(ansm$mshape,ansf$mshape)
[1] 0.05353765
> ansm$rmsd
[1] 0.07872565
> ansf$rmsd
[1] 0.05811237
```

Note that the function procGPA provides many calculations, including the centroid size of each observation ($size), the Riemannian distance to the mean ($rho), the Procrustes rotated figures ($rotated), the sample mean shape ($mshape), and the tangent coordinates ($tan). The particular types of means and tangent coordinates depends on the option choices, but the default is that the full Procrustes mean shape is given with the Procrustes residuals as approximate tangent coordinates.

The R function procGPA contains two tolerances as detailed in Algorithm GPA-Shape-1 (Section 7.4.1) which are set to ensure the accuracy of the algorithm. The option tol1 is the tolerance for optimal rotations in the iterative GPA algorithm (step 2 of GPA-Shape-1), which is the tolerance on the mean sum of squares (divided by size of mean squared) between successive iterations, and the rotation iterations stop when the normalized mean sum of squares is less than tol1. The option tol2 is a tolerance for the scale/rotation steps for GPA for the iterative algorithm, which is tolerance on the mean sum of squares (divided by size of mean squared) between successive iterations. We illustrate the differences in the choices of tolerances for the male macaque data, using the extended output option proc.output=TRUE.

```
ans1<-procGPA(macm.dat,tol1=1e-05,tol2=1e-05,proc.output=TRUE)
Step | Objective function | change
------------------------------------------------------------
Initial objective fn 0.01223246 -
------------------------------------------------------
  Rotation iteration 1 0.008385595 0.003846862
  Rotation iteration 2 0.008278599 0.0001069951
  Rotation iteration 3 0.00827771 8.893455e-07
------------------------------------------------------
Rotation step 0 0.00827771 0.003954746
------------------------------------------------------
  Scaling updated
  Rotation iteration 1 0.006236379 1.663161e-08
------------------------------------------------------
Scale/rotate step 1 0.006236379 0.002041331
------------------------------------------------------
  Scaling updated
  Rotation iteration 1 0.006236379 1.535184e-10
------------------------------------------------------
Scale/rotate step 2 0.006236379 1.535344e-10
------------------------------------------------------
Shape distances and sizes calculation ...
PCA calculation ...
Finished.
```

Each rotation step involves n OPA matchings, where n is the sample size. In this calculation there are three rotation steps, a scaling step, a rotation step, a scaling step and a rotation step. So, here there are 5×9 OPA matchings. For a second choice of tolerances we have:

```
ans2<-procGPA(macm.dat,tol1=1e-08,tol2=1e-08,proc.output=TRUE)
Step | Objective function | change
----------------------------------------------------------
Initial objective fn 0.01223246 -
------------------------------------------------------
  Rotation iteration 1 0.008385595 0.003846862
  Rotation iteration 2 0.008278599 0.0001069951
  Rotation iteration 3 0.00827771 8.893455e-07
  Rotation iteration 4 0.008277695 1.500593e-08
  Rotation iteration 5 0.008277695 1.662341e-10
------------------------------------------------------
Rotation step 0 0.008277695 0.003954761
```

```
 ------------------------------------------------
  Scaling updated
  Rotation iteration 1 0.006236379 1.905269e-09
 ------------------------------------------------
Scale/rotate step 1 0.006236379 0.002041316
 ------------------------------------------------
  Scaling updated
  Rotation iteration 1 0.006236379 2.268505e-11
 ------------------------------------------------
Scale/rotate step 2 0.006236379 2.268794e-11
 ------------------------------------------------
Shape distances and sizes calculation ...
PCA calculation ...
Finished.

riemdist(ans1$mshape,ans2$mshape)
[1] 4.942156e-08
```

Here there are 7×9 OPA matchings, and the resulting mean shapes are almost identical. Using a higher tolerance we have:

```
ans3<-procGPA(macm.dat,tol1=1e-02,tol2=1e-02,proc.output=TRUE)
 Step | Objective function | change
 ------------------------------------------------------------
Initial objective fn 0.01223246 -
 ------------------------------------------------
  Rotation iteration 1 0.008385595 0.003846862
 ------------------------------------------------
Rotation step 0 0.008385595 0.003846862
 ------------------------------------------------
  Scaling updated
 ------------------------------------------------
Scale/rotate step 1 0.006343774 0.00204182
 ------------------------------------------------
Shape distances and sizes calculation ...
PCA calculation ...
Finished.

riemdist(ans2$mshape,ans3$mshape)
[1] 0.0002719985
```

Here there are just n OPA matchings, and there is a non-trivial difference between the mean shape estimators. The default tolerances for GPA in procGPA are 10^{-5} which has worked well in a very large variety of scenarios.

7.5   Generalized Procrustes algorithms for size-and-shape analysis

Partial Procrustes analysis involves just registration by translation and rotation (not scaling) as opposed to full Procrustes analysis which involves the full set of similarity transformations. The terminology was introduced by Kent (1992). The objects must be recorded to the same scale, and so partial Procrustes analysis is appropriate for size-and-shape analysis. We minimize

$$G_p = \inf_{\Gamma_i, \gamma_i, \mu} \sum_{i=1}^{n} \|X_i \Gamma_i + 1_k \gamma_i^{\mathrm{T}} - \mu\|^2$$

$$= \inf_{\mu} \sum_{i=1}^{n} \left\{ \|\mu\|^2 + \|X_i\|^2 - 2\|X_i\| \, \|\mu\| \cos \rho(X_i, \mu) \right\}, \qquad (7.17)$$

which follows since

$$\inf_{\Gamma_i} \operatorname{trace}(\mu^{\mathrm{T}} X_i \Gamma_i) = \|X_i\| \|\mu\| \cos \rho.$$

For size-and-shape analysis there is no need to include an overall size constraint, as was required for the pure shape case in Equation (7.13).

The GPA algorithms to carry out the minimization are simply adapted to the size-and-shape case by not including the scaling steps.

7.5.1   Algorithm: GPA-Size-and-Shape-1

1. **Translations.** Centre the configurations to remove location.

2. **Rotations.** For the ith configuration let

$$\overline{X}_{(i)} = \frac{1}{n-1} \sum_{j \neq i} X_j^P,$$

then the new X_i^P is taken to be the ordinary Procrustes registration, involving only rotation, of the old X_i^P onto $\overline{X}_{(i)}$. The n figures are rotated in turn. This process is repeated until the Procrustes sum of squares of Equation (7.12) cannot be reduced further (i.e. the difference is less than a tolerance parameter $tol1$). Hence, the matrices $(X_i^P)^{\mathrm{T}} \overline{X}_{(i)}$ are symmetric positive semidefinite.

7.5.2   Algorithm: GPA-Size-and-Shape-2

1. **Translations.** Centre the configurations to remove location.

2. Initialize $\hat{\mu}$ for example at $\frac{1}{n} \sum_{i=1}^{n} CX_i$.

3. **Rotations.** For the ith configuration $(i = 1, \ldots, n)$ carry out an ordinary Procrustes match by rotating to $\hat{\mu}$,

$$X_i^P = CX_i\hat{\Gamma}_i.$$

4. Update $\hat{\mu} = \frac{1}{n} \sum_{i=1}^{n} X_i^P$.

5. Repeat steps 3 and 4 until the Procrustes sum of squares of Equation (7.12) cannot be reduced further (i.e. the difference in sum of squares is less than tolerance parameter *tol2*).

7.5.3   Partial GPA in R

Algorithm GPA-Size-and-Shape-1 can be carried out with the R command procGPA(data , scale=FALSE), and so for the Digit 3 data we have:

```
ans1 <- procGPA(digit3.dat, scale=FALSE)

ans1$GSS
[1] 5204.127
```

Here partial GPA is carried out and the resulting Procrustes sum of squares $G_p = 5204$. The resulting estimate of Fréchet mean size-and-shape (see Definition 6.12) is in given in ans1$mshape. Note that this estimate is different from the unit size partial Procrustes mean shape in Section 7.6.2, which requires all configurations to be of unit size in the matching.

The usual full GPA is now carried out for comparison:

```
ans2 <- procGPA(digit3.dat, scale=TRUE)

ans2$GSS
[1] 3851.577

riemdist(ans1$mshape,ans2$mshape)
[1] 0.01165189
```

The full Procrustes sum of squares $G = 3851$, and so the scaling has reduced the sum of squares as expected. The full Procrustes mean shape (up to an arbitrary scaling) is given in ans2$mshape. The Riemannian shape distance between the shapes of the two estimates is 0.01165.

7.5.4   Reflection GPA in R

If matching using orthogonal matrices instead of rotation matrices the method is called reflection Procrustes analysis. This can be carried out using procGPA with option `reflect=TRUE`. For the digit 3 data we consider reflection Procrustes without or with scaling, respectively:

```
ans3 <- procGPA(digit3.dat, scale=FALSE, reflect=TRUE)

ans2$GSS
[1] 5204.127

ans4 <- procGPA(digit3.dat, scale=TRUE, reflect=TRUE)

ans4$GSS
[1] 3851.577
```

Note that reflection GPA in the digit 3 data is the same as GPA without reflection invariance, as all the individuals are very far from their reflected versions.

7.6   Variants of generalized Procrustes analysis

7.6.1   Summary

There are many variants to GPA, and so to help with the terminology we provide a summary in Table 7.1.

7.6.2   Unit size partial Procrustes

If all the configurations have unit size, then it can be seen that the estimator of $[\mu]$ is obtained by minimizing sums of squared chordal distances on the pre-shape sphere,

Table 7.1   Nomenclature for the different types of ordinary Procrustes analysis (OPA) registrations with $n = 2$ objects, and generalized Procrustes analysis (GPA) registrations with $n \geq 2$ objects.

Name	Transformations	Section
Full OPA	Translation, rotation, scale	7.2.1
Partial OPA	Translation, rotation	7.2.3
Reflection OPA	Additional reflection	7.2.4
Full GPA	Translation, rotation, scale	7.3
Partial GPA	Translation, rotation	7.5.3
Reflection GPA	Additional reflection	7.5.4

and the resulting estimator is the partial Procrustes mean (see Section 6.3). In this case

$$G_p = \inf_{\mu} \sum_{i=1}^{n} d_P^2(X_i, \mu)$$

$$= \inf_{\mu} \sum_{i=1}^{n} 2[1 - \cos \rho(X_i, \mu)]. \qquad (7.18)$$

This is the approach used by Ziezold (1994) and Le (1995) and described by Kent (1992). There is some evidence (Ziezold 1989; Stoyan and Frenz 1993) that this approach leads to non-unique solutions, although Le (1995) gave conditions when there is a unique solution. The method has also been called 'Procrustes-without-scaling' by Dryden (1991) and Mardia and Dryden (1994).

In some applications measurement error may be present and Du $et\ al.$ (2015) considered size and shape analysis for error prone landmark data, and conditional score methods were used to provide asymptotically consistent estimators of rotation and/or scale.

7.6.3   Weighted Procrustes analysis

Standard Procrustes methods weight each landmark equally, and effectively treat the landmarks as uncorrelated. In weighted Procrustes analysis the Procrustes methods are adapted by replacing the squared Euclidean norm $\|X\|^2 = \mathrm{trace}(X^T X)$ with the squared Mahalanobis norm

$$\|X\|_{\Sigma}^2 = \mathrm{vec}(X)^T \Sigma^{-1} \mathrm{vec}(X)$$

in Section 7.2.1 for OPA and Section 7.3 for GPA. We write OPA(Σ) and GPA(Σ) for these general approaches, sometimes called weighted Procrustes methods, which are a form of weighted least squares. Some problems with estimating covariance matrices using Procrustes analysis are highlighted by Lele (1993). Lele (1993) recommends estimation of the covariance structure based on inter-landmark distances, and this approach is briefly described later in Section 15.3.

Although estimation of covariance matrices is problematic, working with known covariance matrices is fairly straightforward. In practice, of course, Σ is unknown and has to be estimated. Goodall (1991) gives estimates based on maximum likelihood considerations. If X_i^P are the registered figures by either unweighted or weighted Procrustes and $\bar{X} = \frac{1}{n} \sum X_i^P$ is the resulting Procrustes mean, then

$$\hat{\Sigma} = \frac{1}{n} \sum_{i=1}^{n} \mathrm{vec}(X_i^P - \bar{X}) \mathrm{vec}(X_i^P - \bar{X})^T \qquad (7.19)$$

is an estimate of Σ for unrestricted covariances. This estimate is singular and so generalized inverses could be used in any applications. Principal component analysis of Equation (7.19) provides perhaps the best practical way forward in the case of small variations. This approach is equivalent to carrying out the orthogonal decomposition in a tangent plane to shape space and the approach is discussed in detail in Section 7.7.

Goodall (1991) emphasizes the use of factored covariances of the form (cf. Mardia 1984):

$$\Sigma = \Sigma_k \otimes \Sigma_m, \tag{7.20}$$

where Σ_k measures the covariances between landmarks and Σ_m models the variation identical at each landmark. The choice of factored models leads to a fairly straightforward adaptation of the OPA and GPA algorithms. However, factored covariances can be criticized because it is often unrealistic to assume the same structure of variability at each landmark, and their use can be problematic, as pointed out by Lele (1993) and Glasbey et al. (1995). A refinement was proposed by Goodall (1995) using restricted maximum likelihood estimation. Dimension reduction using eigendecomposition of factored covariance matrices of the form (7.20) has been investigated by Dryden et al. (2009a), with application to face identification using the MLE algorithm of Dutilleul (1999). Other patterned covariance structures include self-similar deflation (Bookstein, 2015a) based on bending energy (see Chapter 12).

In the isotropic case inference is more straightforward albeit unrealistic in practical applications. If $\Sigma = \sigma^2 I_{km}$, then Goodall (1991) takes an estimate of variability as:

$$\hat{\sigma}^2 = \frac{1}{nkm} \sum_{i=1}^{n} \text{vec}(X_i^P - \bar{X})^{\text{T}} \text{vec}(X_i^P - \bar{X}). \tag{7.21}$$

Note that

$$\hat{\sigma}^2 = \frac{1}{nkm} \sum_{i=1}^{n} \|\hat{\mu}\|^2 d_F^2(X_i, \hat{\mu}) \sim \frac{1}{nkm} \sigma^2 \chi^2_{(n-1)q}$$

approximately (for small σ) from Equation (9.7), where q is the dimension of the shape space. Hence, from the approximate chi-squared distribution

$$E(\hat{\sigma}^2) \approx \sigma^2 \frac{(n-1)q}{nkm}$$

so $\hat{\sigma}^2$ can be quite biased under the isotropic normal model, and a less biased estimator of σ^2 is:

$$\hat{\hat{\sigma}}^2 = \frac{1}{(n-1)q} \sum_{i=1}^{n} \text{vec}(X_i^P - \bar{X})^{\text{T}} \text{vec}(X_i^P - \bar{X}).$$

Alternative methods for estimating the covariance structure include the offset normal maximum likelihood approach of Dryden and Mardia (1991b) which is described in Section 11.2. Theobald and Wuttke (2006, 2008) have developed an approximate maximum likelihood procedure which works well in many examples using factored covariance structure for size-and-shape or shape analysis, and is available in Douglas Theobald's Theseus program.

So, one has several choices for GPA with general covariance structure. One possibility is to specify a known Σ and then use GPA(Σ). Alternatively, one could use GPA($\Sigma = I$) and then obtain the estimate $\hat{\Sigma}$ using a suitable technique. A third alternative is to use the following iterative procedure:

1. Use GPA($\Sigma = I$).

2. Obtain the estimate $\hat{\Sigma}$ using a suitable estimator.

3. Carry out GPA($\hat{\Sigma}$).

4. Iteratively cycle between steps 2 and 3 until convergence.

This procedure is not guaranteed to converge.

The development of Procrustes techniques to deal with non-isotropic covariance structures is still a topic of current research. Brignell *et al.* (2005, 2015) and Brignell (2007) have considered covariance weighted Procrustes analysis. Some procedures are available in R in procWOPA and procWGPA. Also see Bennani Dosse *et al.* (2011) for anisotropic GPA.

Brignell *et al.* (2005, 2015) provide an explicit solution for partial covariance weighted OPA. The method of partial covariance weighted OPA involves the least squares matching of one configuration to another using rigid-body transformations. Estimation of the translation and rotation parameters, γ and Γ, is carried out by minimizing the Mahalanobis norm,

$$D^2_{pCWP}(X, \mu; \Sigma) = \| \mu - X\Gamma - 1_k \gamma^T \|^2_\Sigma, \tag{7.22}$$

where Σ ($km \times km$) is a symmetric positive definite matrix, γ is an $m \times 1$ location vector and Γ is an $m \times m$ special orthogonal rotation matrix, and

$$\|X\|^2_\Sigma = \text{vec}(X)^T \Sigma^{-1} \text{vec}(X).$$

The translation which minimizes Equation (7.22) is given by Result 7.3. In general, the minimizing rotation is solved numerically, however when $m = 2$ there is only one rotation angle and a solution is given by Result 7.4.

Result 7.3   *Given two configuration matrices, X and μ, and a symmetric positive definite matrix, Σ, the translation, as a function of rotation, which minimizes the Mahalanobis norm, $D^2_{pCWP}(X, \mu; \Sigma)$ is:*

$$\hat{\gamma} = [(I_m \otimes 1_k)^T \Sigma^{-1} (I_m \otimes 1_k)]^{-1} (I_m \otimes 1_k)^T \Sigma^{-1} \text{vec}(\mu - X\Gamma). \tag{7.23}$$

Result 7.4   *Consider $m = 2$, let $A = [(I_m \otimes 1_k)^T \Sigma^{-1}(I_m \otimes 1_k)]^{-1}(I_m \otimes 1_k)^T \Sigma^{-1}$, and denote the partitioned submatrices as:*

$$A = \begin{bmatrix} A_{11} & A_{12} \\ A_{21} & A_{22} \end{bmatrix}, \qquad X = [X_1 \quad X_2], \qquad \mu = [\mu_1 \quad \mu_2],$$

where the A_{ij} have dimension $(1 \times k)$ and X_i, μ_i have dimension $(k \times 1)$ for $i, j = 1, 2$, then given two configuration matrices, X and μ, and a symmetric positive definite matrix Σ, the rotation which minimizes the Mahalanobis norm, $D^2_{pCWP}(X, \mu; \Sigma)$ is given by:

$$\cos \hat{\theta} = \frac{S(2\lambda - 2Q) + TR}{(2\lambda - 2P)(2\lambda - 2Q) - R^2},$$

$$\sin \hat{\theta} = \frac{T(2\lambda - 2P) + SR}{(2\lambda - 2P)(2\lambda - 2Q) - R^2}, \qquad (7.24)$$

where

$$P = \begin{bmatrix} (X_1 + 1_k\delta_1) \\ (X_2 + 1_k\delta_2) \end{bmatrix}^T \Sigma^{-1} \begin{bmatrix} (X_1 + 1_k\delta_1) \\ (X_2 + 1_k\delta_2) \end{bmatrix},$$

$$Q = \begin{bmatrix} (X_2 - 1_k\zeta_1) \\ -(X_1 + 1_k\zeta_2) \end{bmatrix}^T \Sigma^{-1} \begin{bmatrix} (X_2 - 1_k\zeta_1) \\ -(X_1 + 1_k\zeta_2) \end{bmatrix},$$

$$R = -2 \begin{bmatrix} (X_1 + 1_k\delta_1) \\ (X_2 + 1_k\delta_2) \end{bmatrix}^T \Sigma^{-1} \begin{bmatrix} (X_2 - 1_k\zeta_1) \\ -(X_1 + 1_k\zeta_2) \end{bmatrix},$$

$$S = -2 \begin{bmatrix} (X_1 + 1_k\delta_1) \\ (X_2 + 1_k\delta_2) \end{bmatrix}^T \Sigma^{-1} \begin{bmatrix} (\mu_1 - 1_k\alpha_1) \\ (\mu_2 - 1_k\alpha_2) \end{bmatrix}, \qquad (7.25)$$

$$T = 2 \begin{bmatrix} (X_2 - 1_k\zeta_1) \\ -(X_1 + 1_k\zeta_2) \end{bmatrix}^T \Sigma^{-1} \begin{bmatrix} (\mu_1 - 1_k\alpha_1) \\ (\mu_2 - 1_k\alpha_2) \end{bmatrix},$$

$$\alpha_i = A_{i1}\mu_1 + A_{i2}\mu_2, \quad \delta_i = -A_{i1}X_1 - A_{i2}X_2, \quad \zeta_i = A_{i1}X_2 - A_{i2}X_1,$$

and λ is the real root less than $\frac{1}{2}\left(P + Q - \sqrt{(P - Q)^2 + R^2} \right)$ of the quartic equation:

$$16\lambda^4 - 32(P + Q)\lambda^3 + [16(P^2 + Q^2) + 64PQ - 4(S^2 + T^2) - 8R^2]\lambda^2$$
$$+ [8R^2(P + Q) - 32PQ(P + Q) + 8(QS^2 + PT^2 - STR)]\lambda \qquad (7.26)$$
$$+ 16P^2Q^2 + R^4 - R^2(S^2 + T^2) + 4RST(P + Q)$$
$$- 4P^2T^2 - 4Q^2S^2 - 8PQR^2 = 0.$$

Note that a unique solution of Equation (7.26) that satisfies the constraint may not exist and it may be necessary to evaluate $D^2_{pCWP}(X, \mu; \Sigma)$ for several choices of λ or use numerical methods. Brignell *et al.* (2015) discuss further properties and extensions to general covariance weighted Procrustes with multiple observations.

Other forms of weighted Procrustes methods are when the observations $X_1, \ldots, X_n$ themselves have different weights, or a non-isotropic covariance structure. This situation is quite straightforward to deal with, involving weighted individual terms in the Procrustes sum of squares. The GPA algorithm is simply adapted by replacing the mean at each iteration by a weighted mean. This method is used by Zhou *et al.* (2016) with applications in medical imaging.

7.7   Shape variability: principal component analysis

As well as estimation of mean shape or mean size-and-shape, it is also of greater interest to describe the structure of shape or size-and-shape variability. One such measure is the population covariance matrix of the coordinates in a suitable tangent space to shape space or size-and-shape space, where the pole of the projection μ is a population mean, and the pole is the point where the tangent space touches the manifold. The population tangent covariance matrix is:

$$\Sigma = E[(V - E[V])(V - E[V])^T],$$

and $V \in T_\mu(M)$, where M is the manifold. The choice of tangent coordinates and inverse projection could be any suitable candidates from Section 4.4, for example inverse exponential and the exponential map; or partial tangent coordinates and inverse projection.

Given a random sample of data we can estimate the shape or size-and-shape of μ with a suitable estimator $\hat{\mu}$, and then project the data into the tangent space at $\hat{\mu}$. The sample covariance matrix of the tangent coordinates then provides an estimate of the shape or size-and-shape variability.

Note that we first describe PCA with Procrustes coordinates which is mathematically simple. Shape PCA requires a choice of base metric, called the Procrustes metric here, which is an arbitrary choice (Bookstein, 2015a). PCA with respect to other metrics is considered in Section 12.3.5, including relative warps.

7.7.1   Shape PCA

Cootes *et al.* (1992) and Kent (1994) developed PCA in a tangent space to shape space. In particular Kent (1994) proposed PCA of the partial Procrustes tangent coordinates defined in Equation (4.28), whereas Cootes *et al.* (1992) used PCA of the Procrustes residuals, which are approximate tangent coordinates (see Section 4.4.6).

The general method of sample PCA is the following:

- Choose a pole for the tangent coordinates $\hat{\mu}$.

- Calculate the tangent coordinates $v_i, i = 1, \ldots, n$, where $v_i \in T_{\hat{\mu}}(M)$.

- Compute

$$S_v = \frac{1}{n} \sum_{i=1}^{n} (v_i - \bar{v})(v_i - \bar{v})^T,$$

where $\bar{v} = \frac{1}{n} \sum_{i=1}^{n} v_i$.

- Calculate the eigenvalues $\lambda_j, j = 1, \ldots, p$ and corresponding eigenvectors γ_j of S_v.

- Calculate

$$v(c,j) = \bar{v} + c\lambda_j^{1/2} \gamma_j, j = 1, \ldots, p$$

for a range of values of c.

- Project back from $v(c,j)$ to a suitable icon $X_I(c)$ to examine the structure of the jth principal component (PC).

- The j PC scores are given by:

$$c_{ij} = \lambda_j^{1/2} \gamma_j^T v_i,$$

for $i \in 1, \ldots, n$. The number PCs with non-zero eigenvalues is $p = \min(n - 1, q)$ for shape and $p = \min(n - 1, q + 1)$ for size-and-shape, where $q = km - m - m(m - 1)/2 - 1$.

The structure of the jth PC can be seen by plotting icons $X_I(c,j)$ for a range of values of c. In particular we examine

$$v(c,j) = \bar{v} + c\lambda_j^{1/2} \gamma_j, \quad j = 1, \ldots, p, \tag{7.27}$$

for a range of values of the standardized PC score c and then project back into configuration space using Equation (4.34) and Equation (3.11). The linear transformation to an icon in the configuration space

$$X_I = \text{vec}_m^{-1}(\text{block diag}(H^T, \ldots, H^T)\{v(c,j) + \gamma\}) \tag{7.28}$$

is often a good approximation to the inverse projection from the tangent space to an icon, near the pole.

So, to evaluate the structure of the jth PC for a range of values of c, calculate v from Equation (7.27), project back using the inverse transformation to the pre-shape sphere and then evaluate an icon using, say, Equation (4.34) and Equation (3.11) or the linear approximation (7.28) to give the centred pre-shape.

There are several ways to visualize the structure of each PC:

1. Evaluate and plot an icon for a few values of $c \in [-3, 3]$, where $c = 0$ corresponds to the full Procrustes mean shape. The plots could either be separate or registered.

2. Draw vectors from the mean shape to the shape at $c = +3$ and/or $c = -3$ say to understand the structure of shape variability. The plots should clearly label which directions correspond to positive and negative c if both values are used.

3. Superimpose a square grid on the mean shape and deform the grid to icons in either direction along each PC. The methods of Chapter 12 will be useful for drawing the grids, and for example the thin-plate spline deformation could be used.

4. Animate a sequence of icons backwards and forwards along the range $c \in [-3, 3]$. This dynamic method is perhaps the most effective for displaying each PC.

In datasets where the shape variability is small it is often beneficial to magnify the range of c in order to easily visualize the structure of each PC.

In some datasets only a few PCs may be required to explain a high percentage of shape variability. Some PCs may correspond to interpretable aspects of variability (e.g. thickness, bending, shear) although interpretation is difficult due to the choice of Procrustes metric here (Bookstein, 2016). This can be improved using relative warps (see Chapter 12).

7.7.2   Kent's shape PCA

Following Kent (1992), consider n pre-shapes $Z_1, \ldots, Z_n$ with tangent space shape coordinates given by $v_1, \ldots, v_n$, with a pre-shape $\hat{\mu}$ corresponding to the full Procrustes mean shape as the pole, so

$$v_i = \left[I_{km-m} - \text{vec}(\hat{\mu})\text{vec}(\hat{\mu})^{\mathrm{T}}\right] \text{vec}(Z_i \hat{\Gamma}_i), \tag{7.29}$$

where each v_i is a real vector of length $(k-1)m$, obtained from Equation (4.33). Alternatively we could use the full Procrustes residuals r_i of Equation (8.13), or the inverse exponential map Procrustes tangent coordinates v_{Ei} of Equation (4.35).

Note that $\sum_{i=1}^{n} r_i = 0$ and $\sum_{i=1}^{n} v_i \approx 0$ and $\sum_{i=1}^{n} v_{Ei} \approx 0$.

The PC loadings γ_j are the orthonormal eigenvectors of the sample covariance of the tangent coordinates S_v, corresponding to eigenvalues $\lambda_j, j = 1, \ldots, p = \min(n - 1, q)$ (where q is the dimension of the shape space).

By carrying out PCA in the tangent space we are decomposing variability (the total sum of full Procrustes distances) into orthogonal components, with each PC successively explaining the highest variability in the data, subject to being orthogonal to the higher PCs. If the structure of shape variability is that the points are approximately independent and isotropic with equal variances, then the eigenvalues λ_j of the covariance matrix in tangent space will be approximately equal for 2D data (this property is proved in Section 11.1.6). If there are strong dependencies between landmarks, then only a few PCs may capture a large percentage of the variability.

An alternative decomposition which weights points close together differently from those far apart are relative warps (Bookstein 1991), described in Section 12.3.6. Further types of decomposition are independent components analysis (see Section 7.11) and the non-linear principal nested spheres (see Section 13.4.5).

7.7.3   Shape PCA in R

Carrying out tangent-based PCA is straightforward in the `shapes` library in R. The calculations are carried out in the routine `procGPA` and then plots are given using `shapepca`. To carry out PCA on Kent's partial tangent coordinates for the T2 small data (with $k = 6$ landmarks) we have:

```
ans1 <- procGPA(qset2.dat , tangentcoords="partial")
> ans1$percent
[1] 6.871271e+01 9.721554e+00 7.779814e+00 6.553396e+00 2.657204e+00
[6] 2.362407e+00 1.542515e+00 6.703960e-01 5.842414e-08 8.280442e-30
> ans1$pcasd
[1] 5.425114e-02 2.040601e-02 1.825471e-02 1.675419e-02 1.066848e-02
[6] 1.005930e-02 8.128399e-03 5.358657e-03 1.581927e-06 1.883289e-17
shapepca(ans1,joinline=c(1,6,2,3,4,5,1),type="r")
shapepca(ans1,joinline=c(1,6,2,3,4,5,1),type="v")
shapepca(ans1,joinline=c(1,6,2,3,4,5,1),type="s")
shapepca(ans1,joinline=c(1,6,2,3,4,5,1),type="g")
shapepca(ans1,joinline=c(1,6,2,3,4,5,1),type="m")
```

We see that the percentage of variability explained by the first three PCs are 68.7 9.7 and 7.8%, respectively, and there are $2k - 4 = 8$ PCs with non-zero variance (up to machine error). The PC scores are available in `ans1$scores` and the standard deviations for each PC are given in `ans1$pcasd`, as seen above, and these are the square root of the eigenvalues of the sample covariance matrix. The standardized scores (with sd=1) are in `ans1$stdscores`. In Figure 7.5 we see the first three PCs displayed in two format (using option `type="r"`). The other displays (not shown here) are `type` equal to `"v"`,`"s"`,`"g"` and `"m"` for vector, superposition, grid and movie representations , respectively.

If instead we use the Procrustes residuals we have:

```
ans2<- procGPA(qset2.dat , tangentcoords="residual")
> ans2$percent
[1] 6.843087e+01 9.701803e+00 7.832971e+00 6.554448e+00 2.671126e+00
[6] 2.366390e+00 1.586899e+00 6.752792e-01 1.802141e-01 5.786210e-08
[11] 1.188710e-28 8.399878e-29
> ans2$pcasd
[1] 9.415925e+00 3.545383e+00 3.185666e+00 2.914104e+00 1.860305e+00
[6] 1.750976e+00 1.433876e+00 9.353597e-01 4.832050e-01 2.738004e-04
[11] 1.241009e-14 1.043214e-14
```

and we see now that there are now nine non-zero variances, and the extra dimension of variability arises because the Procrustes residuals do not lie exactly in a tangent space. Note that the Procrustes residuals are on the overall scale of the original data – the figures have not been rescaled to the pre-shape sphere as in the partial Procrustes tangent space approach. However, the percentages of variability and structure of the PCs are very similar, apart from the arbitrary overall scaling in the residual PCA.

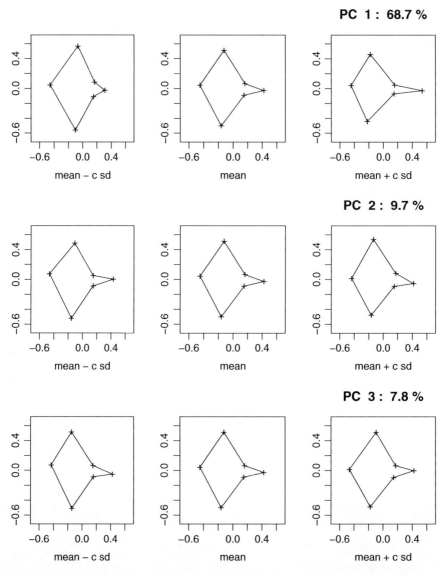

Figure 7.5   Plots of the first three PCs. In the jth row: mean − 3sd PCj, mean, mean + 3sd PCj (where j = 1, 2, 3).

For a 3D example we consider the male macaque data:

```
ans3<-procGPA(macm.dat,tangentcoords="partial")
shapepca(ans3)
ans3$percent
[1] 4.740113e+01 2.083676e+01 1.286306e+01 8.373567e+00 5.866786e+00
```

```
[6] 2.650214e+00 1.292852e+00 7.156279e-01 5.549779e-31
shapepca(ans3)
shapepca(ans3,pcno=1,type="g",zslice=-1)
```

The default plot for the 3D PCA is a plot of the mean shape, with vectors drawn to the mean + 3sd PCj, $j = 1, 2, 3$, as seen in Figure 7.6. Figure 7.6 also shows a deformed grid which gives an indication of the main shape variability in PC1. The first PC explains 47.4% of the variability here.

It can be seen that the main variability in the landmarks is in the 'top most' landmark in the x–y plane (*bregma*), which is difficult to locate in primates. Referring back to Figure 7.3 it can be seen that there are two males with unusual 'top-most' landmarks (*bregma*) and this appears to give rise to the extra variability in these landmarks as seen in the first PC.

Example 7.3   A random sample of 23 T2 mouse vertebral outlines was taken from the Small group of mice introduced in Section 1.4.1. Six mathematical landmarks are located on each outline, and in between each pair of landmarks 9 equally spaced pseudo-landmarks were placed (as in Figure 1.4), giving a total of $k = 60$ landmarks in $m = 2$ dimensions. In Figure 7.7 we see the Procrustes registered outlines.

The sample covariance matrix in the tangent space [using the partial Procrustes coordinates of Equation (4.28)] is evaluated and in Figure 7.8 we see sequences of shapes evaluated along the first two PCs. Alternative representations are given in Figure 7.9, Figure 7.10 and Figure 7.11. The R code for producing the plots is:

```
data(mice)
t2<-mice$outlines[,,mice$group=="s"]
ans<-procGPA(t2,tangentcoords="partial")
x<-ans$rotated
plotshapes(ans$rotated,joinline=c(1:60,1))
shapepca(ans,type="r",mag=2,joinline=c(1:60,1),pcno=c(1:2))
shapepca(ans,type="v",mag=2,joinline=c(1:60,1),pcno=c(1:2))
shapepca(ans,type="s",mag=2,joinline=c(1:60,1),pcno=c(1:2))
shapepca(ans,type="g",mag=2,joinline=c(1:60,1),pcno=c(1:2))
pairs(cbind(ans$size,ans$rho,ans$scores[,1],ans$scores[,2],
    ans$scores[,3]),label=c("size","rho","pc1","pc2","pc3"))
```

Shapes are evaluated in the tangent space and then projected back using the approximate linear inverse transformation of Equation (7.28) for visualization. The percentages of variability captured by the first two PCs are 64.6 and 8.8%, so the first PC is a very strong component here.

The first PC appears to highlight the length of the spinous process (the protrusion on the 'top' of the bone) in contrast to the relative width of the bone. The angle between lines joining landmarks 1 to 5 and 2 to 3 decreases as the height of landmark 4 increases, whereas there is little change in the angles from the lines

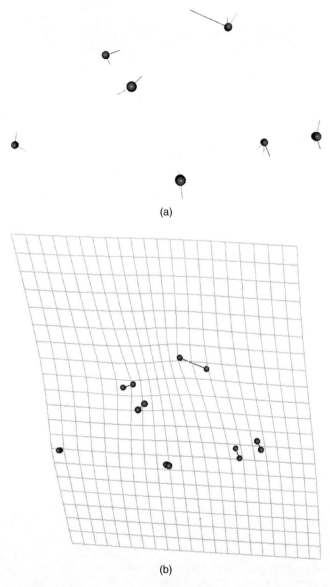

(a)

(b)

Figure 7.6   (a) Plots of the mean (red spheres) with vectors to figures along the first three PCs: (black) mean + 3sd PC1; (red) mean + 3sd PC2; and (green) mean + 3sd PC3. (b) The mean (red) and a figure at mean + 3sd PC1 (blue) with a deformed grid on the blue figure at $z = -1$, which was deformed from being a square grid on the red figure at $z = -1$. For a colour version of this figure, see the colour plate section.

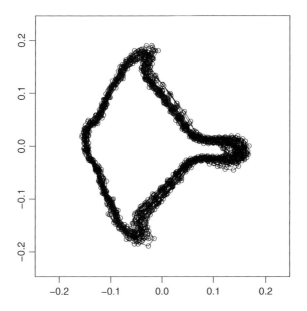

Figure 7.7   Procrustes rotated outlines of T2 Small mouse vertebrae.

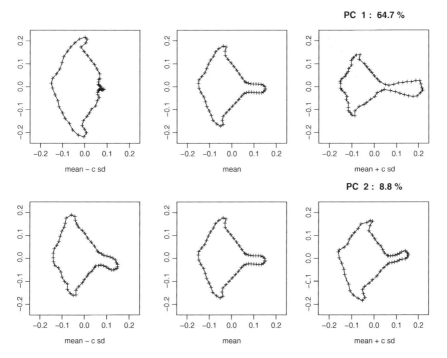

Figure 7.8   Two rows of series of T2 vertebral shapes evaluated along the first two PCs – the ith row shows the shapes at $c \in \{-6, 0, 6\}$ standard deviations along the ith PC. Note that in each row the middle plot ($c = 0$) is the full Procrustes mean shape. By magnifying the usual range of c by 2 the structure of each PC is more clearly illustrated.

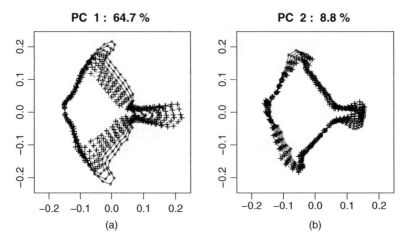

Figure 7.9   The first (a) and second (b) PCs for the T2 Small vertebral data. The plot shows the icons overlaid on the same picture. Each plot shows the shapes at $c \in \{-6, -4, -2\}$ (——), the mean shape at $c = 0$ (circled +) and the shapes at $c \in \{+6, +4, +2\}$ (…+…) standard deviations along each PC.*

joining 1 to 6 and 2 to 6. The second PC highlights the pattern of asymmetry in the end of the spinous process and asymmetry in the rest of the bone.

Pairwise plots of the elements of the vector $(s_i, \rho_i, c_{i1}, c_{i2}, c_{i3})^{\mathrm{T}}$, $i = 1, \ldots, n$, are given in Figure 7.12, where s_i are the centroid sizes, ρ_i are the Riemannian distances to the mean, and c_{i1}, c_{i2} and c_{i3} are the first three standardized PC scores.

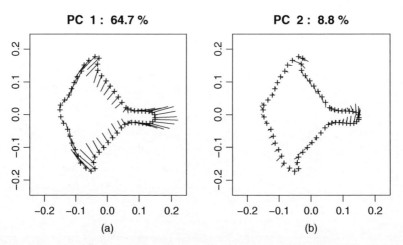

Figure 7.10   The first (a) and second (b) PCs for the T2 Small vertebral outline data. Each plot shows the full Procrustes mean shape with vectors drawn from the mean (+) to an icon which is $c = +6$ standard deviations along each PC from the mean shape.

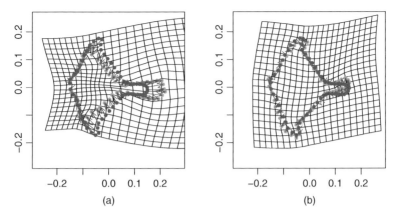

(a)   (b)

Figure 7.11 The first (a) and second (b) PCs for the T2 Small vertebral outline data. A square grid is drawn on the mean shape and deformed using a pair of thin-plate splines (see Chapter 12) to an icon $c = 6$ standard deviations along each PC (indicated by a vector from the mean to the icon). The plots just show the deformed grid at $c = 6$ for each PC and not the starting grids on the mean. For a colour version of this figure, see the colour plate section.

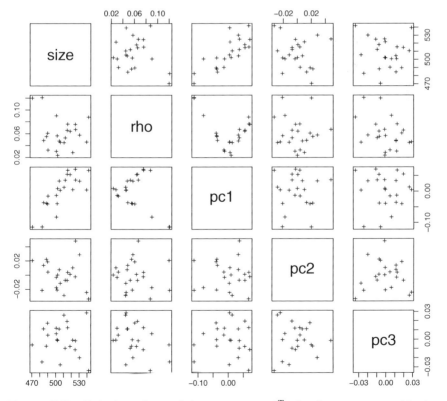

Figure 7.12 Pairwise plots of $(s_i, \rho_i, c_{i1}, c_{i2}, c_{i3})^{\mathrm{T}}$, $i = 1, \ldots, n$, centroid size, Riemannian distance to the mean shape and the first three standardized PC scores, for the T2 Small vertebral outline data.

There appears to be one bone that is much smaller than the rest and it also appears that there is some correlation between the first PC score and the centroid size of the bones.

An overall measure of shape variability is the root mean square of full Procrustes distance $RMS(d_F)$, which here is 0.07, and the shape variability in the data is quite small. ☐

Example 7.4 A random sample was taken of 30 handwritten digit number 3s in the dataset of Section 1.4.2. Thirteen landmarks were located by hand on images of each of the digits, and here $k = 13$ and $m = 2$. The full Procrustes rotated figures are displayed in Figure 7.13. The pairwise plots of $(s_i, \rho_i, c_{i1}, c_{i2}, c_{i3})^T$, $i = 1, \ldots, n$, the centroid size, the Riemannian distance ρ_i to the mean and the first three PC scores are given in Figure 7.14. The first three PCs explain 50.4, 15.4, 12.8, 7.5 and 4.3% of the variability. There is quite a large amount of shape variability in these data – the root mean square of Riemannian distance $RMS(\rho)$ is 0.274.

The PCs are displayed in Figure 7.15 and Figure 7.16. The first PC can be interpreted as partly capturing the amount that the central part (middle prong of the number 3) protrudes in contrast to the degree of curl in the end of the bottom loop and the

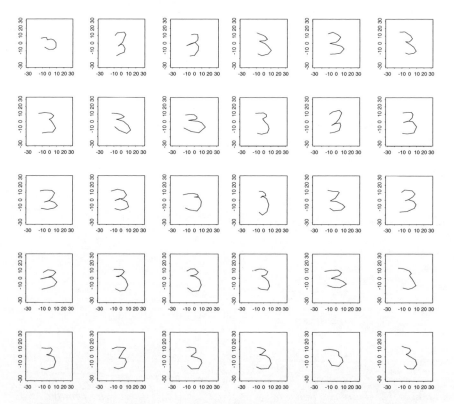

Figure 7.13 The full Procrustes coordinates for all 30 handwritten digits.

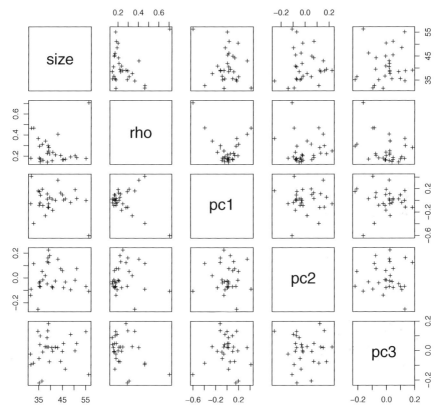

Figure 7.14 Pairwise plots of $(s_i, \rho_i, c_{i1}, c_{i2}, c_{i3})^{\mathrm{T}}$, $i = 1, \ldots, n$, *the centroid size, Riemannian distance* ρ *to the mean shape and the first three PC scores, for the digit 3 data. There appears to be an outlier with a particularly large value of* ρ. *Closer inspection indicates that the first digit may have poorly identified landmarks.*

length of the top loop. The second PC includes measurement of tall thin digits versus short fat digits (vertical/horizontal shear).

In Figure 7.14 there is a positive relationship between the absolute value of the score of PC1 (c_{i1}) and the Riemannian distance ρ_i to the mean. This is not surprising and should be expected in most datasets, as shapes near to the mean will have the small scores along the PCs.

The R code for producing the Digit 3 PCA plots is:

```
data(digit3.dat)
ans<-procGPA(digit3.dat,tangentcoords="partial")
shapepca(ans,type="r",mag=1,joinline=c(1:13),pcno=c(1:2))
shapepca(ans,type="s",mag=1,joinline=c(1:60,1),pcno=c(1:2))
pairs(cbind(ans$size,ans$rho,ans$scores[,1],ans$scores[,2],
    ans$scores[,3]),label=c("size","rho","pc1","pc2","pc3"))
```

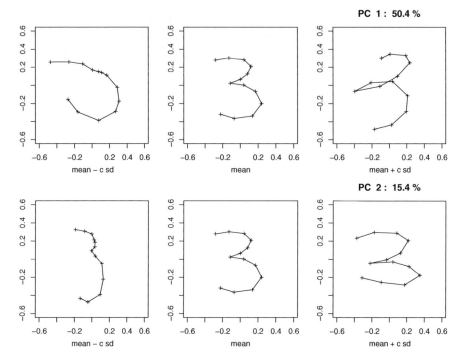

Figure 7.15   *Principal component analysis of the digit number 3s. The ith row represents the ith PC, with configurations evaluated at* $-3, 0, 3$ *standard deviations along each PC from the Procrustes mean. The central figure on each row* $(c = 0)$ *is the Procrustes mean shape.*

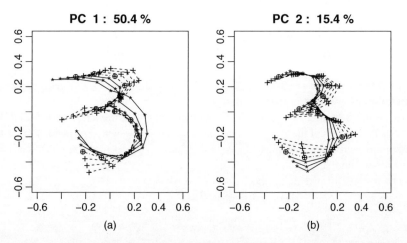

Figure 7.16   *The first three PCs for the digit 3 data from the first (a) and second (b) PCs. Each plot shows the shapes at* $c \in \{-3, -2, -1\}$ *(—*—), the mean shape at* $c = 0$ *(circled +) and the shapes at* $c \in \{+3, +2, +1\}$ *(…+…) standard deviations along each PC.*

There does appear to be one outlier in the dataset with large ρ. On closer inspection, the first digit in the dataset has its landmarks placed very poorly (landmark 7 is placed on the bottom loop where landmark 4 should be, and the other nearby points correspond poorly). The analysis was re-performed without the first digit and the root mean square of Procrustes distance $RMS(\rho)$ is 0.26 and the first five PCs are similar to the first analysis with all the data and the percentages of variability explained by the first five PCs are 43.6, 18.4, 13.2, 9.0 and 4.9%. The similar structure in the first five PCs and the drop in the relative contribution of the first PC is expected, as the outlier digit has a particularly large score on the first PC.   □

There have been numerous applications of PCA in shape analysis, and it can provide a useful way of exploring low dimensional structures in shape data. Some further applications include face identification (Mallet *et al.* 2010; Morecroft *et al.* 2010; Evison *et al.* 2010), computer vision (Dryden 2003), modelling plant roots (Hodgman *et al.* 2006), and examining brain shape in epilepsy (Free *et al.* 2001). An in-depth discussion of shape PCA is given by Bookstein (2015b).

7.7.4 Point distribution models

Principal component analysis with the full Procrustes coordinates of Equation (8.13) has a particularly simple formulation. Cootes *et al.* (1992, 1994) use PCA to develop the **point distribution model** (PDM), which is a PC model for shape and uses Procrustes residuals rather than tangent coordinates. Given n independent configuration matrices $X_1, \ldots, X_n$ the figures are registered to $X_1^P, \ldots, X_n^P$ by full GPA. The estimate of mean shape is taken to be the full Procrustes mean $\hat{\mu}$ which has the same shape as $\bar{X} = \frac{1}{n} \sum_{i=1}^{n} X_i^P$. The sample covariance matrix is:

$$\frac{1}{n} \sum_{i=1}^{n} \text{vec}(X_i^P - \bar{X})(\text{vec}(X_i^P - \bar{X}))^{\text{T}}$$

and the PCs are the eigenvectors of this matrix, $\gamma_j, j = 1, \ldots, \min(n-1, q)$, with corresponding decreasing eigenvalues λ_j and $q = (k-1)m - m(m-1)/2 - 1$ is the dimension of the shape space. Note that PCA using this formulation is the same (up to an overall scaling) as using the full Procrustes tangent coordinates of Equation (8.13), with the tangent coordinates pre-multiplied by H^{T}, the transpose of the Helmert submatrix.

Visualization of the PCs is carried out as in the previous section. In particular, the structure in the jth PC can be viewed through plots of an icon for mean shape $\hat{\mu}$ with displacement vectors

$$\text{vec}(\hat{\mu}) + c\lambda_j^{1/2}\gamma_j$$

for the shapes corresponding to $c \in [-3, 3]$.

The approaches using partial or full Procrustes tangent coordinates or the PDMs are almost identical in practice, for datasets with small variability. The PDM can be thought of as a structural model in the tangent space.

Cootes *et al.* (1994) give several early examples of PDMs, and develop flexible models for describing shape variability in various datasets of images of objects, including hands, resistors and heart ventricles.

Example 7.5   An example of the PDM approach is given in Figure 7.17 taken from Cootes *et al.* (1994), who describe a flexible model for describing shape variability in hands. In Figure 7.17 the first PC highlights the spreading of the fingers, the second PC highlights movement of the thumb relative to the fingers, and the third PC highlights movement of the middle finger. **The shape variability here is complicated because there are multiple patterns.** As well as the biological shape variability of hands, the relative positions of the fingers contributes greatly to shape variability here.   □

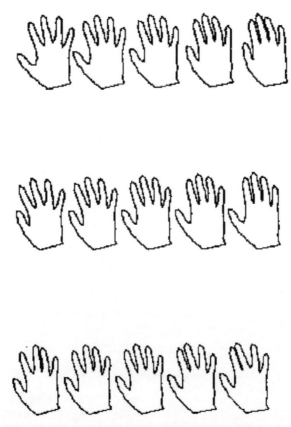

Figure 7.17   Varying hands: the first three PCs with values of $c \in \{-2, -1, 0, 1, 2\}$ here. (Reproduced by permission of Carfax Publishing Ltd.) Source: Cootes et al. 1994.

7.7.4.1   PDM in two dimensions

We can formulate the PDM for a 2D configuration matrix $X(2k \times 1)$ of k landmarks in $\mathbb{R}^2$ as (Mardia 1997):

$$X = \mu + \sum_{j=1}^{p} y_j \gamma_j + \epsilon, \tag{7.30}$$

where $y_j \sim N(0, \lambda_j)$, $\epsilon \sim N_{2k}(0, \sigma^2 I)$, independently and the vectors γ_i satisfy

$$\mu^T \gamma_j = 0, \gamma_j^T \gamma_j = 1, \gamma_i^T \gamma_j = 0, \quad i \neq j,$$

and $\lambda_1 \geq \lambda_2 \geq \cdots \geq \lambda_p$. In addition, for invariance under rotation and for translation, the vectors γ_i satisfy, respectively

$$\gamma_j^T v = 0 \text{ and } \gamma_j^T (1, \ldots, 1, 0, \ldots, 0)^T = 0, \gamma_j^T (0, \ldots, 0, 1, \ldots, 1)^T = 0,$$

where $v = (-\beta_1, \ldots, -\beta_k, \alpha_1, \ldots, \alpha_k)^T$ with $\mu = (\alpha_1, \ldots, \alpha_k, \beta_1, \ldots, \beta_k)^T$. Here $p \leq \min(n - 1, 2k - 4)$ and p is preferably taken to be quite small, for a parsimonious model.

7.7.5   PCA in shape analysis and multivariate analysis

We shall highlight some similarities with PCA in shape analysis with PCA in multivariate analysis. In the standard implementation PCA is used to summarize succinctly the main modes of variation in a dataset, and the principle is the same in shape analysis – we are looking for orthogonal axes of shape variation which summarize large percentages of the shape variability.

In conventional PCA the coefficients with largest absolute value in each PC are interpreted as patterns of interest in the dataset. For example, consider the PCA of a set of length measurements taken from a biological specimen, for example an ape skull. The first PC may give approximately equal weight to each variable; thus, it could be interpreted as a measure of overall size. The second PC might have positive loadings for face measurements and negative loadings for braincase measurements, and so it measures the relative magnitudes of these two regions.

However, in shape PCA it is difficult to interpret the PCs in terms of effects on the original landmarks. Bookstein (2016) argues that coefficients from a PCA from Procrustes registered data cannot be interpreted as effects that have any sense in biological applications, as the Procrustes geometry is entirely a function of the particular selection of landmarks. In Section 12.3.6 we encounter an important variant of PCA relative warps – which provides a different orthogonal basis and typically decomposes shape variability at a variety of scales.

Principal component analysis is also used in multivariate analysis as a dimension reducing technique, and this is also one of the goals of PCA in shape analysis.

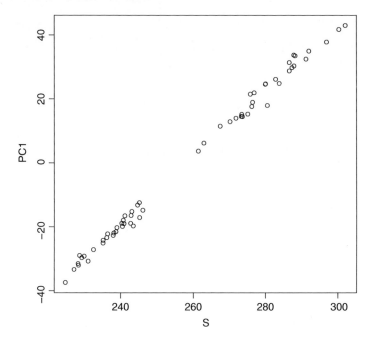

Figure 7.18   Plot of size-and-shape PC1 scores versus centroid size, for the gorilla data. Note the females all have centroid size less than 247 and the males are all greater than 261.

7.8   Principal component analysis for size-and-shape

As for shape analysis we can carry out PCA in the tangent space to size-and-shape space. Because size variability is often a major aspect of the overall variability, PC1 will often involve size. For example we consider a size-and-shape PCA of the combined male and female gorilla data of Section 1.4.8, and we plot PC1 versus centroid size in Figure 7.18. There is clearly an extremely strong correlation between the size-and-shape PC1 scores and centroid size (correlation = 0.9975), and the size-and-shape PC2 scores are strongly correlated with the shape PC1 (correlation 0.6429).

7.9   Canonical variate analysis

As the tangent space is a Euclidean vector space we can carry out the usual techniques from multivariate analysis in this space. For example if we have at least two groups we can consider canonical variate analysis Mardia *et al.* (1979, Section 12.5). The canonical variates are linear combinations of the variables which best separate the groups, and the same components are used in Fisher's linear discriminant

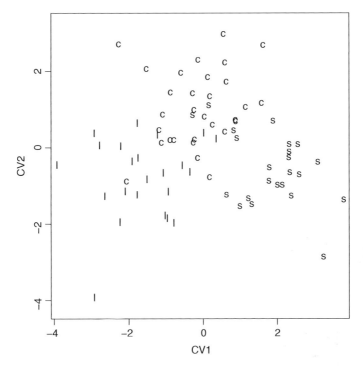

Figure 7.19   Plots of the first two canonical variates for the T2 mouse vertebral data. The three groups are Large (l), Small (s) and Control (c).

analysis. For the mouse vertebral data canonical variate analysis can be carried out in R using:

```
data(mice)
shapes.cva(mice$x,mice$group)
```

The method involves full Procrustes analysis and uses the Procrustes residuals for the shape canonical variate analysis. A plot of the first two canonical variates is given in Figure 7.19. We see that there is good separation in the three groups using the shape canonical variate analysis.

To carry out shape canonical variate analysis on the great apes dataset we have:

```
data(apes)
shapes.cva(apes$x,apes$group)
```

In Figure 7.20 we see there is complete separation of the three species of great ape. The female and male gorillas are very different, and also the orangutans are quite well separated by sex. However, there is a lot of overlap in the male and female groups

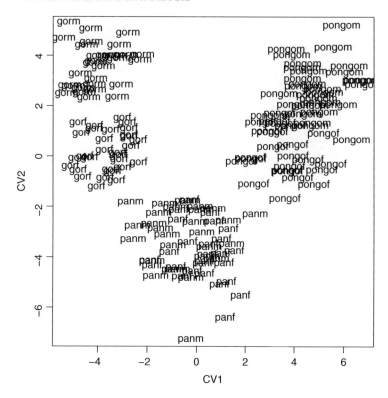

Figure 7.20   Plots of the first two canonical variates for the great ape data. The six groups are male gorillas (gorm), female gorillas (gorf), male chimpanzees (panm), female chimpanzees (panf), male orang utans (pongom) and female orang utans (pongof).

for the chimpanzees, indicating less shape difference between the sexes in this group. Note that the underlying assumption of equal covariance matrices may not hold, as in this example, but nevertheless it is a useful representation.

An alternative to canonical correlation analysis for investigating associations and dimension reduction in groups of shapes was introduced by Iaci *et al.* (2008). The method was based on maximizing Kullback–Leibler information and was applied to a dataset of fly wings and mouse mandible data.

7.10   Discriminant analysis

Fisher's linear discriminant or quadratic discriminant analysis can be carried out on Procrustes tangent coordinates when there are two groups for classification. Mardia *et al.* (2013a) gave an example of quadratic discriminant analysis used in a court case for a prisoner on Death Row, introduced in Section 1.4.6. Quadratic discriminant analysis is often more appropriate as group covariance matrices are usually different.

It was of importance to determine whether the prisoner had suffered brain damage through FASD or not. Using a previous study of controls versus FASD patients there was a difference in the average corpus callosum shape around a particular part on the upper right isthmus region (thin section before the bulge on the right end in Figure 1.16). The FASD patients had a thinner region on average. Using a model based on Procrustes tangent coordinates, the likelihood ratio was very strongly in favour of the prisoner being classified in the FASD group. The resulting testimony in the courtroom, in addition to other expert observations, led to the prisoner being spared the death penalty in favour of life in prison.

Some other methods for discriminant analysis are also available, for example, one based on neural networks Southworth *et al.* (2000). A model-based approach for triangular shape is given in Kent and Mardia (2013).

7.11   Independent component analysis

As a further indication of the type of analysis that can be carried out we consider independent component analysis (ICA) (Hyvärinen *et al.* 2001), which seeks the most non-Gaussian directions of variability. There are many types of ICA and we use `fastICA`, which is available in an R library (Marchini *et al.* 2013). In practice one normally carries out ICA on a reduced set of PC scores if the dimension is high. The commands to carry out ICA on the mouse vertebral outline data, after first reducing to 3 PC scores from partial Procrustes analysis are:

```
library(fastICA)
nic<-3
proc<-procGPA(mice$outlines,scale=FALSE)
ans<-fastICA(proc$scores,nic)
ICscores<-t(ans$S)
par(mfrow=c(2,3))
colgrp<-as.integer(mice$group)+1
plot(ICscores[1,],ICscores[2,],type="n",xlab="ic1",ylab="ic2")
text(ICscores[1,],ICscores[2,],mice$group,col=colgrp)
plot(ICscores[1,],ICscores[3,],type="n",xlab="ic1",ylab="ic3")
text(ICscores[1,],ICscores[3,],mice$group,col=colgrp)
plot(ICscores[2,],ICscores[3,],type="n",xlab="ic2",ylab="ic3")
text(ICscores[2,],ICscores[3,],mice$group,col=colgrp)
PCscores<-t(proc$scores)
plot(PCscores[1,],PCscores[2,],type="n",xlab="pc1",ylab="pc2")
text(PCscores[1,],PCscores[2,],mice$group,col=colgrp)
plot(PCscores[1,],PCscores[3,],type="n",xlab="pc1",ylab="pc3")
text(PCscores[1,],PCscores[3,],mice$group,col=colgrp)
plot(PCscores[2,],PCscores[3,],type="n",xlab="pc2",ylab="pc3")
text(PCscores[2,],PCscores[3,],mice$group,col=colgrp)
```

In Figure 7.21 we see plots of the independent component (IC) scores and PC scores for the mouse vertebral outline data. We see that IC1 versus IC3 give quite good separation of the three groups, as indeed do PC1 versus PC2.

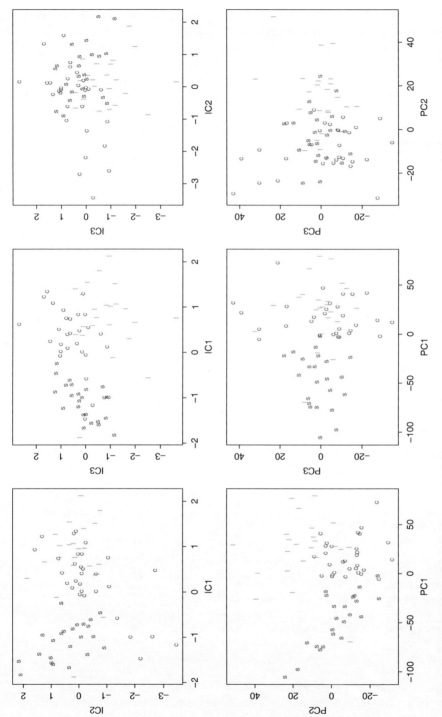

Figure 7.21 Plot of the first three independent components scores (top row), and the first three PC scores (bottom row) for the mouse vertebral outline data. The observations are labelled by group: Control (c); Large (l); and Small (s). The ordering of the ICs is arbitrary. For a colour version of this figure, see the colour plate section.

Brignell *et al.* (2010) gave a further example of using ICA in shape analysis, with an application to brain surface asymmetry in schizophrenia.

7.12   Bilateral symmetry

A particularly important concept in biology is bilateral symmetry, which is symmetry about a plane (as in a mirror image). Kent and Mardia (2001) study the topic in theoretical detail, and in particular provide detailed algebraic decompositions that can be used for a practical investigation of bilateral symmetry. Other investigations include Mardia *et al.* (2000) who provide a statistical assessment of bilateral symmetry of shapes and Bock and Bowman (2006) who consider the measurement and analysis of asymmetry with applications to face modelling, including measuring symmetry after cleft lip surgery in children. Brignell *et al.* (2010) study the application described in Section 1.4.15 where the shape of the cortical surface is investigated. A special type of symmetry was observed called brain torque, where the schizophrenia patients had slightly more symmetric brains on average compared with the control group. Theobald *et al.* (2004) decomposed shape variability into symmetric and asymmetric components, which was also used by Brignell *et al.* (2010). The method is very simple, in that a dataset is supplemented by relabelled mirror images (relabelled reflections) before carrying out PCA.

Example 7.6   Consider the T2 Small data of Section 1.4.1. The percentages explained by the first four PCs after full Procrustes analysis and PCA are 68.4, 9.7, 7.8, and 6.5%, respectively. In Figure 7.22 we see the raw data and the reflected relabelled data, where the y coordinates are made negative and points $\{1, 2, 3, 4, 5, 6\}$ are

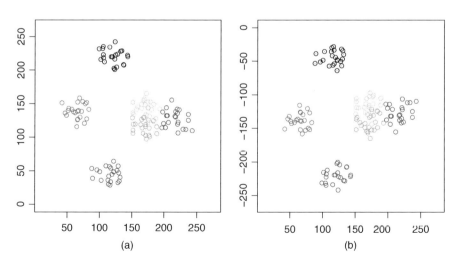

Figure 7.22   The T2 Small landmarks (a) and the reflected relabelled landmarks (b). For a colour version of this figure, see the colour plate section.

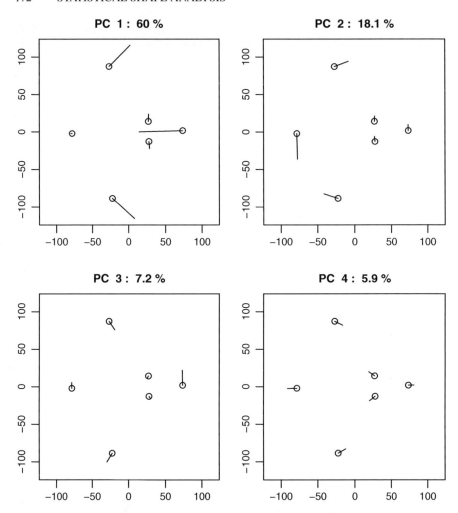

Figure 7.23   The symmetric and asymmetric PCs from the T2 Small vertebrae, with the bilateral symmetric mean and PC vectors magnified three times.

relabelled as {2, 1, 5, 4, 3, 6}. Full Procrustes analysis is carried out and the first four PC vectors are displayed in Figure 7.23. The R code for the analysis is:

```
refx<-qset2.dat
refx[,1,]<- qset2.dat[c(2,1,5,4,3,6),1,]
refx[,2,]<- -qset2.dat[c(2,1,5,4,3,6),2,]
plotshapes(qset2.dat,refx,col=1:6)
xdat<-abind(qset2.dat,refx)
out<-procGPA(xdat)
shapepca(out,type="v",mag=3,pnco=1:4)
```

Note that in this analysis the Procrustes mean shape has bilateral symmetry. We see that PC1 and PC4 are symmetric PCs (with 60.0 and 5.9% variability, respectively) whereas PC2 and PC3 are asymmetric (with 18.1 and 7.2% variability, respectively). The advantage of the symmetry/asymmetry decomposition is that the PCs can be more interpretable. Here the symmetric PC scores are projections of each case into a subspace. Further types of projections (affine components and partial warps) are considered in Chapter 12.   □

Fluctuating asymmetry is a type of deviation from bilateral symmetry, where the deviations are randomly distributed about a bilateral symmetric mean value. Low levels of fluctuating asymmetry often demonstrate only small environmental perturbations of genetic development programmes. Klingenberg and McIntyre (1998) and Klingenberg (2011) discuss appropriate methodology which is implemented in the MorphoJ software (Klingenberg 2011) for investigating fluncuating asymmetry. There are numerous other specialist applications in morphometrics, for example Cardini (2014) discusses the use of 2D images as proxies for 3D shapes, again using MorphoJ. Other software for analysing specialized biological applications includes tpsRelw (Rohlf 2010), geomorph (Adams and Otárola-Castillo 2013), PAST (Hammer et al. 2001) and the Evan toolbox (Phillips et al. 2010).

8

2D Procrustes analysis using complex arithmetic

8.1 Introduction

This chapter provides an overview of the techniques introduced so far when a random sample of landmark configurations is available in two dimensions. In this chapter we make particular use of complex numbers which leads to simple methodology in this important case. Although most of the material for this chapter has been described using general matrix notation in previous chapters, it is often much simpler to use complex vectors in the 2D case. This chapter is designed to be largely self-contained for planar shape analysis using complex notation.

8.2 Shape distance and Procrustes matching

Consider two centred configurations $y = (y_1, \ldots, y_k)^T$ and $w = (w_1, \ldots, w_k)^T$, both in $\mathbb{C}^k$, with $y^* 1_k = 0 = w^* 1_k$, where y^* denotes the transpose of the complex conjugate of y. In order to compare the configurations in shape we need to establish a measure of distance between the two shapes.

A suitable procedure is to match w to y using the similarity transformations and the differences between the fitted and observed y indicate the magnitude of the difference in shape between w and y. Consider the complex regression equation

$$
\begin{aligned}
y &= (a + ib)1_k + \beta e^{i\theta} w + \epsilon \\
&= [1_k, w]A + \epsilon \\
&= X_D A + \epsilon,
\end{aligned}
\tag{8.1}
$$

Statistical Shape Analysis, with Applications in R, Second Edition. Ian L. Dryden and Kanti V. Mardia.
© 2016 John Wiley & Sons, Ltd. Published 2016 by John Wiley & Sons, Ltd.

where $A = (A_1, A_2)^T = (a + ib, \beta e^{i\theta})^T$ are the 2×1 complex parameters with translation $a + ib$, scale $\beta > 0$ and rotation $0 \le \theta < 2\pi$; ϵ is a $k \times 1$ complex error vector; and $X_D = [1_k, w]$ is the $k \times 2$ 'design matrix'. To carry out the registration we could estimate A by minimizing the least squares objective function, the sum of square errors

$$D^2(y, w) = \epsilon^* \epsilon = (y - X_D A)^*(y - X_D A).$$

The full Procrustes registration of w on y is obtained by estimating A with $\hat{A}$, where

$$\hat{A} = (\hat{a} + i\hat{b}, \hat{\beta} e^{i\hat{\theta}})^T = \arginf \epsilon^* \epsilon = \arginf(y - X_D A)^*(y - X_D A).$$

Definition 8.1   *The **full Procrustes fit (registration)** of w onto y is:*

$$w^P = X_D \hat{A} = (\hat{a} + i\hat{b})1_k + \hat{\beta} e^{i\hat{\theta}} w,$$

where $(\hat{\beta}, \hat{\theta}, \hat{a}, \hat{b})$ are chosen to minimize

$$D^2(y, w) = \| y - w\beta e^{i\theta} - (a + ib)1_k \|^2.$$

Result 8.1   *The full Procrustes fit has matching parameters*

$$\hat{a} + i\hat{b} = 0, \tag{8.2}$$
$$\hat{\theta} = \text{Arg}(w^* y) = -\text{Arg}(y^* w), \tag{8.3}$$
$$\hat{\beta} = (w^* y y^* w)^{1/2}/(w^* w). \tag{8.4}$$

Proof: We wish to minimize (over β, θ, a, b) the expression

$$\begin{aligned} D^2 &= \epsilon^* \epsilon \\ &= \| y - w\beta e^{i\theta} - (a + ib)1_k \|^2 \\ &= y^* y + \beta^2 w^* w - y^* w\beta e^{i\theta} - w^* y\beta e^{-i\theta} + k(a^2 + b^2) \end{aligned} \tag{8.5}$$

(remember y and w are centred). Clearly, the minimizing a and b are zero. Let $y^* w = \gamma e^{i\phi}$ ($\gamma \ge 0$) and then

$$\beta(y^* w e^{i\theta} + w^* y e^{-i\theta}) = \beta(\gamma e^{i(\theta+\phi)} + \gamma e^{-i(\theta+\phi)}) = 2\beta\gamma \cos(\theta + \phi).$$

So to minimize $\| y - \beta e^{i\theta} w \|^2$ over θ we need to maximize $2\beta\gamma \cos(\theta + \phi)$. Clearly, a solution for θ is $\hat{\theta} = -\phi = -\text{Arg}(y^* w)$. To find the minimizing scale we solve

$$\frac{\partial D^2}{\partial \beta} = 0 = 2\beta w^* w - 2\gamma,$$

where $\gamma = |y^*w|$. Hence,

$$\hat{\beta} = |y^*w|/(w^*w) = (w^*yy^*w)^{1/2}/(w^*w),$$

as required.   □

The solution is the standard least squares solution (but with complex variables) and we can write the solution in the familiar form

$$\hat{A} = (\hat{A}_1, \hat{A}_2)^{\mathrm{T}} = (X_D^*X_D)^{-1}X_D^*y \Rightarrow \hat{A}_1 = 0, \quad \hat{A}_2 = w^*y/(w^*w). \qquad (8.6)$$

Note that the full Procrustes fit of w onto y is given explicitly by:

$$w^P = X_D\hat{A} = \hat{\beta}e^{i\hat{\theta}}w = w^*yw/(w^*w).$$

The residual vector $r = y - X_D\hat{A}$ is given by:

$$r = [I_k - X_D(X_D^*X_D)^{-1}X_D^*]y = (I_k - H_{hat})y$$

where H_{hat} is the 'hat' matrix for X_D, that is

$$H_{hat} = X_D(X_D^*X_D)^{-1}X_D^*.$$

The minimized value of the objective function is:

$$D^2(r,0) = r^*r = y^*y - (y^*ww^*y)/(w^*w). \qquad (8.7)$$

Now this expression is not symmetric in y and w unless $y^*y = w^*w$. A convenient standardization is to take the configurations to be unit size, that is

$$\sqrt{y^*y} = \sqrt{w^*w} = 1.$$

So, if we include standardization, then we obtain a suitable measure of shape distance.

Definition 8.2 *The **full Procrustes distance** between complex configurations w and y is given by:*

$$d_F(w, y) = \inf_{\beta,\theta,a,b} \left\| \frac{y}{\|y\|} - \frac{w}{\|w\|}\beta e^{i\theta} - 1_k(a + ib) \right\|$$

$$= \left\{ 1 - \frac{y^*ww^*y}{w^*wy^*y} \right\}^{1/2}. \qquad (8.8)$$

The expression for the distance follows from Equation (8.7).

Note that the full Procrustes fit of w onto y is actually obtained by complex linear regression of y on w, which is a simpler procedure than working with minimization over rotation matrices as in Chapter 3.

This is not the only choice of distance between shapes, and further choices of distance were considered in Chapter 3. However, the full Procrustes distance is natural from a statistical point of view, obtained from a least squares criterion and optimizing over the full set of similarity parameters. The squared full Procrustes distance naturally appears exponentiated in the density for many simple probability distributions for shape, as we shall see in Chapter 10.

8.3   Estimation of mean shape

Consider the situation where a random sample of configurations $w_1, \dots, w_n$ is available and we wish to estimate a population mean shape, such as the population full Procrustes mean.

Definition 8.3   *The full Procrustes estimate of mean shape $[\hat{\mu}]$ is obtained by minimizing (over μ) the sum of square full Procrustes distances from each w_i to an unknown unit size mean configuration μ, that is*

$$[\hat{\mu}] = \arg\inf_{\mu} \sum_{i=1}^{n} d_F^2(w_i, \mu),$$

as previously seen in Equation (6.11). Again we us assume that the configurations $w_1, \dots, w_n$ have been centred, so that $w_i^* 1_k = 0$.

Result 8.2   *(Kent 1994) The full Procrustes mean shape $[\hat{\mu}]$ can be found as the eigenvector corresponding to the largest eigenvalue of the **complex sum of squares and products matrix***

$$S = \sum_{i=1}^{n} w_i w_i^* / (w_i^* w_i) = \sum_{i=1}^{n} z_i z_i^*, \qquad (8.9)$$

where the $z_i = w_i / \|w_i\|$, $i = 1, \dots, n$, are the pre-shapes.

Proof: We wish to minimize

$$\sum_{i=1}^{n} d_F^2(w_i, \mu) = \sum_{i=1}^{n} \left\{ 1 - \frac{\mu^* w_i w_i^* \mu}{w_i^* w_i \mu^* \mu} \right\} \qquad (8.10)$$

$$= n - \mu^* S \mu / (\mu^* \mu). \qquad (8.11)$$

Therefore,

$$\hat{\mu} = \arg\sup_{\|\mu\|=1} \mu^* S \mu.$$

Hence, $\hat{\mu}$ is given by the complex eigenvector corresponding to the largest eigenvalue of S [using e.g. Mardia *et al.* 1979, Equation (A.9.11)]. All rotations of $\hat{\mu}$ are also solutions, but these all correspond to the same shape $[\hat{\mu}]$.   □

The eigenvector is unique (up to a rotation) provided there is a single largest eigenvalue of S (which is the case for most practical datasets). We shall see in Section 10.2 that the solution corresponds to the MLE of modal shape under the complex Bingham model. Note that in this special case we do not need to use the iterative GPA algorithm of Section 7.3, and this is a further indication of the 2D case being special using complex arithmetic. In order to estimate the full Procrustes mean shape with this method we can use the R command:

```
procGPA(data,eigen2d=TRUE)
```

and examples are given in Section 8.4.

The **full Procrustes fits** or **full Procrustes coordinates** of $w_1, \ldots, w_n$ are:

$$w_i^P = w_i^* \hat{\mu} w_i / (w_i^* w_i), \quad i = 1, \ldots, n, \tag{8.12}$$

where each w_i^P is the full Procrustes fit of w_i onto $\hat{\mu}$. Calculation of the full Procrustes mean shape can also be obtained by taking the arithmetic mean of the full Procrustes coordinates, that is $\frac{1}{n} \sum_{i=1}^{n} w_i^P$ has the same shape as the Procrustes mean shape $[\hat{\mu}]$ (see Result 8.2). The **Procrustes residuals** are calculated as:

$$r_i = w_i^P - \left(\frac{1}{n} \sum_{i=1}^{n} w_i^P \right), \quad i = 1, \ldots, n, \tag{8.13}$$

and the Procrustes residuals are useful for investigating shape variability (see Section 7.7).

An alternative equivalent procedure to working with centred configurations would be to work with the Helmertized landmarks Hw_i, where H is the sub-Helmert matrix given in Equation (2.10). This procedure was originally used by Kent (1991, 1992, 1994). The least squares estimate of shape is the leading eigenvector $\hat{\mu}_1$ of HSH^T. Note that $H^T \hat{\mu}_1$ is identical to $\hat{\mu}$, up to an arbitrary rotation.

Definition 8.4   *To obtain an overall measure of shape variability we consider the* **root mean square** *$RMS(d_F)$ of full Procrustes distance from each configuration to the full Procrustes mean $[\hat{\mu}]$,*

$$RMS(d_F) = \sqrt{\frac{1}{n} \sum_{i=1}^{n} d_F^2(w_i, \hat{\mu})}. \tag{8.14}$$

Example 8.1   In Figure 8.1 we see the raw digitized data from the female and male gorilla skulls from the dataset described in Section 1.4.8. The landmarks have been recorded by a digitizer to be registered so that *opisthion* is at the origin and the line

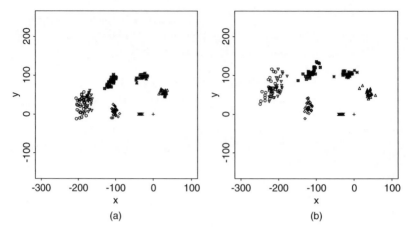

Figure 8.1   (a) The 30 female gorilla skull landmarks registered in the coordinate system as recorded by a digitizer. (b) The original 29 male gorilla skull landmarks.

from *opisthion* to *basion* is horizontal. There are $k = 8$ landmarks in $m = 2$ dimensions. In Figure 8.2 and Figure 8.3 we also see full Procrustes fits of the females and males separately. For each sex the landmarks match up quite closely because the shape variability is small. The full Procrustes mean for each sex is found from the dominant eigenvector of the complex sum of squares and products matrix for each

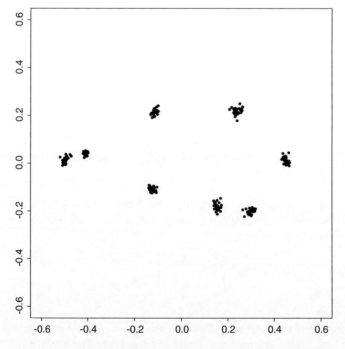

Figure 8.2   The full Procrustes fits of the female gorilla skulls.

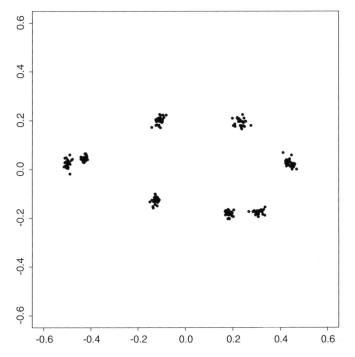

Figure 8.3   The full Procrustes fits of the male gorilla skulls.

sex. In Figure 8.4 we see the full Procrustes registration of the female average shape and the male average shape. It is also of interest to assess whether there is a significant average shape difference between the sexes and, if so, to describe the difference. We consider methods for testing for average shape differences in Chapter 9. The full Procrustes distance d_F between the mean shapes is 0.059, and the within-sample $RMS(d_F)$ is 0.044 for females and 0.050 for males. We see later in Section 9.1.2 that the difference in mean shapes between the sexes is statistically significant.   □

8.4   Planar shape analysis in R

In R we can make explicit use of the complex eigenvector solution to the Procrustes mean by using the option eigen2d=TRUE in function procGPA. This method can be much faster than using the generalized Procrustes algorithm of Section 7.3 if the number of observations n is large.

```
data(apes)
ans1<-procGPA(apes$x,eigen2d=TRUE,tol1=1e-10)
ans2<-procGPA(apes$x,eigen2d=FALSE)
riemdist(ans1$mshape,ans2$mshape)
[1] 1.724934e-07
```

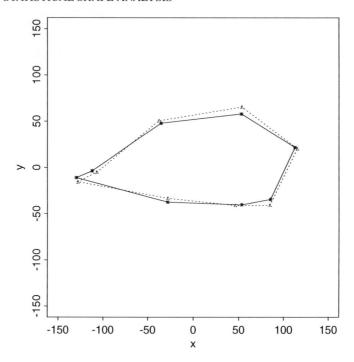

Figure 8.4   The male (—) and female (- - -) full Procrustes mean shapes registered by GPA.

As seen above both the complex eigenvector and GPA give almost identical results. The rotation and scale of both estimates are arbitrary, and the complex eigenvector has centroid size 1, and the first two coordinates are horizontal.

8.5   Shape variability

After having obtained an average configuration we often wish to examine the structure of shape variability in a sample, using tangent space PCA (see Section 7.7). In practice one uses real coordinates in the tangent space to carry out further analysis, as the complex covariance structure (which is isotropic at each landmark) is very restrictive.

We denote the real vectors of the tangent coordinates as $v_i, i = 1, \ldots, n$. These could be the Procrustes residuals r_i of Equation (8.13) or another choice of tangent coordinates which we introduced in Chapter 3.

Example 8.2   Consider the mouse vertebral data described in Section 1.4.1. There are $k = 6$ landmarks in $m = 2$ dimensions. The analysis here is similar to Kent (1994). In Figure 8.5 we have a plot of the Procrustes mean shape obtained from the dominant eigenvector of the complex sum of squares and products matrix. The full Procrustes

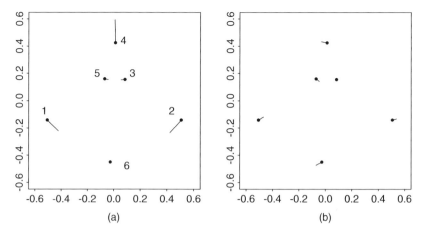

Figure 8.5   The Procrustes mean shape of the T2 vertebal data (landmarks at the dots) and vectors to 6 standard deviations along the first PC (a) and second PC (b).

mean shape is centred, with unit size, and rotated so that the line joining the two farthest apart landmarks is horizontal. Hence, the mean shape has coordinates $(-0.51 - 0.14i, 0.51 - 0.14i, 0.09 + 0.15i, 0.01 + 0.42i, -0.07 + 0.16i, -0.03 - 0.45i)^{\mathrm{T}}$.

In order to examine the structure of variability we examine the eigenstructure of the sample covariance matrix S_v of the Procrustes residuals. The square roots of the eigenvalues of S_v are:

$$0.054, 0.020, 0.018, 0.017, 0.011, 0.010, 0.008, 0.005, 0, 0, 0, 0.$$

Hence, the first two PCs explain 69 and 10% of the variability, respectively. The last four zero eigenvalues are zero due to the four constraints for location, rotation and scale. For each PC, shapes at 6 standard deviations away from the mean are calculated. In Figure 8.5 we see the mean shape with these unit vectors drawn for the first two PCs. The vectors of the first and second PCs in the figure are given by:

$$(0.11 - 0.11i, -0.11 - 0.12i, -0.04, -0.01 + 0.22i, 0.04 - 0.01i, 0.01 + 0.01i)^{\mathrm{T}},$$
$$(0.05 + 0.03i, 0.04 + 0.01i, -0.01, -0.06 + 0.01i, 0.03 - 0.02i, -0.06 - 0.03i)^{\mathrm{T}}.$$

There appears to be a high dependence between certain landmarks, as indicated by the fact that the first PC explains such a large proportion of the variability. The first PC highlights a shift downwards and inwards for landmarks 1 and 2, balanced by an upwards movement for landmark 4. At the same time landmarks 3 and 5 move inwards slightly whereas there is little movement in landmark 6. The second PC is not distinguishable from PC3 and PC4 due to similar eigenvalues. If we had chosen to display the PCs in a different manner (e.g. relative to landmarks 1 and 2 as in Bookstein coordinates), then our interpretation would be different. In particular we

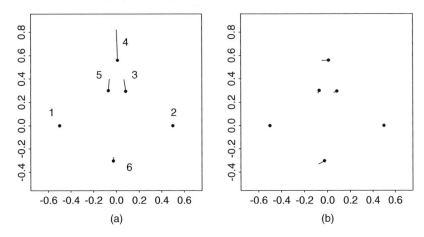

Figure 8.6   The mean shape with vectors to a figure at 6 standard deviations along the first PC (a) and second PC (b), both the same PCs as in Figure 8.5 but the icons are registered on a common baseline 1, 2.

display the PCs in Figure 8.6 where the mean and a figure at 3 standard deviations along the PC have been registered relative to landmarks 1 and 2. Our interpretation would be that the first PC includes the movement of landmarks 3, 4, 5 and 6 upwards relative to points 1 and 2. Landmark 4 shows the largest movement, followed by 3 and 5 together and landmark 6 shows the smallest movement. Both of these interpretations are correct, as they are describing the same PCs.   □

9

Tangent space inference

In this chapter we outline some simple approaches to inference based on tangent space approximations, which are valid in datasets with small variability in shape. We discuss one and two sample Hotelling's T^2 tests for mean shape based on normality assumptions, and then consider non-parametric alternatives. As a special case we consider isotropic covariance structure, although the model is too simple for most applications. We discuss other multivariate inference techniques which work directly on shape coordinates and conclude with the topic of allometry: the relationship between shape and size. Note that any inference is a function of the landmark selection that the investigator designed at the outset.

Statistical models and parametric inference in the pre-shape, shape and size-and-shape space are discussed later in Chapter 10.

9.1   Tangent space small variability inference for mean shapes

The horizontal tangent space to the pre-shape sphere was defined in Section 4.4. A practical approach to analysis is to use the Procrustes tangent space coordinates if the data are concentrated and then perform standard multivariate analysis in this linear space, where the pole is chosen from the data using a consistent estimator of an overall population mean, which is then treated as fixed. For example the pole could be the full Procrustes mean of Equation (6.11).

After the choice of tangent space has been fixed we carry out inference using any convenient linear statistical methods, and we can make use of all the functions and packages in R, or any other statistical package of choice.

Statistical Shape Analysis, with Applications in R, Second Edition. Ian L. Dryden and Kanti V. Mardia.
© 2016 John Wiley & Sons, Ltd. Published 2016 by John Wiley & Sons, Ltd.

9.1.1   One sample Hotelling's T^2 test

Consider carrying out a test on a mean shape of a single population and whether or not the mean shape has a particular special shape $[\mu_0]$, that is test between

$$H_0 : [\mu] = [\mu_0] \quad \text{versus} \quad H_1 : [\mu] \neq [\mu_0].$$

Let $X_1, \ldots, X_n$ be a random sample of configurations with partial Procrustes tangent coordinates (with pole $\hat{\mu}$ from the full Procrustes mean with unit size) given from Equation (4.33) by $v_1, \ldots, v_n$ where

$$v_i = (I_{km-m} - \text{vec}(\hat{\mu})\text{vec}(\hat{\mu})^{\text{T}})\text{vec}(X_i^P / \|X_i^P\|). \tag{9.1}$$

Let the tangent coordinates of μ_0 be γ_0 where

$$\gamma_0 = (I_{km-m} - \text{vec}(\hat{\mu})\text{vec}(\hat{\mu})^{\text{T}})\text{vec}(\mu_0^P / \|\mu_0^P\|)$$

and μ_0^P is the Procrustes fit of μ_0 onto $\hat{\mu}$. Since the dimension of the tangent space is $q = km - m - m(m-1)/2 - 1$ and the length of each vector v_i is $(k-1)m > q$ we have a singular covariance matrix and so we could use generalized inverses.

Definition 9.1   *A **generalized inverse** of a symmetric square matrix A is denoted by A^- and satisfies*

$$A^- A A^- = A.$$

*The **Moore–Penrose generalized inverse** of A is:*

$$A^- = \sum_{j=1}^{p} \lambda_j^{-1} \gamma_j \gamma_j^{\text{T}},$$

where γ_j are the eigenvectors of A corresponding to the p non-zero eigenvalues $\lambda_j, \ j = 1, \ldots, p$.

To obtain a one sample test a standard multivariate analysis approach is carried out on v_i, where a multivariate normal model for v_i is assumed,

$$v_i \sim N(\xi, \Sigma),$$

independently for $i = 1, \ldots, n$. The one sample Hotelling's T^2 test could be used (e.g. Mardia *et al.* 1979, p. 125). We write $\bar{v} = \frac{1}{n} \sum v_i$ for the sample mean and we write $S_v = \frac{1}{n} \sum (v_i - \bar{v})(v_i - \bar{v})^{\text{T}}$, for the sample covariance matrix (with divisor n). The Mahalanobis squared distance between $\bar{v}$ and γ_0 is:

$$D^2 = (\bar{v} - \gamma_0)^{\text{T}} S_v^- (\bar{v} - \gamma_0) \tag{9.2}$$

where S_v^- is the Moore–Penrose generalized inverse of S_v. The rank of S_v is $\min(q, n-1)$ and we assume that the rank of our sample covariance matrices is q in this chapter, although the methods can be extended to $n \leq q$ with appropriate regularization (see Section 9.1.5).

Important point: The test statistic is taken as:

$$F = \frac{(n-q)}{q} D^2 = \frac{(n-q)}{q} \sum_{j=1}^{q} \frac{s_j^2}{\lambda_j},$$

where the $s_j = \gamma_j^{\mathrm{T}}(\bar{v} - \gamma_0)$ is the jth PC score for $(\bar{v} - \gamma_0)$, $j = 1, \dots, q$. The test statistic F has an $F_{q,n-q}$ distribution under H_0. Hence, we reject H_0 for large values of F.

A $100(1 - \alpha)\%$ confidence region for mean shape is given by:

$$\left\{ [\mu] : F = \frac{(n-q)}{q} \sum_{j=1}^{q} \frac{s_j^2}{\lambda_j} \leq F_{q,n-q,1-\alpha} \cdot \right\}$$

where $F_{q,n-q,1-\alpha}$ is the $(1 - \alpha)$-quantile of the F distribution with degrees of freedom $(q, n - q)$. In practice such a confidence region might have limited practical use due to extra uncertainty in the landmark selection or the choice of Procrustes metric.

Example 9.1   Consider the digit 3 data, described in Section 1.4.2, with $k = 13$ points in $m = 2$ dimensions on $n = 30$ objects. We might wish to examine whether the population mean shape could be an idealized template, such as that displayed in Figure 9.1, with equal sized loops, with 12 of the landmarks lying equally spaced on two regular octagons (apart from landmark 7 in the middle). The μ_0 is taken as the template and the data are projected into the tangent plane with the pole at the Procrustes mean $\hat{\mu}$. The $q = 22$ PC scores are retained and the squared Mahalanobis distance from $\bar{v}$ to the pole in the tangent space is $\sum_{j=1}^{q} s_j^2/\lambda_j = 47.727$ and hence $F = 17.356$. Since $P(F_{22,8} > 17.356) \approx 0.0002$ we have very strong evidence that the population mean shape does not have the shape of this template.   □

Of more practical interest is a two sample test.

9.1.2   Two independent sample Hotelling's T^2 test

Consider two independent random samples $X_1, \dots, X_{n_1}$ and $Y_1, \dots, Y_{n_2}$ from independent populations with mean shapes $[\mu_1]$ and $[\mu_2]$. To test between

$$H_0 : [\mu_1] = [\mu_2] \quad \text{versus} \quad H_1 : [\mu_1] \neq [\mu_2],$$

we could carry out a Hotelling's T^2 two sample test in the Procrustes tangent space, where the pole corresponds to the overall pooled full Procrustes mean shape $\hat{\mu}$ (i.e. the full Procrustes mean shape calculated by GPA on all $n_1 + n_2$ individuals). Let

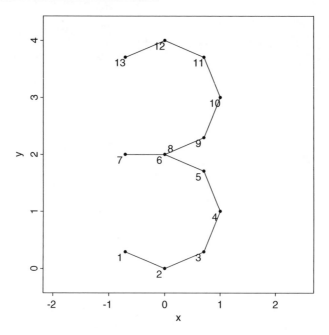

Figure 9.1   A piecewise linear template number 3 digit, with two equal sized arcs, and with 12 landmarks lying on two regular octagons.

$v_1, \ldots, v_{n_1}$ and $w_1, \ldots, w_{n_2}$ be the partial Procrustes tangent coordinates (with pole $\hat{\mu}$). The multivariate normal model is proposed in the tangent space, where

$$v_i \sim N(\xi_1, \Sigma) \ , \quad w_j \sim N(\xi_2, \Sigma), \qquad i = 1, \ldots, n_1; j = 1, \ldots, n_2,$$

and the v_i and w_j are all mutually independent, and common covariance matrices are assumed. We write $\bar{v}, \bar{w}$ and S_v, S_w for the sample means and sample covariance matrices (with divisors n_1 and n_2) in each group. The Mahalanobis distance squared between $\bar{v}$ and $\bar{w}$ is:

$$D^2 = (\bar{v} - \bar{w})^{\mathrm{T}} S_u^{-} (\bar{v} - \bar{w}), \qquad (9.3)$$

where $S_u = (n_1 S_v + n_2 S_w)/(n_1 + n_2 - 2)$, and S_u^{-} is the Moore–Penrose generalized inverse of S_u (see Definition 9.1). Under H_0 we have $\xi_1 = \xi_2$, and we use the test statistic

$$F = \frac{n_1 n_2 (n_1 + n_2 - q - 1)}{(n_1 + n_2)(n_1 + n_2 - 2)q} D^2. \qquad (9.4)$$

The test statistic has an F_{q,n_1+n_2-q-1} distribution under H_0. Hence, we reject H_0 for large values of F.

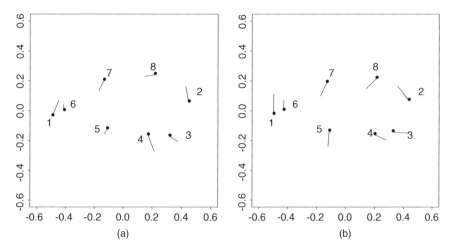

Figure 9.2   The first PC for the gorilla females (a) and males (b). The mean shape is drawn with vectors to an icon +3 (——) standard deviations along the first PC away from the mean.

Example 9.2   Consider the gorilla skull data described in Section 1.4.8. There are $n_1 = 30$ female gorilla skulls and $n_2 = 29$ male gorilla skulls, with $k = 8$ landmarks in two dimensions, and so there are $q = 2k - 4 = 12$ shape dimensions. The first three PCs in each group explain 34.8, 22.9, 11.2% (females) and 42.2, 18.0, 12.4% (males) of the variability in each group.

```
> procGPA(gorf.dat)$percent[1:3]
[1] 34.79297 22.90900 11.25934

> procGPA(gorm.dat)$percent[1:3]
[1] 42.20271 17.96493 12.37879
```

A plot of the first PC for each group is given in Figure 9.2. Although the test requires equal covariance structures this is unreasonable here, as shape variation is invariably different between sexes in biology. However, we continue to illustrate the use of this two sample Hotelling test before considering more reasonable nonparametric tests later in Section 9.1.3.

The percentages of variability explained by the first three within group PCs are 37.3, 16.0, 14.7% and the first three PCs are included in Figure 9.3. In addition, we have no reason to doubt multivariate normality from the pairwise scatters of the standardized PC scores of the data (some of the PC scores are shown in Figure 9.4).

The observed test statistic (9.4) is $F = 26.470$ and since $P(F_{12,46} > 26.47) < 0.0001$ we have very strong evidence that the mean shapes are different. So our

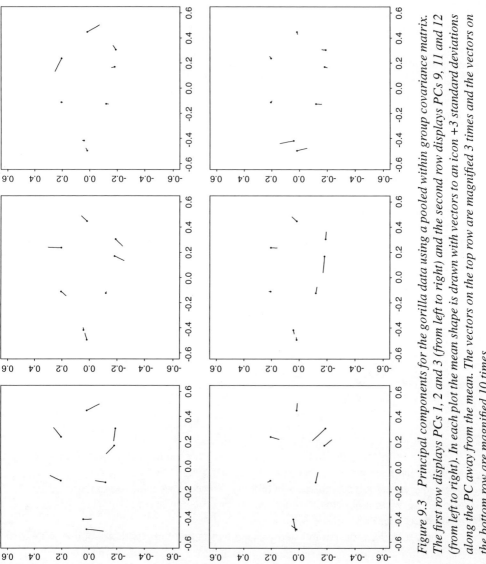

Figure 9.3 Principal components for the gorilla data using a pooled within group covariance matrix. The first row displays PCs 1, 2 and 3 (from left to right) and the second row displays PCs 9, 11 and 12 (from left to right). In each plot the mean shape is drawn with vectors to an icon +3 standard deviations along the PC away from the mean. The vectors on the top row are magnified 3 times and the vectors on the bottom row are magnified 10 times.

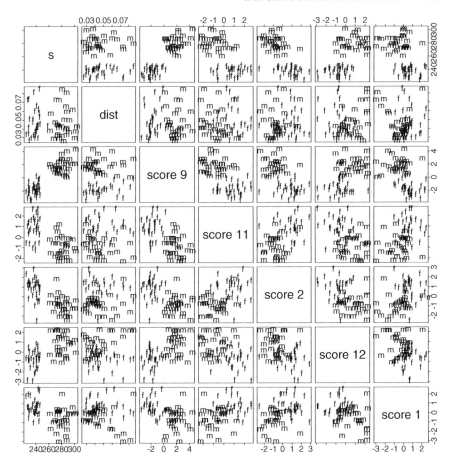

Figure 9.4   Pairwise scatter plots of the centroid sizes, the full Procrustes distances to the pooled mean and PC scores 9, 11, 2, 12, 1 ($s_i, d_{Fi}, c_{i9}, c_{i11}, c_{i2}, c_{i12}, c_{i1}$)^T for the gorilla data: males (m); and females (f). These particular PC scores c_{ij} have the highest correlation with the observed group shape difference.

conclusion would be that there is a significant difference in mean shape between the female and male gorilla skulls in the midline. The test can be carried out in R:

```
> resampletest(gorm.dat,gorf.dat,replace=TRUE)
Resampling...No of resamples = 200
Bootstrap - sampling with replacement
$H
[1] 26.47042
```

```
$H.pvalue
[1] 0.004975124

$H.table.pvalue
[1] 1.110223e-16
```

and here $H.table.pvalue gives the p-value of the test and $H.pvalue corresponds to a bootstrap test (see Section 9.1.3).

We can express the squared Mahalanobis distance of Equation (9.3) as:

$$D^2 = (\bar{v} - \bar{w})^T S_u^-(\bar{v} - \bar{w}) = \sum_{j=1}^{q} s_j^2/\lambda_j$$

where $s_j = (\bar{v} - \bar{w})^T \gamma_j$ are the scores in the direction of the observed group difference (Kent 1997). Large values of s_j^2/λ_j indicate which directions of shape variability are associated with the difference between the groups. In our example the values of $1.0084 s_j^2/\lambda_j$ are:

$$1.53, 2.57, 0.05, 0.07, 0.21, 0.88, 0.59, 1.18, 13.93, 0.22, 2.77, 2.48.$$

Of the first few PCs (which explain most of the shape variability in the dataset) PC2 and, to a lesser extent, PC1 have high scores in the direction of the shape difference, namely 2.57 and 1.53. The later PCs are effectively just arbitrary choices of directions in shape space so no particular meaning or interpretation should be assigned to these.

Figure 9.4 displays some of the PC scores for the data and PC1 and PC2 give a good separation of the two groups. The higher PCs are effectively just arbitrary directions in shape space which explain very little variability. They have been displayed for mathematical curiosity but their utility in any practical analysis is negligible. From Figure 9.4 it is clear that the groups differ substantially in centroid size. The correlations of the PC scores with centroid size are:

$$-0.48, -0.68, -0.21, -0.14, 0.15, 0.42, 0.33, 0.48, 0.88, 0.20, -0.61, 0.60,$$

and so the PCs that have a high contribution in the direction of the shape difference have a high correlation with centroid size, namely PC1, PC2 and PC9 with correlations -0.48, -0.68 and 0.88, respectively. So there is clear allometry here – shape differences are associated with size differences (see Sections 5.7 and 9.5).

As seen above, there is an extremely large contribution from PC9 in terms of s_j^2/λ_j. We plot the first 3 PCs and PCs 9, 11 and 12 in Figure 9.3. There is nothing special about the number 9 here – the lower PCs are measuring very little shape variability and are essentially random directions in shape which are orthogonal to the higher PCs. It just happens to be that PC9 correlates well with the observed shape difference between the two means. Another random sample might easily result in a different PC or several PCs having high correlation with the observed shape difference. The two groups are very different in size and in shape, and PC9 is also the most highly correlated with size.

Now that a significant mean shape difference has been found a biologist would be further interested in how size and shape are related. For example, if using size as a covariate how do the shapes of the gorillas differ after removing the effect of the covariate of size? Note that PC3 is not so highly correlated with size and this PC includes mainly braincase variability, so we might wish to investigate whether the non-size-related shape difference is mainly in the braincase. We would also wish to investigate other non-size related variation in all the PCs. Further methods for describing differences between mean shapes and exploring shape variability using thin-plate splines are given in Chapter 12.   □

9.1.3   Permutation and bootstrap tests

The multivariate normal model in the tangent space may be unreasonable in some applications or the assumption of equal covariance matrices may be doubted, as in Example 9.2. Alternative non-parametric methods are a **permutation test** (Dryden and Mardia 1993; Good 1994; Bookstein 1997) or **bootstrap test** (Amaral *et al.* 2007), with the null hypothesis H_0 that the groups have equal mean shapes. The permutation test requires the assumption that the groups of data are exchangeable under H_0, so for example the covariance matrices must be equal in both groups, but the bootstrap test is less restrictive.

For a two sample permutation test the data are permuted into two groups of the same size as the groups in the data, and the test statistic is evaluated for all possible permutations $T_1, \ldots, T_P$. The ranking r of the observed test statistic T_{obs} is then used to give the p-value of the permutation test:

$$1 - \frac{r-1}{P}.$$

Instead of evaluating all T_i, we can consider a number B (e.g. 200) of random permutations, and the procedure is called a **Monte Carlo test**. The ranking r of the observed test statistic from B random permutations gives a p-value of:

$$1 - \frac{r-1}{B+1}.$$

An alternative non-parametric test is a bootstrap test. In this procedure the data are sampled from the two groups **with replication**, although it is important that the sampling is carried out under H_0, that the mean shapes are equal. To ensure that sampling is carried out under H_0 each group is translated to have a common mean before resampling. The test statistic is evaluated for B Monte Carlo replications $T_1, \ldots, T_B$. One potential advantage of the bootstrap test is that it does not require exchangeability, and so the two groups could have different covariance matrices for example.

Permutation tests and bootstrap tests have been discussed by Amaral *et al.* (2007) in some detail for 2D shape analysis. The methods are implemented in R using `resampletest` and can also be appropriate for large variations provided the sample sizes are large (we return to this issue later in Chapter 13).

Example 9.3   Consider the 2D chimpanzee data of Section 1.4.8. There are $n_1 = 28$ males and $n_2 = 26$ females, each with $k = 8$ landmarks in $m = 2$ dimensions. We wish to test whether the mean population shapes for both sexes are equal. After performing full GPA on the pooled dataset we transform to the tangent space coordinates of Equation (9.1). The dimension of the shape space is $q = 12$ [8×2 (coordinates) -2 (location) -1 (rotation) -1 (size)]. Proceeding with the Hotelling's T^2 test we have statistics $F = 1.53$ and $P(F_{12,41} > 1.53) = 0.153$, and so there is not a significant difference in mean shape.

A permutation test is also carried out. The data are randomly split into two groups each of size 28 and 26. Out of 200 such permutations the observed F statistic of Equation (9.4) had rank 24, giving a p-value of 0.119, and so again there is no evidence for a difference in mean shape. However, for the above two tests the equal covariance assumption may not be reasonable.

Finally a bootstrap test is also carried out which does not require equal covariance matrices. The two full Procrustes means for each group are parallel translated to the common full Procrustes mean, and then samples with replacement are taken of size 28 and 26. Details of the parallel translation are given by Amaral *et al.* (2007) and in Section 4.1.6. Out of 200 such replications the observed F statistic of Equation (9.4) had rank 35, giving a p-value of 0.174, and so again there is no evidence for a difference in mean shape.

The sample sizes are fairly small here and we might expect the Hotelling's T^2 test to be not very powerful.   □

The above example can be carried out in R as follows (with relevant output given):

```
ans <- resampletest(panm.dat,panf.dat,replace=FALSE)
Resampling...No of resamples = 200
Permutations - sampling without replacement

ans$H
[1] 1.530009
ans$$H.pvalue
[1] 0.119403
ans$H.table.pvalue
[1] 0.1526978

ans<-resampletest(panm.dat,panf.dat,replace=TRUE)
Resampling...No of resamples = 200
Bootstrap - sampling with replacement
```

```
ans$H
[1] 1.530009
ans$H.pvalue
[1] 0.1741294
ans$H.table.pvalue
[1] 0.1526978
```

9.1.4   Fast permutation and bootstrap tests

One practical issue with the `resampletest` implementation is that the tangent space and Procrustes registration are computed on each replication, and so instead we describe some fast approximate two sample tests which can be easily carried out using the command `testmeanshapes` in R, which are based on Czogiel (2010). The fast permutation test carries out an initial pooled GPA and then the permutations are carried out without replacement on the Procrustes residuals. The fast bootstrap test carries out an initial pooled GPA and then the bootstrap replications are carried out with replacement on the Procrustes residuals calculated using each group mean. Both methods are fast algorithms as the Procrustes registration is carried out only once.

Example 9.4   Consider the 2D chimpanzee data from Example 9.3. The fast permutation and bootstrap test commands are:

```
> testmeanshapes(panm.dat,panf.dat,replace=FALSE)
Permutations - sampling without replacement: No of permutations = 1000
$H
[1] 1.53001

$H.pvalue
[1] 0.1468531

$H.table.pvalue
[1] 0.1526974

> testmeanshapes(panm.dat,panf.dat,replace=TRUE)
Bootstrap - sampling with replacement within each group under H0:
No of resamples = 1000
$H
[1] 1.53001

$H.pvalue
[1] 0.2177822

$H.table.pvalue
[1] 0.1526974
```

and hence in this example the Hotelling T^2 statistic and tabular p-value are almost identical, and the permutation p-value is 0.146 and the bootstrap p-value is 0.218, which are similar to Example 9.3. Hence, again we conclude there is no shape difference between the male and female chimpanzee means.   □

Example 9.5   We consider the macaques example of Section 1.4.3 for an example in three dimensions.

```
out <- testmeanshapes(macm.dat,macf.dat,replace=FALSE)
out$H
[1] 1.651495

out$H.pvalue
[1] 0.3786214

out$H.table.pvalue
[1] 0.3777721
```

Note that some regularization has been carried out by adding a small constant to the pooled sample covariance matrix, because the dimension is large compared with the small sample sizes. We see that the Hotelling T^2 statistic is 1.65 with tabular p-value (from the F-distribution) of 0.378 and the permutation p-value (from 1000 permutations) of 0.379. Also, rather than a permutation test we can consider a bootstrap test, where the sampling is carried out with replacement and each group is moved to have a common mean before resampling by computing the tangent residuals from each sample mean. Hence sampling is carried out under H_0 that the means are equal. The command in this case is:

```
testmeanshapes(macm.dat,macf.dat,replace=TRUE)
```

and the bootstrap p-value is 0.673. So, for all these tests the conclusion is the same that there is no evidence for a difference in the mean shapes for the male and female macaques. At this point the statistician should consult with the scientist further to enumerate those features that the literature has indicated might differentiate the groups, e.g. biological function, such as diet.   □

Some further studies of morphological variation in primates, humans and hominid fossils include: O'Higgins and Jones (1998); Bookstein *et al.* (1999); O'Higgins (2000); and Mitteroecker *et al.* (2004).

9.1.5   Extensions and regularization

Further inference, such as testing the equality of the mean shapes in several groups, proceeds in a similar manner. An overall pooled full Procrustes mean is taken as the pole and multivariate analysis of variance (MANOVA) (e.g. see Mardia *et al.* 1979, p. 333) is carried out on the Procrustes tangent coordinates. General linear models could be proposed in the tangent space and the full armoury of multivariate data analysis can be used to analyse shape data, provided variations are small.

In some datasets there are few observations and possibly many landmarks on each individual. Although inference can be carried out in a suitable tangent space there is often a problem with the space being over-dimensioned. For example, a Hotelling's T^2 test may not be very powerful unless there are a large number of observations available, or it may not be possible to invert the covariance matrices in the procedures. A solution is to carry out some form of regularization, and there are many possible choices. For example one could perform a PCA on the pooled datasets and retain the first few PC scores, although there are obvious dangers, particularly if a true group difference is orthogonal to the first few PCs. An alternative approach, which is used in the `testmeanshapes` procedure in R, is to add a small multiple of the identity matrix (0.000001 in fact) before taking the inverse. The precise choice of constant will of course make a practical difference. Other types of regularization might include assuming the inverse covariance matrix is a sparse matrix, and methods such as the graphical LASSO would be appropriate in this case (Meinshausen and Bühlmann 2006; Friedman *et al.* 2008). Further discussion of choices of regularization for permutation and bootstrap tests was given by Dryden *et al.* (2014), particularly in the context of MDS estimators (which are discussed in Section 15.3).

9.2   Inference using Procrustes statistics under isotropy

Another simple approach to statistical inference is to work with statistics based on squared Procrustes distances, which is equivalent to assuming that the covariance matrix is proportional to the identity matrix. Goodall (1991) has considered such an approach using approximate chi-squared distributions, following from the work of Sibson (1978, 1979) and Langron and Collins (1985). The underlying model is that configurations are isotropic normal perturbations from mean configurations, and the distributions of the squared Procrustes distances are approximately chi-squared distributions. The procedures require a much more restrictive isotropic model than the previous section, and the assumption often does not hold in practice (e.g. Bookstein, 2014). Non-parametric procedures based on Goodall's procedures can be particularly effective, and do not require such strict assumptions. **Important point:** For a preliminary analysis we will assume isotropy.

9.2.1   One sample Goodall's *F* test and perturbation model

We consider first the case when a random sample of n observations $X_1, \dots, X_n$ (each a $k \times m$ matrix) is taken from an isotropic normal model with mean μ and transformed by an additional location, rotation and scale, that is

$$X_i = \beta_i(\mu + E_i)\Gamma_i + 1_k\gamma_i^{\mathrm{T}}, \quad \mathrm{vec}(E_i) \sim N(0, \sigma^2 I_{km}), \tag{9.5}$$

where $\beta_i > 0$ (scale), $\Gamma_i \in SO(m)$ (rotation) and $\gamma_i \in \mathbb{R}^m$ (translation), and σ is small.

The following approximate analysis of variance (ANOVA) identity holds for $\hat{\mu} \approx \mu$ and small σ:

$$\sum_{i=1}^{n} d_F^2(X_i, \mu) \approx \sum_{i=1}^{n} d_F^2(X_i, \hat{\mu}) + n d_F^2(\mu, \hat{\mu}),$$

where $\hat{\mu}$ is the full Procrustes mean and d_F is the full Procrustes distance of Equation (4.10). The proof can be seen using Taylor series expansions. Note the similarities with ANOVA in classical regression analysis – the left-hand side of the equation is like a total sum of squares and the right-hand side is like the residual sum of squares plus the explained (regression) sum of squares.

Consider testing between $H_0 : [\mu] = [\mu_0]$ and $H_1 : [\mu] \neq [\mu_0]$. Under the null model it can be shown that approximately (to second-order terms in E_i) that

$$d_F^2(X_i, \mu_0) \sim \tau_0^2 \chi_q^2 \tag{9.6}$$

independently for $i = 1, \ldots, n$, where $q = km - m - m(m-1)/2 - 1$ is the dimension of the shape space, $\tau_0 = \sigma/\delta_0$, and $\delta_0 = S(\mu_0) = \|C\mu_0\|$ is the centroid size of μ_0. The proof can be obtained by Taylor series expansions, after Sibson (1978), and the proof for the $m = 2$ dimensional case is seen from Equation (10.14), when discussing the complex Watson distribution.

From the additive property of independent chi-squared distributions,

$$\sum_{i=1}^{n} d_F^2(X_i, \mu_0) \sim \tau_0^2 \chi_{nq}^2.$$

In addition, since q parameters are estimated in $\hat{\mu}$ we have:

$$\sum_{i=1}^{n} d_F^2(X_i, \hat{\mu}) \sim \tau_0^2 \chi_{(n-1)q}^2$$

and $d_F^2(\mu_0, \hat{\mu})$ is approximately independent of $\sum d_F^2(X_i, \hat{\mu})$. Hence, approximately

$$n d_F^2(\mu_0, \hat{\mu}) \sim \tau_0^2 \chi_q^2, \tag{9.7}$$

again using the additive property of independent chi-squared distributions. So, under H_0 we have the approximate result

$$F = (n-1)n \frac{d_F^2(\mu_0, \hat{\mu})}{\sum_{i=1}^{n} d_F^2(X_i, \hat{\mu})} \sim F_{q,(n-1)q}. \tag{9.8}$$

This is valid for small σ and μ_0 close to $\hat{\mu}$, and so we reject H_0 for large values of this test statistic. We call the test the **one sample Goodall's F test**, after Goodall (1991).

If τ_0 is small, E_i is isotropic (but not necessarily normal) and nq is large, then approximately

$$\sum_{i=1}^{n} d_F^2(X_i, \mu) \sim N(\tau_0^2 nq, 2\tau_0^4 nq)$$

by applying the central limit theorem.

The test based on the isotropic model can be seen as a special case of the Hotelling's T^2 procedure of Section 9.1.1. If we replace S_v with $s_v^2 I_{2k-2}$, where s_v^2 is the unbiased estimate of variance, then the Mahalanobis distance of Equation (9.2) becomes

$$D^2 = s_v^{-2} \| \gamma_0 - \bar{v} \|^2 = d_F^2(\mu_0, \hat{\mu})/s_v^2,$$

from Equation (4.30). Now

$$s_v^2 = \frac{1}{n-1} \sum_{i=1}^{n} \| v_i - \bar{v} \|^2 = \frac{1}{n-1} \sum_{i=1}^{n} d_F^2(X_i, \hat{\mu})$$

and hence the test statistic for the one sample Hotelling's T^2 test statistic is proportional to:

$$d_F^2(\mu_0, \hat{\mu}) \Big/ \sum_{i=1}^{n} d_F^2(X_i, \hat{\mu}),$$

and the one sample Hotelling's T^2 test under the isotropic model is identical to using the F statistic of Equation (9.8).

9.2.2   Two independent sample Goodall's F test

Consider independent random samples $X_1, ..., X_{n_1}$ from a population modelled by Equation (9.5) with mean μ_1, and $Y_1, ..., Y_{n_2}$ from Equation (9.5) with mean μ_2. Both populations are assumed to have a common variance for each coordinate σ^2. We wish to test $H_0 : [\mu_1] = [\mu_2](= [\mu_0])$, say, against $H_1 : [\mu_1] \neq [\mu_2]$. Let $[\hat{\mu}_1]$ and $[\hat{\mu}_2]$ be the full Procrustes means of each sample, with icons $\hat{\mu}_1$ and $\hat{\mu}_2$. Under H_0, with σ small, the Procrustes distances are approximately distributed as:

$$\sum_{i=1}^{n_1} d_F^2(X_i, \hat{\mu}_1) \sim \tau_0^2 \chi_{(n_1-1)q}^2,$$

$$\sum_{i=1}^{n_2} d_F^2(Y_i, \hat{\mu}_2) \sim \tau_0^2 \chi_{(n_2-1)q}^2,$$

$$d_F^2(\hat{\mu}_1, \hat{\mu}_2) \sim \tau_0^2 \left(\frac{1}{n_1} + \frac{1}{n_2} \right) \chi_q^2,$$

where $\tau_0 = \sigma/\delta_0$ and $\delta_0 = S(\mu_0)$. Again, proofs of the results can be obtained using Taylor series expansions. In addition these statistics are approximately mutually independent (exactly in the case of the first two expressions). Hence, under H_0 we have the approximate distribution

$$F = \frac{n_1 + n_2 - 2}{n_1^{-1} + n_2^{-1}} \frac{d_F^2(\hat{\mu}_1, \hat{\mu}_2)}{\sum_{i=1}^{n_1} d_F^2(X_i, \hat{\mu}_1) + \sum_{i=1}^{n_2} d_F^2(Y_i, \hat{\mu}_2)} \sim F_{q,(n_1+n_2-2)q}, \quad (9.9)$$

and again this result is valid for small σ. We reject H_0 for large values of this test statistic. We call the test the **two independent sample Goodall's F test** after Goodall (1991).

Example 9.6 Consider the 2D chimpanzee data from Example 9.3. The fast permutation and bootstrap test command testmeanshapes in R give the Goodall statistic with p-values (with the relevant output given only):

```
> testmeanshapes(panm.dat,panf.dat,replace=FALSE)
Permutations - sampling without replacement: No of permutations = 1000
$G
[1] 2.591273

$G.pvalue
[1] 0.02197802

$G.table.pvalue
[1] 0.002276534


> testmeanshapes(panm.dat,panf.dat,replace=TRUE)
Bootstrap - sampling with replacement within each group under H0:
No of resamples = 1000
$G
[1] 2.591273

$G.pvalue
[1] 0.01698302

$G.table.pvalue
[1] 0.002276534
```

From the output using the Goodall test the tabular p-value is $P(G > F_{12,624}) = 0.002$, the permutation p-value is 0.022 and the bootstrap p-value is 0.017. The isotropy assumption is unlikely to be reasonable, so the nonparametric tests should have more weight. The Goodall test is more powerful than the Hotelling T^2 test under either the parametric or nonparametric assumptions as there are many fewer parameters to estimate in the covariance matrix of the tangent coordinates.   □

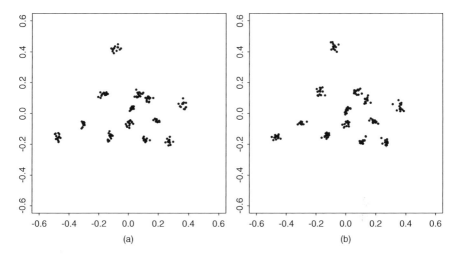

Figure 9.5   The Procrustes rotated brain landmark data for the 14 controls (a) and 14 schizophrenia patients (b).

Example 9.7   Consider the schizophrenia data described in Section 1.4.5. We wish to test whether the mean shapes of brain landmarks are different in the two groups of control subjects and schizophrenia patients. There are $k = 13$ landmarks in $m = 2$ dimensions. The Procrustes rotated data for the groups are displayed in Figure 9.5.

The percentages of variability explained by the first three PCs are 31.6, 21.4, 13.2% for the controls and 27.1, 21.7, 4.8% for the schizophrenia patients. Box's M test was carried out and there is some evidence against equal covariance matrices. The root mean square of d_F in each group is 0.068 in the controls and 0.073 in the schizophrenia group. It is unreasonable to assume isotropy for any practical biological landmark data, and so the nonparametric versions of the tests will be preferred.

The mean configurations are displayed in Figure 9.6. The full Procrustes distance d_F between the mean shapes is 0.038. The sum of squared full Procrustes distances from each configuration to its mean shape is 0.140 and so the F statistic is 1.89. Since $P(F_{22,572} \geq 1.89) \approx 0.01$ we have evidence for a significant difference in shape. However, the assumptions of this test are unreasonable and so we next consider the nonparametric versions.

Following Bookstein (1997) we also consider a Monte Carlo test, as described in Section 9.1.3, based on 999 random permutations. The configurations are randomly assigned into each of the two groups, the F statistic is calculated and the proportion of times that the resulting F statistic exceeds the observed value of 1.89 is the p-value for the test. From 999 random permutations we obtained a p-value of 0.04. Hence, we have some evidence that the mean configurations are different in shape, but with a larger p-value than for the isotropic based tests.

If we carry out a Hotelling's T^2 test in the tangent space we have $F = 0.834$ which is near the centre of the null distribution $[P(F_{22,5} > 0.834) = 0.66]$. So, the

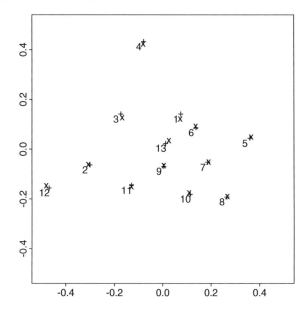

Figure 9.6 The full Procrustes mean shapes of the normal subjects (x) and schizophrenia patients (+) for the brain landmark data, rotated to each other by GPA.

Hotelling's T^2 provides no evidence for a shape difference. We should be aware that the Hotelling T^2 test is expected to be less powerful when the isotropic model holds: power is lost because many degrees of freedom are used in estimating the covariance matrix in the Hotelling's T^2 test.

Consider the permutation test using command `resampletest`. For the schizophrenia data the command for the permutation test is:

```
resampletest(schizophrenia.dat[,,1:14],
        schizophrenia.dat[,,15:28],replace=FALSE)
```

Random samples of size $n_1 = 14$ and $n_2 = 14$ are taken without replacement from the pooled dataset. The new groups are then registered using Procrustes analysis and individual mean shapes are estimated. The observed square full Procrustes distance is then compared with the simulated distribution under H_0, and the p-value is calculated. The bootstrap test is also carried out, as described in Section 9.1.3, and both these tests using the R command `reampletest`. In addition the fast approximations to the procedures are also carried out using `testmeanshapes`.

For the schizophrenia data the permutation p-values for the Goodall statistics are 0.049 (resampletest) and 0.055 (testmeanshapes). For the bootstrap-based test the p-values are 0.079 (resampletest) and 0.067 (testmeanshapes). Hence, this is an interesting example where the evidence is somewhat on the boundary of significance. Later

analysis by Bookstein (2000) using the method of creases found a localized difference in a region corresponding to a known drug effect, with p-value approximately
0.001. □

Note that the test based on the isotropic model can be seen as a special case of the
two sample Hotelling's T^2 procedure of Section 9.1.2. If we replace S_u with $s_u^2 I_{2k-2}$,
where s_u^2 is the unbiased estimate of variance, then the Mahalanobis distance of Equation (9.3) becomes:

$$D^2 = s_u^{-2} \|\bar{v} - \bar{w}\|^2 \approx d_F^2(\hat{\mu}_1, \hat{\mu}_2)/s_u^2,$$

from Equation (4.31). Now

$$s_u^2 = \frac{1}{n_1 + n_2 - 2} \left\{ \sum_{i=1}^{n_1} \|v_i - \bar{v}\|^2 + \sum_{j=1}^{n_2} \|w_j - \bar{w}\|^2 \right\}$$

$$\approx \frac{1}{n_1 + n_2 - 2} \left\{ \sum_{i=1}^{n_1} d_F^2(X_i, \hat{\mu}_1) + \sum_{j=1}^{n_2} d_F^2(Y_j, \hat{\mu}_2) \right\} \qquad (9.10)$$

and so the test statistic for the two sample Hotelling's T^2 test statistic would be proportional to:

$$\frac{d_F^2(\hat{\mu}_1, \hat{\mu}_2)}{\sum_{i=1}^{n_1} d_F^2(X_i, \hat{\mu}_1) + \sum_{j=1}^{n_2} d_F^2(Y_j, \hat{\mu}_2)}.$$

Hence the Hotelling's T^2 test under the isotropic model becomes identical to using
the F statistic of Equation (9.9).

9.2.3   Further two sample tests

A variety of other choices of test statistics are available in the R functions `resam-
pletest`, including the James statistic (James 1954; Seber 1984, p. 115) given by

$$J = (\bar{v} - \bar{w})^T \left(\frac{1}{n_1} S_v + \frac{1}{n_2} S_w \right)^- (\bar{v} - \bar{w}), \qquad (9.11)$$

and the asymptotically pivotal statistic λ_{min} given by Amaral et al. (2007). The λ_{min}
statistic is only appropriate for 2D shapes, but the James statistic is appropriate for
3D data using the fast routine `testmeanshapes`. The tests have performed well in

a range of simulation studies, with the bootstrap test having an advantage over the permutation test when covariance matrices are unequal (Amaral *et al.* 2007).

We consider the 2D chimpanzee data of Example 9.3. We see the relevant results from the `resampletest` command are:

```
> resampletest(panm.dat,panf.dat,replace=FALSE)
Resampling...No of resamples = 200
Permutations - sampling without replacement

$lambda
[1] 30.577

$lambda.pvalue
[1] 0.0199005

$lambda.table.pvalue
[1] 0.002284522

$J
[1] 23.45578

$J.pvalue
[1] 0.1243781

$J.table.pvalue
[1] 0.112

resampletest(panm.dat,panf.dat,replace=TRUE)
Resampling...No of resamples = 200
Bootstrap - sampling with replacement

$lambda
[1] 30.577

$lambda.pvalue
[1] 0.0199005

$lambda.table.pvalue
[1] 0.002284522

$J
[1] 23.45578

$J.pvalue
[1] 0.1691542

$J.table.pvalue
[1] 0.112
```

The James test is also available as a fast permutation and bootstrap routine in `testmeanshapes`:

```
> testmeanshapes(panm.dat,panf.dat,replace=FALSE)
Permutations - sampling without replacement: No of permutations = 1000
100 200 300 400 500 600 700 800 900 1000


$J
[1] 23.45601


$J.pvalue
[1] 0.1448551


$J.table.pvalue
[1] 0.112



> testmeanshapes(panm.dat,panf.dat,replace=TRUE)
Bootstrap - sampling with replacement within each group under H0:
No of resamples = 1000
100 200 300 **400 500 600 700 800 900 1000


$J
[1] 23.45601


$J.pvalue
[1] 0.2127872


$J.table.pvalue
[1] 0.112
```

We see that the results of the James test are similar to those of the Hotelling T^2 test of Example 9.3 and Example 9.4 with no significant difference between the mean shapes. However, the λ_{min} test result is similar to the Goodall test where the conclusion is that there is a significant difference in mean shape. The Goodall test is more powerful than the James test if the assumptions are reasonable as there are many fewer parameters to estimate in the covariance matrix of the tangent coordinates.

9.2.4   One way analysis of variance

Consider a balanced MANOVA with independent random samples $(X_{i1}, ..., X_{in})^{\mathrm{T}}$, $i = 1, ..., n_G$, from n_G groups, each of size n. Let $[\hat{\mu}_j]$ be the group full Procrustes means and $[\hat{\mu}]$ is the overall pooled full Procrustes mean shape. A suitable test statistic Goodall (1991) is:

$$F = n(n-1)n_G \ \frac{\sum_{j=1}^{n_G} d_F^2(\hat{\mu}_j, \hat{\mu})}{(n_G - 1) \sum_{j=1}^{n_G} \sum_{i=1}^{n} d_F^2(X_{ji}, \hat{\mu}_j)}.$$

Under the null hypothesis of equal means the approximate distribution of F is $F_{(n_G-1)q, n_G(n-1)q}$, where $q = (k-1)m - m(m-1)/2$, and the null hypothesis is rejected for large values of the statistic. Since

$$d_F^2(\hat{\mu}_1, \hat{\mu}_2) = 2\left(d_F^2(\hat{\mu}_1, \hat{\mu}) + d_F^2(\hat{\mu}_2, \hat{\mu})\right),$$

the two sample test of the previous section (with $n_1 = n_2$) is a special case.

9.3   Size-and-shape tests

9.3.1   Tests using Procrustes size-and-shape tangent space

Tangent space inference for size-and-shape can proceed in a very similar manner as for shape space. Writing v for the size-and-shape tangent coordinates the test statistics are exactly the same as in the pure shape case for the Hotelling T^2, Goodall and James statistics of Section 9.1 except $q = (k-1)m - m(m-1)/2$ is one larger due to the higher dimension of the size-and-shape space.

We can carry out the tests using the option scale=FALSE in testmeanshapes. In the 2D chimpanzee data of Example 9.3 testing for mean size-and-shape differences between males and females we have:

```
> testmeanshapes(panf.dat,panm.dat,scale=FALSE,replace=TRUE)
Permutations - sampling without replacement: No of permutations = 1000
$H
[1]  2.227968

$H.pvalue
[1]  0.02297702

$H.table.pvalue
[1]  0.02836717

$G
[1]  5.158083

$G.pvalue
[1]  0.001998002

$G.table.pvalue
[1]  2.802277e-08

$J
[1]  34.3913

$J.pvalue
[1]  0.01898102
```

```
$J.table.pvalue
[1] 0.014

> testmeanshapes(panf.dat,panm.dat,scale=FALSE,replace=TRUE)
Bootstrap - sampling with replacement within each group under H0:
No of resamples = 1000
$H
[1] 2.227968

$H.pvalue
[1] 0.05194805

$H.table.pvalue
[1] 0.02836717

$G
[1] 5.158083

$G.pvalue
[1] 0.001998002

$G.table.pvalue
[1] 2.802277e-08

$J
[1] 34.3913

$J.pvalue
[1] 0.04995005

$J.table.pvalue
[1] 0.014
```

Hence for all tests there is evidence for a difference between the mean size-and-shapes of the female and male chimpanzees. If we look specifically at centroid size we see that there is a significant difference in mean centroid size:

```
> t.test(centroid.size(panf.dat),centroid.size(panm.dat))

Welch Two Sample t-test

data: centroid.size(panf.dat) and centroid.size(panm.dat)
t = -3.8176, df = 50.704, p-value = 0.0003682
alternative hypothesis: true difference in means is not equal to 0
95 percent confidence interval:
-10.887324 -3.382205
sample estimates:
mean of x mean of y
196.0349 203.1697
```

with the females being smaller on average than the males. The next stage of the analysis would be to study whether the sexual dimorphism of shape aligns with any suggested allometry.

Also, for analysis of variance for size-and-shape we can use the statistic:

$$F_{SS} = n(n-1)n_G \; \frac{\sum_{j=1}^{n_G} d_S^2(\hat{\mu}_j, \hat{\mu})}{(n_G - 1)\sum_{j=1}^{n_G}\sum_{i=1}^{n} d_S^2(X_{ji}, \hat{\mu}_j)} \tag{9.12}$$

where $\hat{\mu}_i$ and $\hat{\mu}$ are the respective size-and-shape estimators and $q = (k-1)m - m(m-1)/2$. We provide an example of using this test in Section 14.

9.3.2   Case-study: Size-and-shape analysis and mutation

Here we describe a case study to illustrate the size-and-shape tests. This application was described originally by Mardia (2013b) and illustrates the use of a method called TorusDBN (Boomsma *et al.* 2008) in mutation studies. In particular the application is from a study by Airoldi *et al.* (2010) who investigated some aspects of evolution by conducting a mutation experiment on two test species of flowering plants. A basic step in the mutation process involves the changing of amino acids in a sequence through addition or removal of an amino acid which are referred to as gaps (Durbin *et al.* 1998, p. 130). Airoldi *et al.* (2010) have obtained in their experiment five protein sequences, two original sequences to be called AG and FAR sequences, and their three mutant sequences AG+Q, AG+R, FAR–Q with insertion or deletion of one amino acid. The study involves five fragments displayed below, one from each of these five protein sequences, that are of 12 or 13 amino acids in length (the hyphen in each sequence denotes a gap).

```
AG fragment sequences:
 DYMQKR-EVDLHN  (AG)   Original DYMQKREVDLHN
 DYMQKRQEVDLHN  (AG+Q) Insertion of Q
 DYMQKRREVDLHN  (AG+R) Insertion of R.


FAR fragment sequences:
 EYMQKRQEIDLHH  (FAR)   Original
 EYMQKR-EIDLHH  (FAR-Q) Deletion of Q.
```

For brevity, we will denote these five sequences as $1, \dots, 5$, respectively. The inference based on the sequences by Airoldi *et al.* (2010) is that the sequences 1 and 5 have the same functionality and so do each pair: 2 and 3; 2 and 4; 3 and 4. So the key functional pairs are:

$$(1,5), (2,3), (2,4), (3,4).$$

The conclusions of Airoldi *et al.* (2010) were based only on the sequences as there is no knowledge of their true structures. There are many sequences of amino acids where the structures are still unknown. In fact, the site UniParc http://www.uniprot .org/uniparc/ has about 32 million protein sequences but on the other hand there

Table 9.1   The RMSD between Procrustes means forms in Set 1.

	1	2	3	4	5
1	–	0.004	0.0076	0.0139	0.0112
2		–	0.0044	0.0166	0.0157
3			–	0.0291	0.0221
4				–	0.0047
5					–

are only about 85 000 known structures in the Protein Data Bank (PDB). However, the function of a protein is usually determined by its structure rather than its sequence (e.g. see Tramontano 2006, p. 1) and we use TorusDBN to explore inference based on the structure. Mardia (2013b) simulated 100 structures (3D coordinates) from TorusDBN (Boomsma *et al.* 2008) given each of the five fragments, and these 500 simulated configurations are available in the Supplementary Material of Mardia (2013b). Note that these fragments have local functional information as they are selected around the gap region in the protein sequences which gave rise to physical changes in the plants (Airoldi *et al.* 2010).

The sequences 1 and 5 have 12 amino acids but the sequences 2, 3 and 4 have 13 amino acids. One plausible way to carry out inference on functionality through these structures is to treat each point of their configuration with the label number given by its position (termed a landmark so there is correspondence across the configurations). Also not all fragments have the same length, therefore, for a meaningful comparison we compare the N and C terminal portions on both sides of the point mutation, that is, use the two sets of six landmarks: Set 1 the first six, Set 2 the last six. (N-terminus refers to the 'start' of a protein and C-terminus to the 'end' of a protein.)

So, there are 5 groups of $n = 100$ configurations, $j = 1, \ldots, 5$ for each of two sets of data (Set 1: N-terminus; Set 2: C-terminus). The observations each have $k = 6$ landmarks in $m = 3$ dimensions. It is of interest to compare the size-and-shape (form) of the molecules, and examine which pairs are closer.

First, we assess their similarities by using the RMSD as in the following. We calculate the Procrustes mean size-and-shape of each group by GPA for size-and-shape, as in Section 7.5.1. Then we obtain the RMSD between each mean pair leading to the 5×5 form distance matrix; Table 9.1 and Table 9.2 give these distance matrices for the two sets and Figure 9.7 gives, for each set, a corresponding dendrogram (using single linkage clustering). For Set 1, the dendrogram shows no clusters whereas for Set 2 the dendrogram highlights the relative clustering of the pairs (2,3), (2,4), (3,4) and (1,5) under discussion.

Next, for each set, we formulate a plausible perturbation model. We can proceed one step further by defining and testing the null hypothesis as follows. The plausible model for this ANOVA situation is:

$$X_{ij} = (\mu + \mu_j + E_{ij})\Gamma_{ij} + 1_k \gamma_{ij}^T, \quad i = 1, \ldots, n, \quad j = 1, \ldots, n_G, \qquad (9.13)$$

Table 9.2   The RMSD between Procrustes means forms in Set 2.

	1	2	3	4	5
1	–	0.0905	0.0901	0.1595	0.0181
2		–	0.0046	0.0095	0.1062
3			–	0.0200	0.1114
4				–	0.2067
5					–

where X_{ij} are the configuration matrices ($k \times m$), μ is an overall mean configuration, μ_j is the jth mean configuration, E_{ij} are independent Gaussian with zero means, Γ_{ij} are rotations and γ_{ij} are translations, 1_k is the vector $(1, 1, \dots, 1)^T$; Γ_{ij}, γ_{ij} are the nuisance parameters. We test the null hypothesis that the size-and-shape of the means are equal:

$$H_0 : [\mu_1]_S = [\mu_2]_S = \dots = [\mu_{n_G}]_S.$$

MANOVA can then be applied to these preprocessed configurations where each configuration matrix is stacked as a vector having $3k - 6$ elements. Note that $3k - 6$ denotes the number of variables left after removing the 3 degrees of freedom for translation and 3 for rotation. Furthermore, we use the Goodall statistic for size-and-shape (9.12); that is, we have the F statistic given by:

$$F_{SS} = \text{BetweenMSS}/\text{Within MSS with} f_1 = (n_G - 1)q, \ f_2 = n_G(n - 1)q,$$

where $q = 3k - 6$, $n = 100$, $n_G = 5$ so $q = 12$, $f_1 = 48$, $f_2 = 5940$.

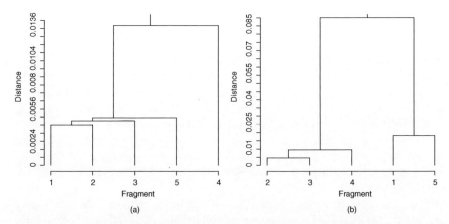

Figure 9.7   Dendrograms for the five fragments based on the RMSD for (a) Set 1 and (b) Set 2: 1, AG; 2, AG + Q; 3, AG + R; 4, FAR; and 5, FAR − Q. Source: Mardia 2013. Reproduced with permission from Taylor & Francis.

Table 9.3   Set 1: Pairwise F values corresponding to Hotelling's T^2 statistics for the five structures with the bootstrap p-values.

Pair	F	p-value	Pair	F	p-value
1,2	1.09	0.40	2,4	1.57	0.14
1,3	1.37	0.22	2,5	2.11	0.03
1,4	1.99	0.04	3,4	2.37	0.02
1,5	1.30	0.25	3,5	1.69	0.12
2,3	0.92	0.56	4,5	1.76	0.09

It is found that:

Set 1 : Between MSS $= 0.0483$, Within MSS $= 0.0208$, $F = 2.3$, p-value $= 0$,

Set 2 : Between MSS $= 0.2982$, Within MSS $= 0.0384$, $F = 7.8$, p-value $= 0$.

Hence, the null hypothesis of the equal means has almost zero p-value, but relatively Set 2 has a larger value of F so this again indicates that these fragments have larger mean differences.

We now examine the Hotelling T^2 statistic in the Procrustes tangent space for all the pairs, and Table 9.3 and Table 9.4 give the p-values from the bootstrap for Set 1 and Set 2, respectively, under the null hypothesis. The p-values from the bootstrap are computed in R; 1000 bootstrap samples were used. For Set 2, only the same four key pairs (1,5), (2,3), (2,4) and (3,4) have large p-values in testing the same forms, but the rest of the combinations have almost zero p-values. Again, there is no clear pattern in Set 1. So we arrive at the the same evidence as through using the RMSD. Note that the RMSD in Table 9.1 implicitly assumes isotropic errors, whereas the T^2 statistic allows for correlated errors so the tests are more stringent.

In passing, we note that from TorusDBN, we find that the predicted secondary structures of the fragment sequences are α-helices, which is consistent with prediction from established deterministic software such as PHYRES (Kelley *et al.* 2015).

Table 9.4   Set 2: Pairwise F values corresponding to Hotelling's T^2 statistics for the five structures with the bootstrap p-values.

Pair	F	p-value	Pair	F	p-value
1,2	4.98	0.00	2,4	0.78	0.65
1,3	4.38	0.00	2,5	8.00	0.00
1,4	8.38	0.00	3,4	1.83	0.06
1,5	1.32	0.20	3,5	5.95	0.00
2,3	0.49	0.91	4,5	12.96	0.00

However, the analysis given here through the structures themselves reveals more about their shape differences. To conclude, all the analysis indicates that there is no pattern in Set 1 (N-terminus) but Set 2 (C-terminus) shows a strong effect. Also for Set 2, the dendrogram and the results from the Hotelling's T^2 tests are in accordance with the conclusions drawn by Airoldi *et al.* (2010). The mutation affects primarily the C-terminus and, as these results are similar to sequence analysis, the structural effects are local rather than global.

9.4   Edge-based shape coordinates

An alternative method to Procrustes registration is to work directly with the edge registration coordinates such as Bookstein or Kendall or QR coordinates, and then use standard multivariate techniques on the vectors of coordinates. Kent (1994) showed that there exists an approximate linear transformation between Bookstein shape coordinates and the partial Procrustes tangent coordinates, and hence multivariate normal-based inference will be similar using either method for small variability.

However, edge registration in Bookstein coordinates induces correlations between landmarks, as we will see in Corollary 11.3. Hence it is not advisable to use Bookstein coordinates for exploring the structure of shape variability through PCA, as this can be misleading (Kent 1994). For example, if the original landmark coordinates are isotropic then the Bookstein shape variables may be far from isotropic. However, the Procrustes tangent coordinates do not suffer from this problem for 2D data.

Examples of using edge based shape coordinates for inference include Bookstein (1991, p. 282) and Bookstein and Sampson (1990), who describe Hotelling's T^2 tests for shape difference using Bookstein coordinates (see Section 9.4), and they also consider testing for affine or uniform shape changes (see Section 13.5.1). Also, O'Higgins and Dryden (1993) use Hotelling's T^2 tests for investigating sexual dimporphism in great apes. One advantage of the approach is that localized shape differences relative to a baseline can be explored, for example O'Higgins and Dryden (1993) investigate the differences between the face and braincase regions in chimpanzees, gorillas and orangutans, and take the baseline as the line between *nasion* and *basion* in the data described in Section 1.4.8.

9.5   Investigating allometry

Allometry can also be investigated in our geometrical framework, using regression of shape on size, and an initial discussion was given in Section 5.7. In biology allometry is a far more detailed topic than we can study here. Allometry often involves regressions of shape on size and other predictors and their interactions, such as sex.

Example 9.8   Consider the T2 Small mouse vertebra data of Section 1.4.1 as a simplified, illustrative example. In Figure 9.8 we see plots of centroid size versus the first three PC scores. There appears to be a positive correlation between the first PC score and size, and the correlation is 0.74. When examining the plot of Figure 8.5

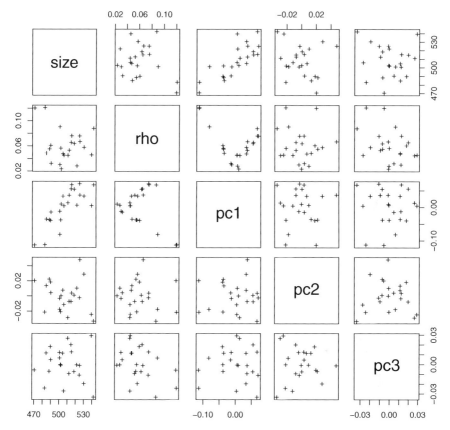

Figure 9.8   Pairwise scatter plots for centroid size and the first three PC scores of shape. There is a positive correlation between the first PC score and size.

it can be seen that the first PC partly measures the length of the spinous process (top protrusion). Therefore a shape measure capturing this effect is associated with centroid size in the dataset. From a biological point of view the association between size of the spinous process and the overall size of the bone is reasonable, because the spinous process bears important muscles.

In the following R code we carry out Procrustes analysis, provide pairwise scatterplots, and fit a linear regression model of centroid size on the PC scores:

```
> data<-qset2.dat
> n<-dim(data)[3]
> k<-dim(data)[1]
> mice<-procGPA(data)
> pairs(cbind(mice$size,mice$scores[,1:3]),
                labels=c("size","PC1","PC2","PC3"))
> Y<-mice$size
```

```
> X<-mice$scores[,1:(2*k-4)]
> ans0<-lm(Y~X)
> summary(ans0)


Call:
lm(formula = Y ~ X)

Residuals:
    Min      1Q  Median      3Q     Max
-4.4085 -2.0525 -0.3534  1.8196  7.7677

Coefficients:
             Estimate Std. Error t value Pr(>|t|)
(Intercept) 173.82121    0.81607 212.997 < 2e-16 ***
XPC1          0.46823    0.08862   5.284 0.000116 ***
XPC2         -0.34271    0.23535  -1.456 0.167412
XPC3          0.09082    0.26193   0.347 0.733937
XPC4         -0.47672    0.28634  -1.665 0.118143
XPC5          0.11087    0.44854   0.247 0.808357
XPC6          0.64310    0.47654   1.350 0.198591
XPC7         -0.48876    0.58193  -0.840 0.415081
XPC8         -1.07219    0.89208  -1.202 0.249338
-
Signif. codes: 0 '***' 0.001 '**' 0.01 '*' 0.05 '.' 0.1 ' ' 1

Residual standard error: 3.914 on 14 degrees of freedom
Multiple R-squared: 0.7253, Adjusted R-squared: 0.5683
F-statistic: 4.62 on 8 and 14 DF, p-value: 0.00624
```

We see that the only PC that is significantly related to centroid size is PC1. We can also fit other types of regression models, exactly as we would do for Euclidean data. For example, we can fit a least squares, ridge regression (Hoerl and Kennard 1970), LASSO (Tibshirani 1996) and Elastic Net (Zou and Hastie 2005) model of centroid size versus PC scores, and calculate the within sample mean square prediction error. We use the glmnet library in R as follows:

```
> library(glmnet)
> #LS
> ans<-as.double(coef(glmnet(X,Y,alpha=0,lambda=0)))
> print(ans)
[1] 173.82120674 0.46823144 -0.34270710 0.09082315 -0.47672340
[6] 0.11086931 0.64310197 -0.48875932 -1.07219378
> out.ls<-glmnet(X,Y,alpha=0)
> ls.pred<-predict(out.ls,s=0,newx=Xtest)
> mean((ls.pred-Ytest)^2)
[1] 9.440591
>
> #Ridge
> out.ridge<-glmnet(X,Y,alpha=0)
```

```
> out<-cv.glmnet(X,Y,alpha=0,grouped=FALSE)
> ans<-as.double(coef(glmnet(X,Y,alpha=0,lambda=out$lambda.min)))
> print(ans)
[1] 173.82120674 0.36051172 -0.26386508 0.06992869 -0.36705005
[6] 0.08536310 0.49515214 -0.37631703 -0.82552856
>
> ridge.pred<-predict(out.ridge,s=out$lambda.min,newx=Xtest)
> mean((ridge.pred-Ytest)^2)
[1] 10.62653
>
> #Lasso
> out.lasso<-glmnet(X,Y,alpha=1)
> plot(out.lasso)
> out<-cv.glmnet(X,Y,alpha=1,grouped=FALSE)
> plot(out)
> ans<-as.double(coef(glmnet(X,Y,alpha=1,lambda=out$lambda.min)))
> print(ans)
[1] 173.8212067 0.3149068 0.0000000 0.0000000 0.0000000 0.0000000
[7] 0.0000000 0.0000000 0.0000000
>
> lasso.pred<-predict(out.lasso,s=out$lambda.min,newx=Xtest)
> mean((lasso.pred-Ytest)^2)
[1] 17.34093
>
> #Elastic Net
> out.en<-glmnet(X,Y,alpha=0.5)
> out<-cv.glmnet(X,Y,alpha=0.5,grouped=FALSE)
> ans<-as.double(coef(glmnet(X,Y,alpha=0.5,lambda=lam2)))
>
> print(ans)
[1] 173.82120674 0.31065524 -0.05315727 0.00000000 -0.11589235
[6] 0.00000000 0.06409357 0.00000000 -0.00714971
> enet.pred<-predict(out.en,s=out$lambda.min,newx=Xtest)
> mean((enet.pred-Ytest)^2)
[1] 12.51252
```

We use cross-validation to choose the values of the regularization parameters. The full regularization path for the LASSO is given in Figure 9.9.

We see that least squares has the lowest mean square prection error as expected, the coefficient of PC1 is the largest in the two sparse methods (LASSO and Elastic Net). The sparse models again point towards PC1 having the strongest association with centroid size.

Finally we consider out of sample prediction by taking 1000 random subsets of 17 of the $n = 23$ vertebrae as training data and the remaining 6 observations as test data, for example using:

```
train<-sample(1:n,trunc(3*n/4))
Y<-mice$size[train]
X<-mice$scores[train,1:(2*k-4)]
Ytest<-mice$size[-train]
```

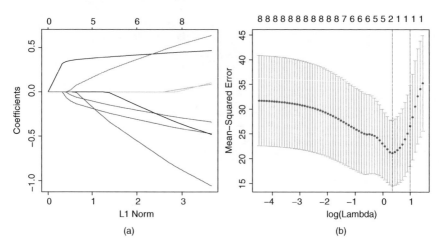

Figure 9.9   Regularization path for the LASSO (a) and cross-validation fitting (b). For a colour version of this figure, see the colour plate section.

```
Xtest<-mice$scores[-train,1:(2*k-4)]
#Lasso
out.lasso<-glmnet(X,Y,alpha=1)
out<-cv.glmnet(X,Y,alpha=1,grouped=FALSE)
plot(out)
lasso.pred<-predict(out.lasso,s=out$lambda.min,newx=Xtest)
mean((lasso.pred-Ytest)^2)
```

The test mean square prediction errors are: least squares 30.57; ridge 29.48; LASSO 26.00; and Elastic Net 27.28. Hence here LASSO regression has the best out of sample mean square prediction error. Given that the only non-zero coefficient fitted for the full dataset is for PC1, this again outlines the importance of the allometric relationship between size and PC1.   □

10

Shape and size-and-shape distributions

Of fundamental interest are probability distributions in shape spaces and pre-shape spaces, which provide models for statistical shape analysis. We shall also consider the joint distribution of size and shape later, but first we concentrate on shape. Since the shape space is non-Euclidean, special care is required. There are several approaches we could consider for stochastic modelling, for example we could specify a probability distribution in:

1. the configuration space

2. the pre-shape space

3. the shape space (directly)

4. a tangent space.

If the probability distribution also includes certain transformation variables (as in 1. and 2.) which are not of interest, then we could either integrate them out (a **marginal approach**) or condition on them (a **conditional approach**). Note that 2. is an ambient space model whereas 3. is a quotient space model (see Section 3.2.1).

The work in this chapter is primarily for $m = 2$ dimensional landmarks, and some extensions to higher dimensions will be considered in Section 11.6. The main inference procedure in this chapter is the method of maximum likelihood.

10.1 The uniform distribution

We gave the volume measure in pre-shape space in Equation (4.24) and the volume measure in shape space in Equation (4.26), using Kent's polar coordinates of

Statistical Shape Analysis, with Applications in R, Second Edition. Ian L. Dryden and Kanti V. Mardia.
© 2016 John Wiley & Sons, Ltd. Published 2016 by John Wiley & Sons, Ltd.

Section 4.3.3. We normalize the volume measure in shape space to give the uniform measure $d\gamma$ in shape space. Given original landmarks z^o recall that Kent's polar coordinates on the pre-shape sphere are obtained from the complex pre-shape $z = (z_1, \dots, z_{k-1})^T = Hz^o/\|Hz^o\|$ by:

$$\text{Re}(z_j) = s_j^{1/2} \cos \theta_j, \ \text{Im}(z_j) = s_j^{1/2} \sin \theta_j,$$

for $j = 1, \dots, k-1$, $s_j \geq 0$, $0 \leq \theta_j < 2\pi$. Kent's polar coordinates on shape space are $s_j, \phi_j = (\theta_j - \theta_{k-1}) \text{mod}(2\pi)$, $j = 1, \dots, k-2$.

Result 10.1   *The **uniform shape measure** in Kent's polar coordinates is:*

$$d\gamma = \frac{(k-2)!}{(2\pi)^{k-2}} ds_1 \dots ds_{k-2} d\phi_1 \dots d\phi_{k-2}$$

which has the property that $\int d\gamma = 1$.

Result 10.2   *Transforming to Kendall coordinates $U^K = (U_3^K, \dots, U_k^K, V_3^K, \dots, V_k^K)^T$ we have the uniform measure on the shape space given by:*

$$d\gamma = f_\infty(u^K) dU_3^K \cdots dU_k^K dV_3^K \cdots dV_k^K, \tag{10.1}$$

where

$$f_\infty(u) = \frac{(k-2)!\pi}{\{\pi(1 + u^T u)\}^{k-1}}. \tag{10.2}$$

Proof: Consider the coordinate system (Kendall 1984) $z_j/z_{k-1} = r_j e^{i\phi_j}$, where $r_j > 0$ and $0 \leq \phi_j < 2\pi$, $j = 1, \dots, k-2$. On substituting $z_j = s_j^{1/2} e^{i\theta_j}$ we have:

$$r_j^2 = s_j \bigg/ \left(1 - \sum_{j=1}^{k-2} s_j \right), \quad j = 1, \dots, k-2,$$

and $\phi_j = (\theta_j - \theta_{k-1}) \text{mod} 2\pi$, $j = 1, \dots, k-2$. We see that

$$s_j = r_j^2 \bigg/ \left(1 + \sum_{j=1}^{k-2} r_j^2 \right)$$

and the Jacobian $J = |\partial s_i/\partial r_j|$ is given by:

$$J = 2^{k-2} \left(\prod_{i=1}^{k-2} r_i \right) \bigg/ \left(1 + \sum_{j=1}^{k-2} r_j^2 \right)^{k-1}.$$

Hence the uniform measure is given by:

$$\frac{(k-2)!}{\pi^{k-2}}\left(1+\sum_{j=1}^{k-2}r_j^2\right)^{1-k}\prod_{j=1}^{k-2}(r_j dr_j d\phi_j).$$

Furthermore, using the coordinate system $u_j + iv_j = r_j e^{i\phi_j}, j = 1,\dots,k-2$, the uniform measure is:

$$\frac{(k-2)!}{\pi^{k-2}}\left(1+\sum_{j=1}^{k-2}u_j^2+v_j^2\right)^{1-k}\prod_{j=1}^{k-2}(du_j dv_j).$$

By suitably permuting the labels of the original pre-shape coordinates $z_1,\dots,z_{k-1}$ we arrive at the expression for the uniform density in terms of Kendall coordinates $U_j^K + iV_j^K = z_{j-1}/z_1, j = 3,\dots,k$. $\square$

When using Kent's polar coordinates the density is constant and hence Kent's coordinate system is a natural way to represent the uniform shape distribution.

We shall see in Section 11.1.1 that if the original landmarks are i.i.d. with a rotationally symmetric distribution, then the resulting marginal shape distribution is uniform in the shape space. Mardia (1980) gave the uniform shape distribution for triangles using angles to measure shape [see Equation (11.5)].

10.2 Complex Bingham distribution

10.2.1 The density

We consider the case where we have a probability distribution on the pre-shape sphere $S^{m(k-1)-1}$, corresponding to k points in m dimensions. For the $m = 2$ dimensional case and using complex notation we have seen in Section 4.3 that $S^{2(k-1)-1} \equiv \mathbb{C}S^{k-2}$, where $\mathbb{C}S^{k-2} = [z : (k-1) - \text{vector}, z^*z = 1]$ is the unit complex sphere in $k-1$ complex dimensions. In our particular application to shape analysis consider the k original landmarks in $m = 2$ dimensions with complex coordinates written as z^o ($k \times 1$ complex vector). Pre-multiply by the $(k-1) \times k$ Helmert submatrix H of Equation (2.10) to give the $k-1$ Helmertized complex landmarks z_H [a $(k-1)$-vector]. Rescaling by $\|z_H\|$ it follows that the pre-shape is:

$$z = (z_1, z_2, \dots, z_{k-1})^{\mathrm{T}} = z_H/\|z_H\| \in \mathbb{C}S^{k-2} \ (\equiv S^{2k-3}).$$

Definition 10.1 *The **complex Bingham distribution** on $\mathbb{C}S^{k-2}$, denoted $\mathbb{C}B_{k-2}(A)$, has probability density function (p.d.f.)*

$$f(z) = c(A)^{-1}\exp(z^*Az), \quad z \in \mathbb{C}S^{k-2},$$

where z^* represents the complex conjugate of the transpose of z, the matrix A is $(k-1) \times (k-1)$ Hermitian (i.e. $A = A^*$), and $c(A)$ is the normalizing constant.

The complex Bingham distribution is analogous to the real Bingham distribution (e.g. see Mardia *et al.* 1979, p. 433) and in fact, it is a particular case (see Section 10.2.5). The distribution has the property that

$$f(e^{i\theta}z) = f(z)$$

and is thus invariant under rotations of the pre-shape z. So, if an object is rotated, then it has the same density and will contribute in an identical way to inference as the same object in the original rotation. This property therefore makes the distribution appropriate for shape analysis (location and scale were previously removed because z is on the pre-shape sphere). The complex Bingham distribution provides a very elegant framework for the analysis of 2D shape data. The distribution was proposed by Kent (1994) and the properties that we describe are from his paper.

Since $z^*z = 1$ for $z \in \mathbb{C}S^{k-2}$, we can see that the parameter matrices A and $A + \alpha I$ define the same complex Bingham distribution, with $c(A + \alpha I) = c(A) \exp \alpha$, where α is a complex number. It is convenient to remove this non-identifiability by setting $\lambda_{\max}(A) = 0$, where $\lambda_{\max}(A)$ denotes the largest eigenvalue of A.

10.2.2   Relation to the complex normal distribution

In order to understand the covariance structure of a complex distribution we describe the complex multivariate normal distribution. If $z_j = x_j + iy_j$ have a joint complex normal distribution with means $\xi_j = \mu_j + iv_j$, $j = 1, \ldots, p$, and a $p \times p$ Hermitian covariance matrix $\Sigma = \Sigma_1 + i\Sigma_2$, then writing $x = (x_1, \ldots, x_p, y_1, \ldots, y_p)^T$ and $\mu = (\mu_1, \ldots, \mu_p, v_1, \ldots, v_p)^T$ we have

$$x \sim N_{2p}\left(\mu, \frac{1}{2}\begin{bmatrix} \Sigma_1 & \Sigma_2 \\ -\Sigma_2 & \Sigma_1 \end{bmatrix}\right), \tag{10.3}$$

where $\Sigma_2 = -\Sigma_2^T$ is skew-symmetric and Σ_1 is symmetric positive definite. In particular, $\text{var}(x_j) = \text{var}(y_j)$ and $\text{cov}(x_j, y_j) = 0$, and so for each j the covariance structure is isotropic. Writing $z = (z_1, \ldots, z_p)^T$ and $\xi = (\xi_1, \ldots, \xi_p)^T$ the density of the complex normal distribution is:

$$f(z) = \frac{1}{\pi^p |\Sigma|} e^{-(z-\xi)^* \Sigma^{-1}(z-\xi)} \tag{10.4}$$

and we write $z \sim \mathbb{C}N(\xi, \Sigma)$; see, for example, Goodman (1963) and Andersen *et al.* (1995).

The complex Bingham distribution can be obtained through conditioning a zero mean complex multivariate normal distribution to have unit norm. In particular, if $w \sim \mathbb{C}N_{k-1}(0, \Sigma)$ then,

$$w|\{\|w\| = 1\} \sim \mathbb{C}B_{k-2}(-\Sigma^{-1}).$$

So, an interpretation of the Hermitian matrix $-A$ is that of the inverse covariance matrix of a zero mean complex normal random variable, which is conditioned to have unit norm to give the complex Bingham distribution. Hence, the complex Bingham distribution is an example of a conditional approach.

10.2.3   Relation to real Bingham distribution

The $(k-2)$ dimensional complex Bingham distribution can be regarded as a special case of a $(2k-3)$ dimensional real Bingham distribution as in the following. If the jth element of z is $(z)_j = x_j + iy_j$, define a $(2k-2)$ dimensional vector $u = (x_1, y_1, \ldots, x_{k-1}, y_{k-1})^T$, say, by splitting each complex number into its real and imaginary parts. Also, if $A = (a_{hj})$ has entries $a_{hj} = \alpha_{hj} \exp(i\theta_{hj})$ with $\theta_{jh} = -\theta_{hj}$, $-\pi < \theta_{jh} \leq \pi$, define a $(2k-2) \times (2k-2)$ matrix B made up of $(k-1)^2$ blocks of size (2×2) given by:

$$B_{hj} = \alpha_{hj} \begin{pmatrix} \cos\theta_{hj} & -\sin\theta_{hj} \\ \sin\theta_{hj} & \cos\theta_{hj} \end{pmatrix}.$$

Then $z^*Az = u^TBu$ so that a complex Bingham distribution for z is equivalent to a real Bingham distribution for u (Kent 1994).

10.2.4   The normalizing constant

Result 10.3   (Kent 1994) *The normalizing constant of the complex Bingham distribution is:*

$$c(A) = 2\pi^{k-1} \sum_{j=1}^{k-1} a_j \exp\lambda_j, \quad a_j^{-1} = \prod_{i \neq j}(\lambda_j - \lambda_i), \tag{10.5}$$

where $\lambda_1 < \lambda_2 < \cdots < \lambda_{k-1} = 0$ denote the eigenvalues of A. Note that $c(A) = c(\Lambda)$ depends only on the eigenvalues of A, with $\Lambda = \mathrm{diag}(\lambda_1, \ldots, \lambda_{k-1})$.

The result and proof were given by Kent (1994). If equalities exist between some λ_js then $c(A)$ can be integrated using repeated applications of L'Hôpital's rule. For example, for $\lambda_{k-2} = \lambda_{k-1}$ but all other λ_js distinct, we get (Bingham *et al.* 1992):

$$c(A) = (k-2)! \left[\sum_{i=1}^{k-3} \frac{e^{-\lambda_i}}{\prod_{j \neq i}(\lambda_j - \lambda_i)} + \frac{e^{-\lambda_{k-1}}\{-\sum_{i=1}^{k-3}(\lambda_i - \lambda_{k-1})^{-1}\}}{\prod_{j \leq k-3}(\lambda_j - \lambda_{k-1})} \right].$$

A particularly useful and simple case is the complex Watson distribution when there are just two distinct eigenvalues, as described in Section 10.3.

10.2.5   Properties

We give various properties of the complex Bingham distribution. Let $\gamma_1, \ldots, \gamma_{k-1}$ denote the standardized eigenvectors of A so that $\gamma_j^* \gamma_j = 1$, $\gamma_i^* \gamma_j = 0$, $i \neq j$, and $A\gamma_j = \lambda_j \gamma_j$, $j = 1, \ldots, k-1$. Each γ_j is defined only up to rotation by a unit complex scalar. Provided that $\lambda_{k-2} < 0$, $z = \gamma_{k-1}$ maximizes the density and γ_{k-1} is unique up to a scalar rotation $\exp(i\theta)\gamma_{k-1}$. Furthermore, if $\lambda_1, \ldots, \lambda_{k-2}$ are far below 0, the distribution becomes highly concentrated about this modal axis.

10.2.5.1   High concentrations

To model high concentration, replace A by κA transforming to Kent's polar coordinates and integrating out $\theta_1, \ldots, \theta_{k-1}$

$$f(s) = 2\pi^{k-1} c(\kappa\Lambda)^{-1} \exp\left(\sum_j \kappa \lambda_j s_j \right), \quad s \in S_{k-1}.$$

As $\kappa \to \infty$, $u_j = \kappa s_j$, $j = 1, \ldots, k-2$, tend to independent exponential random variables with densities

$$\lambda_j^{-1} \exp(\lambda_j u_j), \quad u_j > 0, \ \lambda_j < 0.$$

Thus for large κ,

$$c(\kappa\Lambda) = 2\pi^{k-1} \prod_{j=1}^{k-2} (-\kappa\lambda_j)^{-1}. \tag{10.6}$$

We assume that there is a unique largest eigenvalue $\lambda_{k-1} = 0 > \lambda_{k-2}$. Another way to look at the high concentration about the axis γ_{k-1} is to rotate z to γ_{k-1} and then to project z onto the tangent plane at γ_{k-1}. Given $z \in \mathbb{C}S^{k-2}$,

$$v = \exp(i\theta)(I_{k-1} - \gamma_{k-1}\gamma_{k-1}^*)z, \quad z \in \mathbb{C}^{k-1}, \quad \theta = -\text{Arg}(\gamma_{k-1}^* z),$$

and it can be proved that θ and v are independently distributed with θ uniform on $[0, 2\pi)$. Writing $A = \sum_{j=1}^{k-1} \lambda_j \gamma_j \gamma_j^* = \sum_{j=1}^{k-2} \lambda_j \gamma_j \gamma_j^*$, we have

$$\kappa z^* A z = \kappa v^* A v$$

and therefore v is asymptotically complex multivariate normal in $k - 1$ complex dimensions with zero mean and covariance matrix

$$\Sigma = (-A)^- = -\sum_{j=1}^{k-2} (-\lambda_j)^{-1} \gamma_j \gamma_j^*,$$

where the last expression is the Moore–Penrose generalized inverse of $-A$ (see Definition 9.1). Here Σ is singular because v lies in the tangent plane $\gamma_{k-1}^* v = 0$. Thus z is approximately distributed as $\mathbb{C}N_{k-1}(\gamma_{k-1}, \Sigma)$ so that Σ is determined by the eigenvectors orthogonal to γ_{k-1}.

The asymptotic distribution shows that the real and imaginary parts of v_i are jointly distributed as $N_2(0, I_2 \sigma_i^2)$. Hence the complex Bingham distribution imposes an isotropic distribution on the marginal distributions. However, it does allow some intercorrelation between landmarks, that is that of a complex normal covariance matrix as seen in Equation (10.3).

10.2.5.2 Low concentrations

As $\lambda_j \to 0$ for all j, the complex Bingham distribution tends to the uniform distribution on $\mathbb{C}S^{k-2}$. Note that $c(0) = 2\pi^{k-1}/(k-2)!$.

10.2.5.3 High dimensions

As the number of landmarks k increases one can study the asymptotic distribution of the complex Bingham distribution. Dryden (2005b) considers statistical analysis on high-dimensional spheres and shape spaces, and in particular the high-dimensional complex Bingham distribution tends to a multivariate normal distribution for fixed rank parameter matrix A.

10.2.6 Inference

Let $z_1, \ldots, z_n$ be a random sample from a population modelled by a complex Bingham distribution, $n \geq k - 1$. Set

$$S = \sum_{i=1}^{n} z_i z_i^*$$

to be the $(k - 1) \times (k - 1)$ complex sum of squares and products matrix. Suppose that the eigenvalues of S are positive and distinct, $0 < l_1 < \cdots < l_{k-1}$ and let $g_1, \ldots, g_{k-1}$ denote the corresponding eigenvectors. Note that $\sum_j l_j = n$. Maximum likelihood estimation is carried out as for the real Bingham distribution.

Result 10.4   (Kent, 1994) *Under the complex Bingham distribution the MLE of* $\gamma_1, \ldots, \gamma_{k-1}$ *and* $\Lambda = diag(\lambda_1, \ldots, \lambda_{k-1})$ *are given by:*

$$\hat{\gamma}_j = g_j, \quad j = 1, \ldots, k-1,$$

and the solution to

$$\frac{\partial \log c(\Lambda)}{\partial \lambda_j} = \frac{1}{n} l_j, \quad j = 1, \ldots, k-2.$$

Under high concentrations

$$\hat{\lambda}_j \cong -n/l_j, \quad j = 1, \ldots, k-2.$$

Proof: The log-likelihood for the data reduces to:

$$L = \sum_{i=1}^{n} z_i^* A z_i - n \log c(A) = \text{trace}\left(\sum_{i=1}^{n} z_i z_i^* A\right) - n \log c(\Lambda)$$

$$= \text{trace}(SA) - n \log c(\Lambda) = \sum_j \lambda_j \gamma_j^* S \gamma_j - n \log c(\Lambda). \qquad (10.7)$$

Holding $\lambda_1 < \ldots < \lambda_{k-2} < \lambda_{k-1} = 0$ constant, it can be seen that

$$\hat{\gamma}_j = g_j, \quad j = 1, \ldots, k-1,$$

and so

$$L = \sum_{j=1}^{k-1} l_j \lambda_j - n \log c(\Lambda).$$

The MLEs of the eigenvalues are found by solving

$$\frac{\partial \log c(\Lambda)}{\partial \lambda_j} = \frac{1}{n} l_j, \quad j = 1, \ldots, k-2.$$

Under high concentration, from Equation (10.6),

$$\log c(\Lambda) \simeq \text{constant} - \sum_{j=1}^{k-2} \log(-\lambda_j)$$

giving $\hat{\lambda}_j \cong -n/l_j, \quad j = 1, \ldots, k-2.$ □

The dominant eigenvector $\hat{\gamma}_{k-1}$ can be regarded as the average axis of the data: an estimate of modal shape. Hence, the average axis from the complex Bingham MLE is the same as the full Procrustes estimator of mean shape for $m = 2$ dimensional data, which was also given by the dominant eigenvector of S as shown in Result 8.2.

Example 10.1   Consider the mouse vertebral data described in Section 1.4.1. There are $k = 6$ landmarks in $m = 2$ dimensions here. Using the complex Bingham distribution the MLE of the population modal shape is:

$$g_{k-1} = (0.04 + 0.25i, 0.40 + 0.07i, -0.49 - 0.72i, -0.06 + 0.04i, 0.09)^{\mathrm{T}}$$

corresponding to the eigenvalue $l_{k-1} = 22.9$. Since $l_{k-1} \approx n$ the data are highly concentrated here. In Figure 8.5 we saw a plot of the MLE of the modal shape (which is the full Procrustes mean) as a centred pre-shape.

In order to examine the structure of variability we could examine $g_1, \ldots, g_{k-2}$ although the restrictions of a complex covariance structure are in force. In particular, we necessarily have isotropic perturbations at each landmark. So, a practical method for inference would be to examine the shape variability in the tangent space; details of this example were given in Example 8.2.   □

10.2.7   Approximations and computation

In order to carry out more accurate estimation of the complex Bingham parameters Kume and Wood (2007) study the derivatives of the normalizing constant of the Bingham distribution, and Kume and Wood (2005) consider highly accurate saddlepoint approximations for the Bingham normalizing constant (as well as the Fisher–Bingham). Their methodology enables more accurate maximum likelihood estimation for a whole range of parameter values, including when the data are quite dispersed.

10.2.8   Relationship with the Fisher–von Mises distribution

Result 10.5   *For triangles ($k = 3$) modelling with the complex Bingham in pre-shape space is equivalent to using the Fisher–von Mises distribution on the spherical shape space $S^2(\frac{1}{2})$.*

Proof: If $k = 3$, then there is just one non-zero eigenvalue of A. Hence we can re-parameterize A as:

$$A = -2\kappa(I_2 - \mu\mu^*).$$

So, we have

$$z^*Az = -2\kappa + 2\kappa z^*\mu\mu^*z = \kappa \cos 2\rho(H^{\mathrm{T}}z, H^{\mathrm{T}}\mu),$$

where $\rho(\cdot)$ is the Riemannian distance. We can obtain the density in shape space by using Kent's polar shape coordinates and integrating over the independent uniform θ_2 component (Mardia 1996b). So, the density in shape space is proportional to:

$$\exp\{\kappa \cos 2\rho(H^{\mathrm{T}}z, H^{\mathrm{T}}\mu)\},$$

where $H^{\mathrm{T}}z$ is the centred icon corresponding to z. Since the shape space can be regarded as a sphere $S^2(\frac{1}{2})$ with great circle metric ρ, it follows that, using the direction vector $l = (l_x, l_y, l_z)^{\mathrm{T}}$, the probability element of l is:

$$f(l) = \frac{\kappa}{4\pi^2 \sinh(\kappa)} e^{\kappa l^{\mathrm{T}} v} \mathrm{d}S, \tag{10.8}$$

where $v = (v_x, v_y, v_z)^{\mathrm{T}}$ is the modal direction and $\mathrm{d}S$ is the probability element. The density (10.8) is the density of the Fisher–von Mises distribution on $S^2(\frac{1}{2})$. $\square$

Mardia (1989b) provided motivation to use the Fisher–von Mises distribution for shape analysis of triangles.

Example 10.2   Consider the microfossil data described in Section 1.4.16. Using the complex Bingham distribution on the pre-shape sphere will be equivalent to using the Fisher–von Mises distribution on $S^2(\frac{1}{2})$. Let $l_i = (l_{xi}, l_{yi}, l_{zi})$, $i = 1, \ldots, n$, be the directions on $S^2(\frac{1}{2})$ corresponding to each triangle shape, and we assume that $l_1, \ldots, l_n$ are a random sample from the Fisher–von Mises distribution. From the data we find that

$$\bar{R}_x = \frac{1}{n}\sum_{i=1}^{n} l_{xi} = 0.145, \quad \bar{R}_y = \frac{1}{n}\sum_{i=1}^{n} l_{yi} = 0.075, \quad \bar{R}_z = \frac{1}{n}\sum_{i=1}^{n} l_{zi} = 0.468.$$

Thus the resultant mean length is $\bar{R} = (\bar{R}_x^2 + \bar{R}_y^2 + \bar{R}_z^2)^{1/2} = 0.495$. The mean directional vector is $(0.146, 0.076, 0.472)$ and the resulting estimate of κ from the Fisher–von Mises distribution is $\hat{\kappa} = 419.2$, so we have a large value of the concentration here. When mapped to Bookstein coordinates the mean shape estimate is $u^B = -0.101$ and $v^B = 0.632$. $\square$

10.2.9   Simulation

There are several different methods available for simulation from the complex Bingham distribution, and Kent *et al.* (2004) provide several algorithms, which have rather different efficiencies depending on the amount of concentration. A detailed description of simulation methods for a variety of directional distributions is given by Kent *et al.* (2013). Also, Kume and Walker (2009) discuss simulation for the Fisher–Bingham distribution.

10.3   Complex Watson distribution

10.3.1   The density

Definition 10.2   *The* **complex Watson distribution** *on the pre-shape sphere* $\mathbb{C}S^{k-2}$, *denoted by* $\mathbb{C}W_{k-2}(\mu, \xi)$, *has p.d.f.*

$$f(z) = c_1(\xi)^{-1} \exp\{\xi |z^* \mu|^2\}. \tag{10.9}$$

Properties of the distribution were discussed by Mardia and Dryden (1999). In shape analysis we are primarily interested in $\xi \geq 0$, although the concentration parameter ξ can be negative in some circumstances. For $\xi > 0$, the modal pre-shape direction is $\mu e^{i\theta}$, where $0 \leq \theta < 2\pi$ is an arbitrary rotation angle. The complex Watson distribution is an important special case of the complex Bingham distribution, when there are just two distinct eigenvalues in A (a single distinct largest eigenvalue and all other eigenvalues being equal). In this case all directions orthogonal to the modal axis have equal weight, and so this model implicitly assumes a spherical error distribution, as would be obtained from independent isotropic landmarks with equal variances. In this case the complex Bingham distribution can be re-parameterized so that $A = -2\kappa(I_{k-1} - \mu\mu^*)$, where μ is the modal vector on the pre-shape sphere. The density on the pre-shape sphere can be written as:

$$\begin{aligned}
f(z) &= c_1(\xi)^{-1} \exp\{\xi z^* \mu\mu^* z\} = c_1(2\kappa)^{-1} \exp\{2\kappa z^* \mu\mu^* z\} \\
&= c_1(2\kappa)^{-1} \exp\{2\kappa \cos^2 \rho(H^T z, H^T \mu)\} \\
&= c_1(2\kappa)^{-1} e^{\kappa} \exp\{\kappa \cos 2\rho(H^T z, H^T \mu)\},
\end{aligned} \tag{10.10}$$

where $\rho(H^T z, H^T \mu)$ is the Procrustes distance between the shapes of $H^T z$ and $H^T \mu$. Note that $H^T z$ is the centred icon corresponding to the pre-shape z. The parameters of the distribution are the population mode pre-shape direction μ and the concentration parameter $\xi = 2\kappa$. We have chosen to re-parameterize the complex Watson distribution with the parameter $\kappa = \xi/2$ in order to make connections with the offset normal distributions below in Section 11.1.

We call this distribution the complex Watson distribution due to the similar form of the density to the Dimroth–Watson distribution on the real sphere, see Watson (1983, 1965), who called it the Scheiddegger–Watson distribution.

For triangles ($k = 3$) the complex Watson distribution is the same as the complex Bingham distribution, as there are at most two distinct eigenvalues in that case. An alternative expression for the density is

$$f(z) = c_1(2\kappa)^{-1} \exp\left\{ 2\kappa \left[1 - d_F^2(H^T z, H^T \mu) \right] \right\},$$

where $d_F(\cdot)$ is the full Procrustes distance of Equation (4.10).

This distribution was briefly discussed by Dryden (1991), Kent (1994) and Prentice and Mardia (1995). Full details and further developments can be found in Mardia and Dryden (1999), on which subsequent details are based.

Result 10.6   *The integrating constant for the complex Watson distribution is:*

$$c_1(2\kappa) = 2\pi^{k-1}(2\kappa)^{2-k}\left\{ e^{2\kappa} - \sum_{r=0}^{k-3}\frac{(2\kappa)^r}{r!}\right\} = \frac{2\pi^{k-1}}{(k-2)!}{}_1F_1(1;k-1;2\kappa),$$

where

$${}_1F_1(a;b;x) = 1 + \frac{a}{b}\frac{x}{1!} + \frac{a(a+1)}{b(b+1)}\frac{x^2}{2!} + \frac{a(a+1)(a+2)}{b(b+1)(b+2)}\frac{x^3}{3!} + \cdots \quad (10.11)$$

is the confluent hypergeometric function (e.g. see Abramowitz and Stegun 1970).

See Mardia and Dryden (1999) for a proof. Note that the integrating constant has a relatively simple form, and so inference is quite straightforward.

10.3.2   Inference

Let $z_1, \ldots, z_n$ be a random sample from a population modelled by a complex Watson distribution, with $n \geq k - 1$. Set

$$S = \sum_{i=1}^{n} z_i z_i^*$$

to be the $(k-1) \times (k-1)$ complex sum of squares and products matrix. Suppose that the eigenvalues of S are positive with eigenvalues $l_{k-1} > l_{k-2} \geq \ldots \geq l_1 > 0$ and let $g_{k-1}, \ldots, g_1$ denote the corresponding eigenvectors.

Result 10.7   *Under the complex Watson distribution the MLEs of μ and κ are given by:*

$$\hat{\mu} = g_{k-1},$$

and the solution to

$$\frac{c_1'(\kappa)}{c_1(\kappa)} = \frac{l_{k-1}}{n},$$

where $c_1'(\kappa) = \frac{\partial c_1(\kappa)}{\partial \kappa}$. Under high concentrations

$$\hat{\kappa} \approx \frac{n(k-2)}{n - l_{k-1}}. \quad (10.12)$$

Proof: The log-likelihood for the data reduces to:

$$\log L = \kappa \sum_{i=1}^{n} z_i^* \mu \mu^* z_i - n \log c_1(\kappa) = \kappa \, \text{trace}(S\mu\mu^*) - n \log c_1(\kappa)$$

$$= \kappa \, \text{trace}\left(\left(\sum_{j=1}^{k-1} l_j g_j g_j^* \right) \mu\mu^* \right) - n \log c_1(\kappa).$$

Holding κ constant, it can be seen that $\hat{\mu} = e^{i\alpha} g_{k-1}$ provides the maximum, where α is an arbitrary rotation angle ($0 \le \alpha < 2\pi$), and so

$$\log L = \kappa l_{k-1} - n \log c_1(\kappa).$$

The MLE of κ is found by solving

$$\frac{\partial \log c_1(\kappa)}{\partial \kappa} = \frac{c_1'(\kappa)}{c_1(\kappa)} = \frac{1}{n} l_{k-1}.$$

Under high concentration

$$c_1(\kappa) \approx 2\pi^{k-1} \kappa^{2-k} e^{\kappa}$$

and so

$$c_1'(\kappa) \approx 2\pi^{k-1} \kappa^{2-k} e^{\kappa} \left(1 - \frac{k-2}{\kappa} \right).$$

Therefore

$$\frac{l_{k-1}}{n} = \frac{c_1'(\kappa)}{c_1(\kappa)} \approx 1 - \frac{k-2}{\kappa}$$

and so the high concentration approximate MLE for κ follows.   □

As for the complex Bingham distribution the dominant eigenvector $\hat{\mu} = g_{k-1}$ can be regarded as the average axis of the data (up to a rotation), and it has the same shape as the sample full Procrustes mean of Equation (6.11).

Since the complex Watson distribution is a complex Bingham distribution with all but the largest eigenvalue equal, then under the complex Watson distribution we should expect $l_1, \ldots, l_{k-2}$ to be approximately equal. In other words, the shape variability for the complex Watson distribution is isotropic and all principal components of shape variability have equal weightings. This observation provides us with a model checking procedure.

10.3.3   Large concentrations

We now consider the case of large κ.

Result 10.8   *If κ is large, then the distribution tends to a complex multivariate normal distribution in the tangent space with mean 0 and covariance matrix $\frac{1}{2\kappa}(I_{k-1} - \mu\mu^*)$, which is a generalized inverse of $2\kappa(I_{k-1} - \mu\mu^*)$.*

Proof: We show that the following expression is true:

$$(z - \mu)^*(I - \mu\mu^*)^-(z - \mu) = 1 - z^*\mu\mu^*z, \tag{10.13}$$

where $z^*z = \mu^*\mu = 1$. Furthermore, $A^- = A$ where $A = I - \mu\mu^*$, and A^- is a generalized inverse. First we prove the generalized inverse, that is $AA^-A = A$. It is easily verified that A is idempotent and therefore $A^3 = A$. Hence, a generalized inverse of A is also A. To prove Equation (10.13) expand the left-hand side:

$$LHS = z^*(I - \mu\mu^*)z + \mu^*(I - \mu\mu^*)\mu - z^*(I - \mu\mu^*)\mu - \mu^*(I - \mu\mu^*)z.$$

Since $z^*z = 1$, we have

$$z^*(I - \mu\mu^*)z = z^*z - z^*\mu\mu^*z = 1 - z^*\mu\mu^*z.$$

Furthermore, $(I - \mu\mu^*)\mu = \mu - \mu(\mu^*\mu) = \mu - \mu = 0$. Hence Equation (10.13) is proved. Thus, the exponential term in the complex Watson density reduces to:

$$\exp\{-2\kappa(z - \mu)^*(I - \mu\mu^*)^-(z - \mu)\} \approx \exp\{-2\kappa v^*(I - \mu\mu^*)^-v\},$$

where v are the tangent plane coordinates, and z is close to μ. Hence,

$$v \approx \mathbb{C}N_{k-1}\left(0, \frac{1}{2\kappa}(I_{k-1} - \mu\mu^*)\right),$$

for large concentrations.   □

For large concentrations the distribution is also very similar to the offset isotropic normal distribution described in Section 11.1.2. It follows from the normal approximation that for large κ

$$4\kappa\{(z - \mu)^*(I - \mu\mu^*)^-(z - \mu)\} = 4\kappa\{1 - z^*\mu\mu^*z\}$$

is approximately distributed as χ^2_{2k-4}, since the complex rank of $(I - \mu\mu^*)$ is $k - 2$. Furthermore, we have

$$\kappa z^*\mu\mu^*z = \kappa \cos^2\rho,$$

where ρ is the Riemannian distance between the shapes of $H^T z$ and $H^T \mu$, so that

$$4\kappa(1 - z^*\mu\mu^* z) = 4\kappa \sin^2 \rho = 4\kappa d_F^2 \simeq \chi^2_{2k-4}. \tag{10.14}$$

This result agrees with the approximate distribution for the squared full Procrustes distance under isotropic normal errors given in Equation (9.6), with

$$\kappa = S(H^T \mu)^2 / (4\sigma^2) = \|\mu\|^2 / (4\sigma^2),$$

where $S(H^T \mu)$ is the centroid size of the icon $H^T \mu$ corresponding to μ.

Important point: The central role that the Fisher–von Mises distribution plays in directional data analysis is played by the complex Watson distribution for 2D shape analysis. For triangles we have seen that the two distributions are equivalent. For more than three points there are also similarities. Both distributions are very tractable, and each can be viewed as a conditional normal distribution.

10.4 Complex angular central Gaussian distribution

Kent (1994, 1997) suggested the complex angular central Gaussian distribution for shape analysis, which has the density

$$f(z) = \frac{(k-2)!}{2\pi^{k-1}} |\Sigma|^{-1} (z^* \Sigma^{-1} z)^{-(k-1)}, \quad z \in \mathbb{C}S^{k-2}. \tag{10.15}$$

The parameter matrix Σ is a positive definite Hermitian matrix. The density in Equation (10.15) is invariant under rescaling $\Sigma \to c\Sigma, c > 0$. Also, $f(z) = f(e^{i\theta} z)$ for all θ, and so the distribution is appropriate for shape analysis. Kent (1997) gives an expectation–maximization (EM) algorithm for computing the MLE of Σ, and derives various properties. In particular, the distribution is more resistant to outliers than the complex Bingham or Watson distributions.

10.5 Complex Bingham quartic distribution

We have seen that complex distributions can be of great benefit for planar shape analysis, but the complex structure has been restrictive imposing isotropy at each landmark. An extension which has approximate elliptical structure at each landmark is the complex Bingham quartic distribution (Kent et al. 2006) which includes an extra quartic term in the exponent of the density. The density is:

$$f(z) \propto \exp(z^* A z + z^* B z z^* C z)$$

and Kent et al. (2006) carefully described the identifiable parameters. Note that if the distribution is concentrated the approximate distribution in tangent space is a multivariate normal distribution with general elliptic covariance structure. For k landmarks in two dimensions there are $2k - 4$ mean parameters and $(2k - 4)(2k - 3)/2$ covariance parameters that are identifiable.

10.6   A rotationally symmetric shape family

A rotationally symmetric family of shape distributions can be obtained with densities as functions of shape distance to a mean configuration. The class of densities is given by Mardia and Dryden (1999):

$$c_\phi(\kappa)^{-1} \exp\left[-\kappa\phi\left(d_F^2\right)\right],$$

where $\phi(\cdot) \geq 0$ is a suitable penalty function, and it is assumed that $\phi(d_F)$ is an increasing function of d_F and $\phi(0) = 0$. This distribution is more difficult to work with in $m > 2$ dimensions, as the integrating constant then depends on the mean.

A particular subclass is given by the densities with

$$\phi(d_F^2) = \left[1 - \left(1 - d_F^2\right)^h\right]\big/h, \tag{10.16}$$

which give the same MLE as a class of shape estimators proposed by Kent (1992) and discussed in Section 6.3. The estimators become more resistant to outliers as h increases.

A shape distribution can be obtained from a pre-shape distribution by transforming to Kent's polar shape coordinates and dropping the independent uniform rotation parameter θ_{k-1}. For example, the shape distribution obtained from the complex Watson distribution is a member of the family with $h = 1$ in Equation (10.16). The density in this form was introduced by Dryden (1991) and is given, with respect to the uniform measure, by:

$$c_1(2\kappa)^{-1}2\pi e^\kappa \exp(\kappa \cos 2\rho), \tag{10.17}$$

where ρ is the Procrustes distance from the observed shape to the mean shape. The MLE of the mean shape is the same as the full Procrustes estimator of the mean shape. If $k = 3$, then the density reduces to:

$$\frac{\kappa}{4\pi^2 \sinh(\kappa)} \exp(\kappa \cos 2\rho)$$

which is the Fisher–von Mises distribution on the shape sphere.

An alternative distribution with $h = \frac{1}{2}$ is the distribution with density

$$c_2(\kappa)^{-1} \exp(4\kappa \cos \rho), \tag{10.18}$$

where the mean shape is the same as the partial Procrustes mean (Dryden 1991). We call this distribution the partial Procrustes shape distribution. The integrating constant is given by:

$$c_2(\kappa) = 1 + [\sqrt{\pi}(2\kappa)^{5/2-k}(k-2)!\{I_{k-3/2}(4\kappa) + \mathbf{L}_{k-3/2}(4\kappa)\}],$$

where $I_v(\cdot)$ and $\mathbf{L}_v(\cdot)$ are the modified Bessel function of the first kind and the modified Struve function. If $k = 3$, then $c_2(\kappa) = 8\kappa^2/[1 - \exp(4\kappa) + 4\kappa \exp(4\kappa)]$.

Both of the distributions of Equation (10.17) and Equation (10.18) are asymptotically normal as $\kappa \to \infty$ and uniform if $\kappa = 0$. The density in Equation (10.18) has particularly light tails and so the MLE of μ under the partial Procrustes distribution will be more affected by outliers than that of the complex Watson shape distribution of Equation (10.17).

10.7   Other distributions

Another family of shape distributions for $m = 2$ was given by Micheas *et al.* (2006) who considered complex elliptical distributions, and related work of Micheas and Dey (2005) for modelling shape distributions and inference for assessing differences in shapes.

Pennec (2006) described an intrinsic Gaussian distribution for symmetric Riemannian manifolds, defined as the distribution with fixed mean and covariance matrix Σ which maximizes entropy. The curvature of the manifold appears in the concentration matrix, and for small variations the concentration matrix is approximately

$$\Sigma^{-1} - \text{Ric}/3,$$

where Ric is the Ricci curvature tensor of Equation (4.15). Also, see Pennec (1996, 1999) for earlier, related work.

10.8   Bayesian inference

As an alternative to maximum likelihood estimation we could consider a Bayesian approach to inference. Let $u_1, \ldots, u_n$ be a random sample of shapes, and let Θ be the location shape parameters and Σ the shape variability parameters. After specifying a prior distribution for (Θ, Σ) with density $\pi(\Theta, \Sigma)$, the posterior density is given by:

$$\pi(\Theta, \Sigma | u_1, \ldots, u_n) = \frac{L(u_1, \ldots, u_n; \Theta, \Sigma)\pi(\Theta, \Sigma)}{\int L(u_1, \ldots, u_n; \Theta, \Sigma)\pi(\Theta, \Sigma)\mathrm{d}\Theta\mathrm{d}\Sigma},$$

where $L(u_1, \ldots, u_n; \Theta, \Sigma)$ is the likelihood. We could then use the posterior mean of the parameters or the posterior mode [also known as the maximum a posteriori (MAP) estimator].

A Bayesian estimator of mean shape is the posterior mean shape

$$E[\Theta|u_1, \ldots, u_n] = \int_{\Theta, \Sigma} \Theta \pi(\Theta, \Sigma|u_1, \ldots, u_n)d\Theta d\Sigma,$$

and the posterior mean shape variance is:

$$E[\Sigma|u_1, \ldots, u_n] = \int_{\Theta, \Sigma} \Sigma \pi(\Theta, \Sigma|u_1, \ldots, u_n)d\Theta d\Sigma.$$

Another point estimator is the MAP estimator given by:

$$(\tilde{\Theta}, \tilde{\Sigma}) = \arg\sup_{\Theta, \Sigma} \pi(\Theta, \Sigma|u_1, \ldots, u_n),$$

that is the posterior mode.

A $100(1 - \alpha)\%$ credibility region for shape and covariance is given by A where

$$\int_A \pi(\Theta, \Sigma|u_1, \ldots, u_n)d\Theta d\Sigma = 1 - \alpha.$$

In some situations the posterior density is very complicated and inference can then be carried out using Markov chain Monte Carlo (MCMC) simulation (e.g. Smith and Roberts 1993; Besag *et al.* 1995; Gilks *et al.* 1996). If the prior density for (Θ, Σ) is uniform, then the MAP estimator is identical to the maximum likelihood estimator, and this is the main approach that we have considered in this chapter for ease of exposition.

The Bayesian approach has been particularly useful for unlabelled shape analysis, for example in matching molecules (Green and Mardia, 2006; Dryden *et al.* 2007), where the correspondence between landmarks needs to be modelled. A comparison of approaches is given by Kenobi and Dryden (2012), and we give more detail in Chapter 14.

The choice of prior for the shape Θ (or pre-shape) is quite straightforward, for example we could choose a complex Watson or a complex Bingham prior. One particular case of interest is a prior which encourages smooth outlines in the case of close neighbouring landmarks and this is discussed in the next example.

More care needs to be taken with a prior distribution for Σ, although Wishart or inverse-Wishart distributions are obvious choices. If we were considering a simple model such as the complex Watson, then a suitable prior distribution for the single concentration parameter κ is a gamma distribution.

Example 10.3   Consider modelling a random sample of pre-shapes $z_1, \ldots, z_n$ with a complex Watson distribution with mode μ and concentration parameter κ. The concentration parameter κ is assumed to be **known**. Let the prior distribution of μ be complex Bingham with known parameter matrix A. So, the posterior density of μ is given by:

$$\pi(\mu|z_1, \ldots, z_n) \propto \pi(\mu)L(z_1, \ldots, z_n)$$

$$\propto \exp\left\{\mu^*A\mu + \kappa \sum_{i=1}^{n} z_i^* \mu\mu^* z_i\right\}$$

$$= \exp\{\mu^*(\kappa S + A)\mu\},$$

where $S = \sum_{i=1}^{n} z_i z_i^*$ is the complex sum of squares and products matrix. Hence, we have a conjugate prior here, because the posterior is also a complex Bingham distribution, but with parameter matrix $\kappa S + A$. The posterior mode μ_{MAP} is the dominant eigenvector of $\kappa S + A$.   □

If we have a series of landmarks on outlines of objects, then a sutiable prior might encourage smoothness of the outline. It would be convenient to label the landmarks sequentially around the outline, so that landmarks i and j are neighbours if $|(i - j)\text{mod}k| = 1$. We include the modulo k term so that landmarks 1 and k are considered neighbours. In particular, in this periodic case $A = -FF^T$, where

$$(F)_{ij} = \begin{cases} 2, & \text{if } i = j, \\ -1, & \text{if } |(i - j) \text{ mod } k| = 1. \end{cases}$$

The choice of smoothing parameter λ is clearly of vital importance, as illustrated in the following example. We shall consider a complex Bingham prior with parameter matrix λA, and data distributed as complex Watson with parameter κ. Hence in this case the posterior distribution is also a complex Bingham with parameter matrix $\kappa S + \lambda A$, and the MAP estimate is the principal eigenvector of this matrix.

Example 10.4 Consider the Small group of T2 mouse vertebrae in the dataset of Section 1.4.1. There are $k = 6$ landmarks on $n = 23$ bones. We re-order the landmarks by swapping landmarks 1 and 2, so that the landmarks follow round anti-clockwise. In Figure 10.1 we see various smoothed estimates of the mean shape given by the posterior mode, which we write as μ_{MAP}. This example is used for illustration only to see what would happen if the smoothing parameter is taken to be too large. In practice this method is most useful when a large number of noisy landmarks are available around an outline. In Figure 10.1a we see the Procrustes mean ($\lambda/\kappa = 0$). In Figure 10.1b we see little difference in the MAP estimate ($\lambda/\kappa = 0.1$) but in Figure 10.1c ($\lambda/\kappa = 1$) there is a visible difference, with landmarks being brought in more to a discrete circle. We also consider what happens when a much too large smoothing parameter is used. In Figure 10.1d with a very large value of $\lambda/\kappa = 100$ the mean is very nearly a discrete circle. For such large λ/κ the integrated square derivative matrix dominates, which has the trigonometric eigenvectors for a circulant matrix (the eigenvectors form the discrete Fourier transform matrix). An automatic choice of λ/κ could be obtained by cross-validation in a similar manner to that described in Section 12.3.3.   □

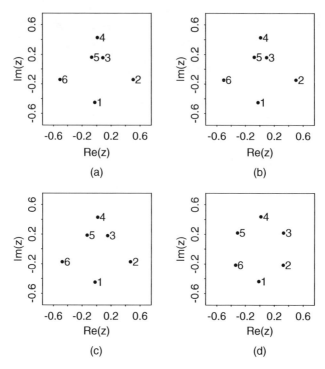

Figure 10.1   The smoothed Procrustes mean of the T2 Small data: (a) $\lambda/\kappa = 0$; (b) $\lambda/\kappa = 0.1$; (c) $\lambda/\kappa = 1.0$; and (d) $\lambda/\kappa = 100$.

There may be occasions when smoothed PCA is appropriate, particularly for outline data. After a smoothed mean μ_{MAP} has been obtained, the figures are transformed to μ_{MAP} by Procrustes analysis leading to the sample covariance matrix S. The first smoothed PC is obtained by finding the dominant eigenvector of $\kappa S + \lambda A$ and practical details are very similar to the procedure of Rice and Silverman (1991). The first few scores from a smoothed PCA of the spine lines were used by Mardia *et al.* (1994) to summarize the shape information. Although it was a small pilot study it did appear that the twin spine shapes showed higher familial correlation than those of non-twin siblings. Mardia *et al.* (1996a) and Dryden *et al.* (2008b) also give some summary measures for measuring spinal shape.

Many authors have considered Bayesian shape analysis, especially for unlabelled shape as discussed in Chapter 14. Some other examples include Brignell (2007) and Mardia *et al.* (2013b) using Markov chain Monte Carlo simulation; Leu and Damien (2014) provide a detailed discussion of Bayesian shape analysis of the complex Bingham distribution; Micheas and Peng (2010) discuss a Bayesian version of Procrustes analysis with applications to hydrology; and Theobald and Mardia (2011) provide a framework for Bayesian analysis of the generalized non-isotropic Procrustes problem with scaling.

10.9   Size-and-shape distributions

10.9.1   Rotationally symmetric size-and-shape family

We could specify size-and-shape distributions directly in size-and-shape space $S\Sigma_m^k$. A suitable candidate could be that based on Riemannian distance, with density proportional to:

$$\exp\{-\kappa\phi(d_S(X, \mu))\},$$

where $d_S(\cdot)$ was given in Equation (5.5) and $\phi(\cdot) \geq 0$ is a suitable penalty function. In the case $\phi(x) = x^2$ the MLE of size-and-shape will be the same as those using least squares partial Procrustes analysis, provided the integrating constant does not depend on μ, that is for the planar case. As for the shape case in Section 10.6 we can choose different functions for estimators with different properties. Further size-and-shape distributions based on normally distributed data are discussed in Section 11.5.

10.9.2   Central complex Gaussian distribution

For 2D data we could exploit the complex representation and, in an analogous manner to using the complex Bingham distribution in pre-shape space for shape analysis, we could specify a distribution for the Helmertized landmarks $z_H = Hz^o \in \mathbb{C}^{k-1} \setminus \{0\}$ which is invariant under rotations. A suitable candidate is the **central complex Gaussian distribution**,

$$z_H \sim \mathbb{C}N_{k-1}(0, \Sigma),$$

where Σ is Hermitian. The distribution of z_H is identical to that of $e^{i\theta}z_H$ for any arbitrary rotation θ. The mode of the distribution is given by the dominant eigenvector of Σ, hence the MLE of size-and-shape under this model is the dominant eigenvector of

$$\frac{1}{n}\sum_{i=1}^{n} z_{Hi}z_{Hi}^*,$$

given a random sample of Helmertized observations $z_{H1}, \dots, z_{Hn}$.

10.10   Size-and-shape versus shape

If both size and shape are available, then we have two strategies in our analysis: either analyse shape and size separately (shape in shape space Σ_m^k and size on the positive real line) or analyse size and shape jointly in size-and-shape space $S\Sigma_m^k$. Which method is preferable depends on the aim of the study. If we are really interested in shape alone, then one would expect that more robust inference will be carried out using the first approach, even if the joint size-and-shape information is available, as less modelling assumptions need to be made. However, in many applications in biology one is interested in investigating joint size and shape information or in using size as a covariate, and then the size-and-shape space plays an important rôle.

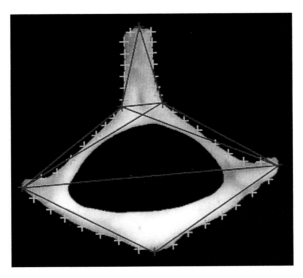

Figure 1.4 Image of a T2 mouse vertebra with six mathematical landmarks on the outline joined by lines (red +) and 42 pseudo-landmarks (yellow +). Source: Dryden & Mardia 1998. Reproduced with permission from John Wiley & Sons. See page 5.

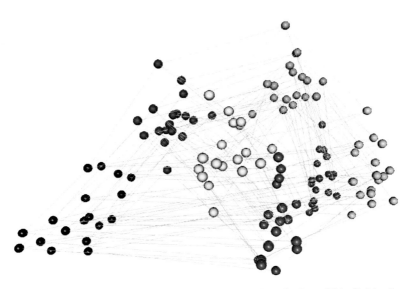

Figure 1.12 The macaque skull data with seven landmarks from 18 individuals, with each landmark displayed by a different colour. See page 15.

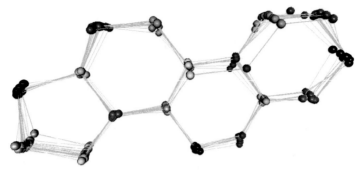

Figure 1.13 The first 17 carbon atoms in the 31 steroid molecules. See page 15.

Figure 1.17 A small dataset of 22 phosphorous atoms from a DNA molecule at n = 30 time points. See page 18.

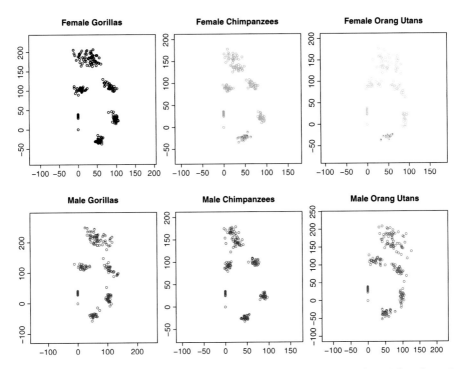

Figure 1.19 The six groups of great ape skull landmarks: (left column) female and male gorillas; (middle column) female and male chimpanzees; and (right column) female and male orangutans. See page 20.

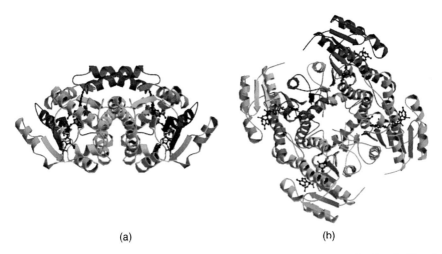

Figure 1.20 The proteins 1a27 (a) and 1cyd (b) from the PDB databank (Tanaka et al. 1996; Mazza 1997; Berman et al. 2000). See page 21.

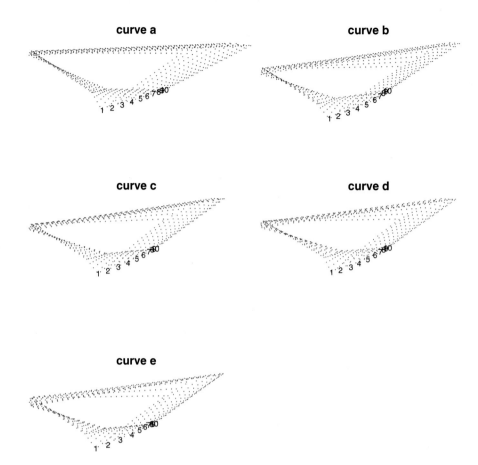

Figure 1.25   The five series of configurations projected into the plane of the table. Each series consists of 10 quadrilaterals observed at equal fractions of the time taken to carry out the pointing movement. Source: Kume et al. 2007. Reproduced with permission of Oxford University Press. See page 26.

Figure 1.26   A set of 62 501 cortical surface points in three dimensions. The colouring indicates the computer ordering of the points on a rainbow scale: red (lower right) to light blue (top of the surface) to violet (lower left). See page 27.

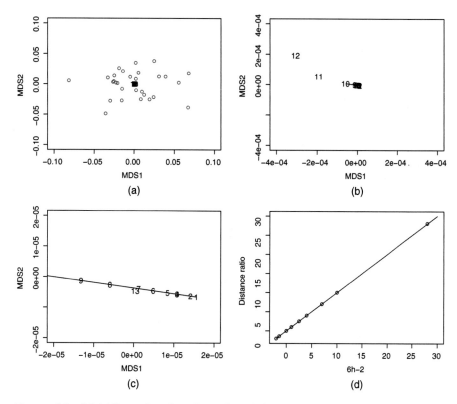

Figure 6.1 Multidimensional scaling plot of the male gorilla data and 13 means using Riemannian distance to form the dissimilarity matrix. (a) The principal coordinates of the 29 skulls (red circles) and the 13 means (black numbers). (b) Zoomed in version of (a) near the origin. (c) Another zoomed in version of (a) near the origin. In each of (a)–(c) the mean shapes are indicated by their number label, and the first ten means are joined by straight lines. (d) A plot of the observed ratios and expected approximate ratios of Riemannian distances for different h. See page 116.

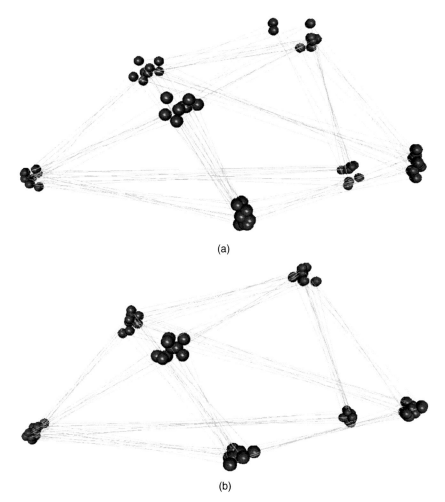

(a)

(b)

Figure 7.3   The male (a) and female (b) macaque skulls registered by full GPA. See page 137.

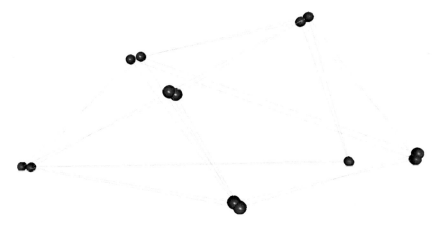

Figure 7.4   The male (red) mean shape registered to the female (blue) mean shape of the macaque skulls registered by OPA. See page 137.

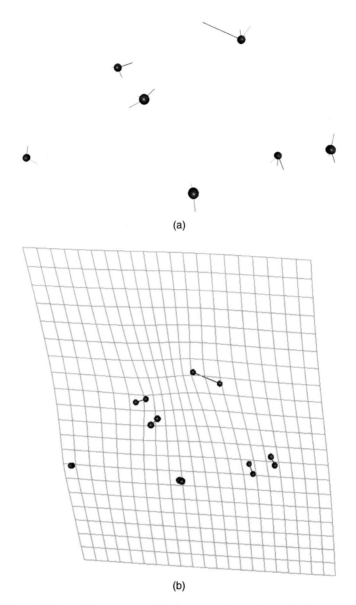

(a)

(b)

Figure 7.6 (a) Plots of the mean (red spheres) with vectors to figures along the first three PCs: (black) mean + 3sd PC1; (red) mean + 3sd PC2; and (green) mean + 3sd PC3. (b) The mean (red) and a figure at mean + 3sd PC1 (blue) with a deformed green grid on the blue figure at $z = -1$, which was deformed from being a square grid on the red figure at $z = -1$. See page 156.

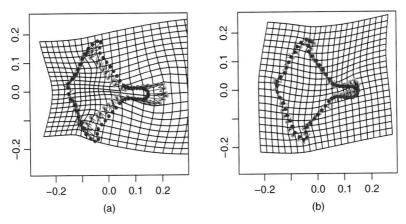

Figure 7.11 The first (a) and second (b) PCs for the T2 Small vertebral outline data. A square grid is drawn on the mean shape (red points) and deformed using a pair of thin-plate splines (see Chapter 12) to an icon (green points) c = 6 standard deviations along each PC (indicated by a blue vector from the mean to the icon). The plots just show the deformed grid at c = 6 for each PC and not the starting grids on the mean. See page 159.

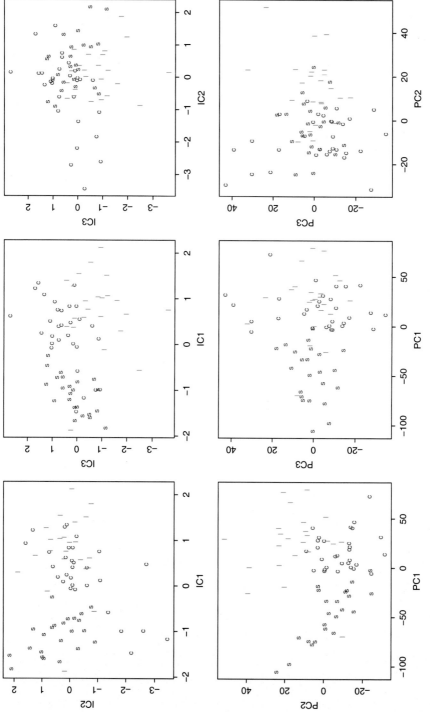

Figure 7.21 Plot of the first three independent components scores (top row), and the first three PC scores (bottom row) for the mouse vertebral outline data. The observations are labelled by group: Control (red c); Large (green l); and Small (blue s). The ordering of the ICs is arbitrary. See page 170.

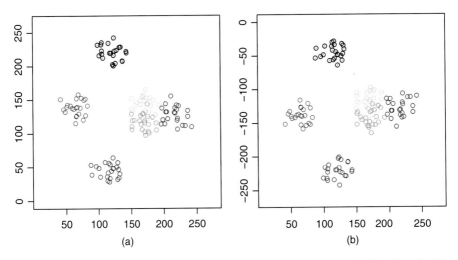

Figure 7.22   The T2 Small landmarks (a) and the reflected relabelled landmarks (b). Each landmark has a different colour. See page 171.

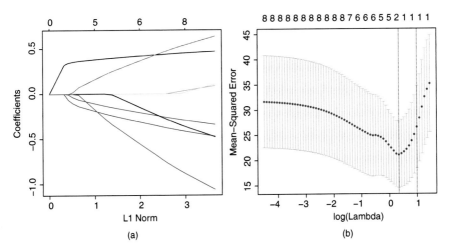

Figure 9.9   Regularization path for the LASSO (a) and cross-validation fitting (b). See page 216.

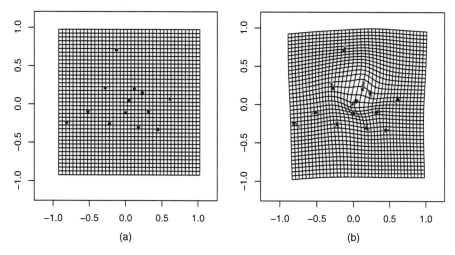

Figure 12.7 A thin-plate spline transformation grid between the control mean shape estimate and the schizophrenia mean shape estimate, from the schizophrenia study. The square grid is placed on the estimated mean control shape (a) and the curved grid is pictured on the estimated mean shape of the schizophrenic patients (b), magnified three times, with arrows drawn from the control mean (red) to patient mean (green). See page 285.

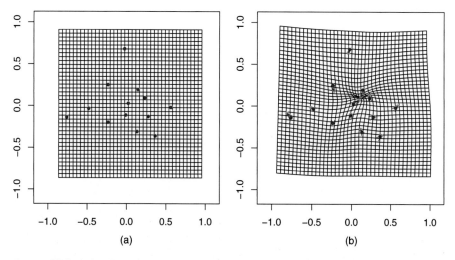

Figure 12.8 The thin-plate spline transformation grid from the patient mean (a) to the control mean (b), with arrows drawn from the patient mean (red) to control mean (green). The shape change has been magnified three times. See page 286.

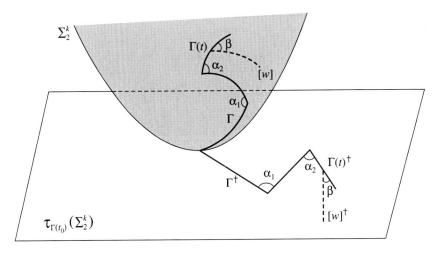

Figure 13.2 A diagrammatic view of unrolling and unwrapping with respect to a piecewise geodesic curve. The blue solid line is the piecewise linear path in the base tangent space. Source: Kume et al. 2007. Reproduced with permission of Oxford University Press. See page 329.

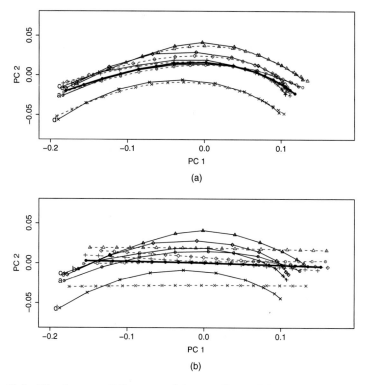

Figure 13.3 The first two PC scores of the unrolling of the human movement data paths with respect to the fitted mean path. In (a) fitted smoothing splines are shown in solid black ($\lambda = 0.00013$) with the projected data points joined by red dashed lines. In (b) fitted approximate geodesics ($\lambda = 60658.8$) are shown in solid black, with the projected data points joined by red dashed lines. In both plots the encircled points are knots of the mean path (bold black). Source: Kume et al. 2007. Reproduced with permission of Oxford University Press. See page 331.

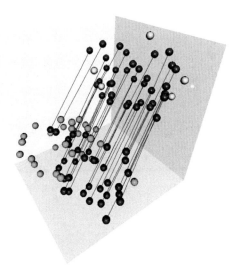

Figure 14.1   The active site locations of proteins `1cyd` *(red and yellow) and* `1a27` *(blue and green) from the Protein Data Bank. The lines connect the top 35 estimated aligned active sites from Green and Mardia (2006), with the red points of 1cyd matching the blue points of 1a27. The non-matched points are yellow and green. See page 345.*

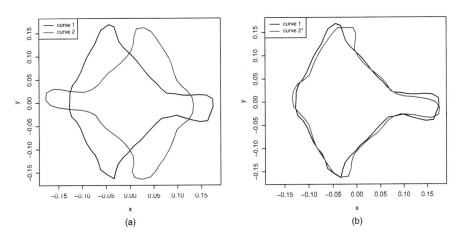

Figure 16.2   (a) Unregistered curves and (b) registration through $\hat{\gamma}(t)_A$*. Source: Cheng* et al. *2016. See page 372.*

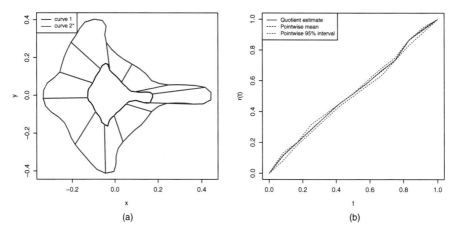

Figure 16.3   (a) Correspondence based on $\hat{\gamma}(t)_A$ and (b) 95% credibility interval for $\gamma(t)$ (dotted red lines). In (a) one of the bones is drawn artificially smaller in order to better illustrate the correspondence. Source: Cheng et al. 2016. See page 373.

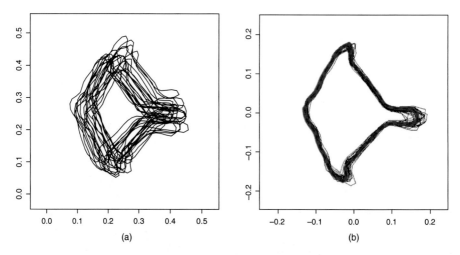

Figure 16.4   The original curves from the Small group, (a) without and (b) with registration. The dashed blue curve in (b) is the estimated μ_A and red colour shows the credible region given by 10 000 samples of the posterior mean. Source: Cheng et al. 2016. See page 373.

11

Offset normal shape distributions

11.1 Introduction

Rather than conditioning we could consider the marginal distribution of shape after integrating out the similarity transformations. These distributions will be called **offset shape distributions**. We consider multivariate normal configurations of k-points in $\mathbb{R}^m$, that is the model for the landmarks is:

$$x = \text{vec}(X) \sim N_{km}(\mu, \Omega), \qquad (11.1)$$

where Ω is symmetric positive definite. We shall concentrate initially on the important $m = 2$ dimensional case in this chapter. Kendall (1984) introduced the case where μ consists of coincident points and Ω was proportional to the identity matrix. Mardia and Dryden (1989a,b) and Dryden and Mardia (1991b) further developed the shape distributions for general μ and Ω, in two dimensions. These distributions have been called 'Mardia–Dryden' distributions in the literature (Bookstein 1991, 2014; Lele and Richtsmeier 1991; Kendall 1991b; Stuart and Ord 1994).

11.1.1 Equal mean case in two dimensions

11.1.1.1 Isotropic case

Result 11.1 *Consider the $k \geq 3$ landmarks in $\mathbb{R}^2$ to be i.i.d. as bivariate normal with equal means and covariance matrix proportional to the identity, that is*

$$(X_j, Y_j)^{\text{T}} \sim N_2\{(\mu, \nu)^{\text{T}}, \sigma^2 I_2\}, \qquad j = 1, \dots, k. \qquad (11.2)$$

The resulting shape distribution is uniform in Kent's polar shape coordinates.

Statistical Shape Analysis, with Applications in R, Second Edition. Ian L. Dryden and Kanti V. Mardia.
© 2016 John Wiley & Sons, Ltd. Published 2016 by John Wiley & Sons, Ltd.

Proof: If the original landmarks have the normal distribution of Equation (11.2), then the Helmertized landmarks $X^* = \text{vec}(HX)$ have a zero mean $2k - 2$ dimensional multivariate normal with covariance matrix $\sigma^2 I_{2k-2}$. Hence the density of X^* is:

$$(2\pi\sigma^2)^{1-k} \exp\left\{-\frac{1}{2\sigma^2}\|X^*\|^2\right\}.$$

We transform from X^* to $(R, s_1, \ldots, s_{k-2}, \theta_1, \ldots, \theta_{k-1})$, where $R = \|X^*\|$ and $(s_1, \ldots, s_{k-2}, \theta, \ldots, \theta_{k-2})^T$ are Kent's polar pre-shape coordinates of Section 4.3.3. The Jacobian of the inverse transformation is $R^{2k-3}2^{2-k}$ and so the joint density of $(R, s_1, \ldots, s_{k-2}, \theta_1, \ldots, \theta_{k-1})$ is:

$$(2\sigma^2\pi)^{1-k}2^{2-k}R^{2k-3} \exp\left\{-\frac{1}{\sigma^2}R^2\right\}.$$

We now transform from R to $A = R^2/(2\sigma^2)$ and the Jacobian of the inverse transformation is σ^2/R. Hence the joint density of $(A, s_1, \ldots, s_{k-2}, \theta_1, \ldots, \theta_{k-1})$ is:

$$(2\pi)^{1-k}A^{k-2}e^{-A}.$$

Integrating out the rotation θ_{k-1} (which is independently uniform) and the scale A by using:

$$\int_0^\infty A^{k-2}e^{-A}dA = (k-2)!$$

we have the density of Kent's polar shape coordinates as:

$$\frac{(k-2)!}{(2\pi)^{k-2}},$$

which is the uniform density.   □

The shape distribution will also be uniform for non-normal i.i.d. symmetric distributions (subject to certain regularity conditions), that is where the landmarks are independent and coordinates (X_j, Y_j) have joint p.d.f. given by:

$$f(x_j, y_j) \propto g(r_j^2),$$

where $r_j^2 = x_j^2 + y_j^2$ for $j = 1, \ldots, k$.

Although not as simple as Kent's polar shape coordinates we shall also use the Kendall coordinates of Equation (2.11) and **for convenience for the rest of this chapter we write U instead of U^K.** Under the i.i.d. normal model of Equation (11.2) the p.d.f. of Kendall coordinates $U = (U_3, \dots, U_k, V_3, \dots, V_k)^T$, with respect to the Lebesgue measure, is given by Equation (10.2), and repeated here:

$$f_\infty(u) = \frac{(k-2)!\pi}{\{\pi(1 + u^T u)\}^{k-1}}.$$

Corollary 11.1   *In the triangle case* $(k = 3)$ *the shape distribution under the i.i.d. normal model of Equation* (11.2) *is uniform on the shape sphere* $S^2(\frac{1}{2})$. *Using Kendall's spherical coordinates* (θ, ϕ) [*given by Equation* (2.13)] *the density, with respect to the Lebesgue measure, is:*

$$f(\theta, \phi)d\theta d\phi = (4\pi)^{-1} \sin\theta d\theta d\phi$$

and $0 \leq \theta \leq \pi, 0 \leq \phi < 2\pi$.

Proof: Transforming to Kendall's spherical coordinates using Equation (2.15) we have:

$$f(\theta, \phi)d\theta d\phi = \frac{1}{\pi(1 + u_3^2 + v_3^2)^2} \frac{\partial(U_3, V_3)}{\partial(\theta, \phi)}. \tag{11.3}$$

The Jacobian of the inverse transformation is given by:

$$\frac{\partial(U_3, V_3)}{\partial(\theta, \phi)} = \frac{\sin\theta}{(1 + \sin\theta\cos\phi)^2} \tag{11.4}$$

and so the result follows.   □

Mardia (1980), describing joint work with Robert Edwards, gave the offset normal distribution for the reflection shape of triangles using internal angles α_1 and α_2. The joint density of α_1 and α_2 is given by:

$$f(\alpha_1, \alpha_2) = \frac{6S}{\pi(3 - C)^2}, \quad 0 \leq \alpha_1, \alpha_2 \leq \pi, \tag{11.5}$$

and $\alpha_3 = \pi - \alpha_1 - \alpha_2 \geq 0$, where $S = \sum_{j=1}^3 \sin 2\alpha_j$ and $C = \sum_{j=1}^3 \cos 2\alpha_j$.

The connection with the uniform distribution in terms of Bookstein coordinates (U^B, V^B) is as in the following. The uniform density using Bookstein coordinates for the triangle case is:

$$f(u^B, v^B) = \frac{1}{3\pi(\frac{1}{2} + \frac{2}{3}(u^B)^2 + \frac{2}{3}(v^B)^2)^2}.$$

The transformation between (U^B, V^B) and (α_1, α_2), $V^B > 0, 0 \le \alpha_1, \alpha_2 \le \pi$, is obtained from

$$\tan \alpha_1 = \frac{V^B}{U^B + \frac{1}{2}}, \quad \tan \alpha_2 = \frac{V^B}{\frac{1}{2} - U^B}, \quad (V_B \ge 0, U^B \ne \pm\frac{1}{2})$$

and the Jacobian is:

$$\left| \frac{\partial(U^B, V^B)}{\partial(\alpha_1, \alpha_2)} \right| = \frac{|\sin \alpha_1 \sin \alpha_2 \sin \alpha_3|}{(\sin \alpha_3)^4}.$$

Using the fact that

$$\frac{1}{2} + \frac{2}{3}(U^B)^2 + \frac{2}{3}(V^B)^2 = \frac{(3 - C)}{4 \sin^2 \alpha_3}$$

and

$$\sin \alpha_1 \sin \alpha_2 \sin \alpha_3 = S/4,$$

the uniform density corresponding to $V^B \ge 0$ is:

$$f_1(\alpha_1, \alpha_2) = \frac{3S}{\pi(3 - C)^2}.$$

There is an equal contribution to the density from $V^B \le 0$ and so the uniform shape density is given by $2f_1$ as required.

11.1.1.2 Distribution of the Riemannian distance

Result 11.2   *When the points are distributed according to the i.i.d. normal model of Equation (11.2) then the marginal distribution of the Riemannian distance ρ between the observed shape u and any fixed shape, has density*

$$2(k - 2) \sin^{2k-5} \rho \cos \rho. \tag{11.6}$$

Proof: Assume the original landmarks z^o are i.i.d. normal with zero mean and unit variance. Transforming to the Helmertized coordinates $z_H = (z_{H1}, \dots, z_{H(k-1)})^T = Hz^o$ it follows that $|z_{Hj}|^2 \sim \chi_2^2$ independently. Without loss of generality we consider the Riemannian distance ρ from z^o to the special rank 1 configuration $\mu = (1, 0, \dots, 0)^T$. Using the formula for the Riemannian distance of Equation (4.21) we have:

$$\tan^2 \rho = \frac{|z_{H2}|^2 + \dots + |z_{H(k-1)}|^2}{|z_{H1}|^2},$$

which is the ratio of independent chi-squared random variables with $2k - 4$ and 2 degrees of freedom, respectively. So,

$$\frac{1}{k-2} \tan^2 \rho \sim F_{2k-4,2}$$

a scaled F distribution with $2k - 4$ and 2 degrees of freedom. Hence, using the connection between the F and beta distributions (e.g. see Mardia *et al.* 1979, Appendix B),

$$S = \sin^2 \rho \sim \mathrm{beta}(k - 2, 1),$$

with density $(k - 2)s^{k-3}$. Transforming to ρ with Jacobian

$$\left| \frac{\mathrm{d}S}{\mathrm{d}\rho} \right| = 2 \sin \rho \cos \rho$$

leads to the required density of Equation (11.6).   □

In Figure 11.1 we see how the shape of the density changes as the number of points k increases. The mode increases as the number of points increases.

This density was first communicated to us by David Kendall (1989, personal communication); see also Goodall and Mardia (1991) and Le (1991b).

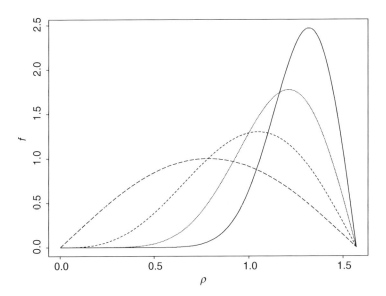

Figure 11.1   The density (f) of the Riemannian distance ρ from any fixed shape for different numbers of points k, in the uniform case. The plot shows the densities for values of k = 3, 4, 6, 10, and the mode of the distribution increases as k increases.

If one uses Goodall and Mardia polar shape coordinates described at the end of Section 5.6.2, then the uniform shape distribution is given by the product of marginal density of ρ from Equation (11.6) and the uniform density of the spherical coordinates ϕ.

11.1.1.3   Anisotropic case

If the covariance matrix in Equation (11.2) is replaced by $\sigma^2\text{diag}(s, 1/s)$, $s > 0$, the p.d.f. of shape (Kendall 1984), with respect to the uniform measure in the shape space, is:

$$f_\infty(u) \left(\frac{2\lambda}{s + s^{-1}} \right)^{k-1} \mathcal{P}_{k-2}(\lambda), \tag{11.7}$$

where

$$1 - \lambda^{-2} = \frac{(s^2 - 1)^2 |(z^+)^{\mathrm{T}} z^+|^2}{(s^2 + 1)^2 (1 + u^{\mathrm{T}} u)^2}, \tag{11.8}$$

with $z^+ = (1, u_3 + iv_3, \ldots, u_k + iv_k)^{\mathrm{T}}$ and

$$\mathcal{P}_n(x) = \sum_{k=0}^{[n/2]} \frac{(-1)^k (\frac{1}{2})_{n-k} (2x)^{n-2k}}{k!(n - 2k)!} \tag{11.9}$$

is the Legendre polynomial of degree n (e.g. see Rainville 1960).

The maximum value of λ is $(s + s^{-1})/2$, which has been called the 'Broadbent factor' by Kendall (1984). The model was motivated by the archaeological example of the standing stones of Land's End described in Section 1.4.18, which was first studied by Broadbent (1980).

11.1.2   The isotropic case in two dimensions

An i.i.d. model (with the same mean for each point) is often not appropriate. For example, in many applications one would be interested in a model with different means at each landmark. The first extension that we consider is when the landmarks $(X_j, Y_j)^{\mathrm{T}}, j = 1, \ldots, k$, are independent isotropic bivariate normal, with different means but the same variance at each landmark. The isotropic normal model is:

$$X = (X_1, \ldots, X_k, Y_1, \ldots, Y_k)^{\mathrm{T}} \sim N_{2k}(\mu, \sigma^2 I_{2k}), \tag{11.10}$$

where $\mu = (\mu_1, \ldots, \mu_k, \nu_1, \ldots, \nu_k)^{\mathrm{T}}$. This model is particularly appropriate when the major source of variability is measurement error at each landmark. For example, consider Figure 11.2 where the points of a triangle are independently perturbed by isotropic normal errors. We wish to find the resulting perturbed shape distribution

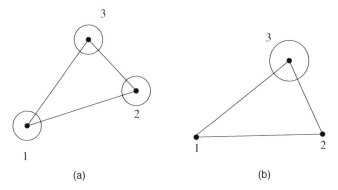

Figure 11.2   The isotropic model is appropriate for independent isotropically per-turbed landmarks (a). The resulting shape distribution can be thought of as the distribution of the landmarks after translating, rotating and rescaling so that the baseline is sent to a fixed position (b).

using say Bookstein or Kendall coordinates after translating, rotating and rescaling two points to a fixed baseline.

Result 11.3   *The **offset normal** shape density under the isotropic normal model of Equation* (11.10), *with respect to the uniform measure in the shape space, is:*

$$_1F_1\{2 - k; 1; -\kappa[1 + \cos 2\rho(X, \mu)]\} \exp\{-\kappa[1 - \cos 2\rho(X, \mu)]\}, \qquad (11.11)$$

where

$$\kappa = S^2(\mu)/(4\sigma^2)$$

is a concentration parameter, $S(\mu)$ is the population centroid size, ρ is the Riemannian distance between the observed shape $[X]$ and the population shape $[\mu]$, and the confluent hypergeometric function $_1F_1(\cdot)$ is defined in Equation (10.11).

Proof: Transform first to the Helmertized landmarks,

$$X_H = (X_{H2}, \ldots, X_{Hk}, Y_{H2}, \ldots, Y_{Hk})^{\mathrm{T}} = LX \sim N_{2k-2}(\mu_H, \sigma^2 I_{2k-2}), \qquad (11.12)$$

where $L = I_2 \otimes H$ is a $(2k - 2) \times 2k$ block diagonal matrix with $LL^{\mathrm{T}} = I_{2k-2}$, $\mu_H = L\mu$, and H is the Helmert submatrix of Equation (2.10). We partition into scale/rotation and shape by transforming to $(W_1 = X_{H2}, W_2 = Y_{H2}, U^{\mathrm{T}})^{\mathrm{T}}$, where $U = (U_3, \ldots, U_k, V_3, \ldots, V_k)^{\mathrm{T}}$ are Kendall coordinates and W_1 and W_2 contain both rotation and scale information. Note that

$$U_j = (X_{Hj}W_1 + Y_{Hj}W_2)/(W_1^2 + W_2^2),$$
$$V_j = (Y_{Hj}W_1 - X_{Hj}W_2)/(W_1^2 + W_2^2), \qquad (11.13)$$

$j = 3, \ldots, k$ and under the model $P(W_1^2 + W_2^2 = 0) = 0$. Defining the $(2k - 2)$ vectors u^+ and the orthogonal vector v^+ by:

$$u^+ = (1, u_3, \ldots, u_k, 0, v_3, \ldots, v_k)^{\mathrm{T}}, \quad v^+ = (0, -v_3, \ldots, -v_k, 1, u_3, \ldots, u_k)^{\mathrm{T}}, \quad (11.14)$$

we can write the transformation from X_H to (W_1, W_2, U) as:

$$X_H = W_1 u^+ + W_2 v^+,$$

and the Jacobian of the inverse transformation is:

$$(W_1^2 + W_2^2)^{k-2}.$$

The p.d.f. of (W_1, W_2, U) is therefore

$$(2\pi\sigma)^{1-k}(w_1^2 + w_2^2)^{k-2} \exp(-G/2),$$

where

$$G = (w_1 u^+ + w_2 v^+ - \mu_H)^{\mathrm{T}}(w_1 u^+ + w_2 v^+ - \mu_H)/\sigma^2.$$

In order to obtain the marginal shape distribution we integrate out W_1 and W_2, that is

$$f(u) = \int_{-\infty}^{\infty}\int_{-\infty}^{\infty} (2\pi\sigma)^{1-k}(w_1^2 + w_2^2)^{k-2} \exp(-G/2) dw_1 dw_2.$$

Keeping U fixed for the moment let us suppose that $(W_1, W_2)|U$ has a bivariate normal distribution with mean

$$\begin{bmatrix} b_1 \\ b_2 \end{bmatrix} = \begin{bmatrix} \mu_H^{\mathrm{T}} u^+ \\ \mu_H^{\mathrm{T}} v^+ \end{bmatrix},$$

and covariance matrix $\sigma^2\{(u^+)^{\mathrm{T}}u^+\}I_2$. We write the density of this distribution as $f^*(w_1, w_2)$. It can be seen that

$$f(u) = (2\pi)^{2-k}\{(u^+)^{\mathrm{T}}u^+\}^{-1}e^{-g/2}\int_{-\infty}^{\infty}\int_{-\infty}^{\infty}(w_1^2 + w_2^2)^{k-2}f^*(w_1, w_2)dw_1 dw_2,$$

where

$$\sigma^2 g = \mu_H^{\mathrm{T}}\mu_H - \frac{(\mu_H^{\mathrm{T}}u^+)^2 + (\mu_H^{\mathrm{T}}v^+)^2}{(u^+)^{\mathrm{T}}u^+}.$$

Now, from standard results of the normal distribution, if W_i, $i = 1, 2$, are independently normal with means b_i and common variance ψ^2, then

$$W_1{}^2 + W_2{}^2 \sim \chi_2^2(\lambda), \quad \lambda = (b_1^2 + b_2^2)/\psi^2,$$

where $\chi_2^2(\lambda)$ is the non-central chi-squared distribution with two degrees of freedom and non-centrality parameter λ. Hence,

$$\int_{-\infty}^{\infty} \int_{-\infty}^{\infty} (w_1{}^2 + w_2{}^2)^{k-2} f^*(w_1, w_2) dw_1 dw_2 = E\{[\chi_2^2(\lambda)]^{k-2}\}.$$

Important point: Note that the integral reduces to calculation of the $(k-2)$th moment of the non-central chi-squared distribution. The rth moment for the non-central $\chi_2^2(\lambda)$ distribution is:

$$E\{[\chi_2^2(\lambda)]^r\} = 2^r r!_1 F_1(-r; 1; -\lambda/2), \quad r = 1, 2, 3, \ldots .$$

Simple algebra shows that

$$\cos^2 \rho(X, \mu) = \frac{b_1^2 + b_2^2}{\|u^+\|^2 \|\mu_H\|^2}$$

and so

$$\lambda = \frac{S(\mu)^2}{\sigma^2} \cos^2 \rho(X, \mu) = 4\kappa \cos^2 \rho(X, \mu).$$

Using simple trigonometric identities we have:

$$f(u) = f_\infty(u)_1 F_1\{2 - k; 1; -\kappa(1 + \cos 2\rho)\} \exp\{-\kappa(1 - \cos 2\rho)\}.$$

and hence the result is proved by recognizing the uniform measure of Equation (10.1). $\qquad\square$

This distribution was introduced by Mardia and Dryden (1989a,b). Kendall (1991b) and Le (1991b) verified the result using stochastic calculus and made connections with Brownian motion (see Section 14.4.3). The density is also described in Stuart and Ord (1994, p. 179), for the triangle case.

The distribution has $2k - 3$ parameters. There are $2k - 4$ parameters in the mean shape $[\mu]$ and a concentration parameter κ.

The confluent hypergeometric function is a finite series in the density since

$$_1F_1\{-r; 1; -x\} = \mathcal{L}_r(-x) = \sum_{j=0}^{r} \binom{r}{j} \frac{x^j}{j!}, \quad r = 0, 1, 2, \ldots , \qquad (11.15)$$

where $\mathcal{L}_r(\cdot)$ is the Laguerre polynomial of degree r; see Abramowitz and Stegun (1970, pp. 773–802). Using Kummer's relation,

$$_1F_1\{a; b; x\} = e^x {}_1F_1\{b - a; b; -x\}$$

we can also rewrite the density, with respect to the uniform measure, as:

$$e^{2\kappa} {}_1F_1\{k - 1; 1; \kappa[1 + \cos 2\rho(X, \mu)]\}.$$

A convenient parameterization for the mean shape uses Kendall coordinates $\Theta = (\theta_3, \ldots, \theta_k, \phi_3, \ldots, \phi_k)^T$ obtained from μ and a coefficient of variation $\tau^2 = \sigma^2 / \|\mu_2 - \mu_1\|^2$. There are $2k - 4$ shape parameters Θ and one variation parameter τ. The density can be written, with respect to the uniform measure, as:

$$_1F_1\{2 - k; 1; -(1 + \Theta^T\Theta)\frac{1 + \cos 2\rho}{4\tau^2}\} \exp\{-(1 + \Theta^T\Theta)\frac{1 - \cos 2\rho}{4\tau^2}\}, \quad (11.16)$$

where

$$\cos^2 \rho = \{[(\Theta^+)^T u^+]^2 + [(\Theta^+)^T v^+]^2\} / \{(u^+)^T u^+ (\Theta^+)^T \Theta^+\}, \quad (11.17)$$

with $\Theta^+ = (1, \theta_3, \ldots, \theta_k, 0, \phi_3, \ldots, \phi_k)^T$, and ρ is the Riemannian distance between the shapes u and Θ given earlier in Equation (4.21).

11.1.3   The triangle case

The triangle case ($k = 3$) is particularly simple. The offset normal shape density is given by:

$$\{1 + \kappa[1 + \cos 2\rho(X, \mu)]\} \exp\{-\kappa[1 - \cos 2\rho(X, \mu)]\}$$

and transforming to Kendall's shape sphere using Equation (2.13) and Equation (2.14), the probability element of $l = (l_x, l_y, l_z)^T$, is:

$$\{1 + \kappa(1 + l^T v)\} \exp\{-\kappa(1 - l^T v)\} dS,$$

where v is the direction on the shape sphere corresponding to the mean shape and dS is the probability element of $S^2(\frac{1}{2})$. This result can be seen by using the same argument that led to Equation (10.8). Mardia (1989b) introduced this result and explored various properties of the distribution, including an approximation based on the Fisher–von Mises distribution.

   Mardia (1980) gave the result for the reflection shape for triangles where the mean is equilateral, using two internal angles α_1 and α_2 to measure the reflection shape.

11.1.4   Approximations: Large and small variations

11.1.4.1   Large variations

As $\tau \to \infty$ in the shape density of Equation (11.16) (or equivalently as the concentration $\kappa \to 0$) we clearly see that the shape distribution tends to the uniform distribution, since $_1F_1(2-k; 1; x) \to 1$ as $x \to 0$.

11.1.4.2   Small variations

Result 11.4   *As $\tau \to 0$ it can be seen that*

$$(U - \Theta)/\tau \overset{D}{\to} N_{2k-4}\{0, 2(I_{2k-4} + \Theta\Theta^{\mathrm{T}} + \Phi\Phi^{\mathrm{T}})\}, \qquad (11.18)$$

where $\overset{D}{\to}$ means 'converges in distribution to' and $\Phi = (-\phi_3, \ldots, -\phi_k, \theta_3, \ldots, \theta_k)^{\mathrm{T}}$ is orthogonal to the mean shape $\Theta = (\theta_3, \ldots, \theta_k, \phi_3, \ldots, \phi_k)^{\mathrm{T}}$.

Proof: The result can be proved by using a Taylor series expansion of U about Θ. Specifically if we write $X_H = \mu + \epsilon$ where, without loss of generality, we have rescaled X by $\delta_{12} = \{(\mu_2 - \mu_1)^2 + (\nu_2 - \nu_1)^2\}^{1/2}$ and

$$\epsilon = (\epsilon_1, \ldots, \epsilon_{k-1}, \eta_1, \ldots, \eta_{k-1})^{\mathrm{T}} \sim N_{2k-2}(0, \tau^2 I_{2k-2}),$$

where $\tau = \sigma/\delta_{12}$. The Taylor series expansion is:

$$U = \Theta(1 - \sqrt{2}\epsilon_1) - \sqrt{2}\eta_1\Phi + \sqrt{2}Z + \sum_j \sum_k O(\epsilon_j\epsilon_k),$$

where $Z = (\epsilon_2, \ldots, \epsilon_{k-1}, \nu_2, \ldots, \nu_{k-1})^{T}$. Hence for small variations we consider up to first-order terms and so the result follows.   □

This result was first given by Bookstein (1984, 1986) using an approximation argument independent of the exact distribution. Bookstein suggested using normal models for shape variables for approximate inference, as described in Section 9.4.

Note that the approximate covariance in Equation (11.18) is not proportional to the identity matrix except for the $k = 3$ triangle case. Hence, correlations have been induced in general into the shape variables by the edge registration procedure.

If using Bookstein coordinates the approximate distribution, for small τ, is:

$$U^B \sim N_{2k-4}(\Theta^B, \tau^2(D^{-1} + 2\Theta^B(\Theta^B)^{\mathrm{T}} + 2\Phi^B(\Phi^B)^{\mathrm{T}})), \qquad (11.19)$$

where $D = I_2 \otimes (H_1^{\mathrm{T}}H_1)$ and

$$H_1^{\mathrm{T}}H_1 = I_{k-2} - \frac{1}{k}1_{k-2}1_{k-2}^{\mathrm{T}}, \quad (H_1^{\mathrm{T}}H_1)^{-1} = I_{k-2} + \frac{1}{2}1_{k-2}1_{k-2}^{\mathrm{T}},$$

where H_1 is the lower right $(k-2) \times (k-2)$ partition matrix of the Helmert submatrix (see Section 2.5), and $\Theta^B = (\theta_3^B, \dots, \theta_k^B, \phi_3^B, \dots, \phi_k^B)^T$ are the Bookstein coordinates of the mean μ, with $\Phi^B = (-\phi_3^B, \dots, -\phi_k^B, \theta_3^B, \dots, \theta_k^B)^T$.

Corollary 11.2   *For $k = 3$ the approximate normal distribution of Bookstein coordinates is isotropic, and in particular*

$$\text{var}(u_3^B) = \text{var}(v_3^B) = 3\tau^2 \left(\frac{1}{2} + \frac{2}{3}(\theta_3^B)^2 + \frac{2}{3}(\phi_3^B)^2 \right), \tag{11.20}$$

$$\text{cov}(u_3^B, v_3^B) = 0,$$

where

$$\theta_3^B + i\phi_3^B = \frac{\mu_3 + iv_3 - \frac{1}{2}(\mu_1 + iv_1 + \mu_2 + iv_2)}{\mu_2 + iv_2 - \mu_1 - iv_1} \tag{11.21}$$

and $\tau = \sigma / \{(\mu_2 - \mu_1)^2 + (v_2 - v_1)^2\}^{1/2}$.

Also, the form of the covariance leads to large variability if the mean location of the corresponding landmark is far away from the mid-point of the baseline.

Corollary 11.3   *For $k = 4$ we have the following approximate covariances between the shape variables:*

$$\text{cov}(u_3^B, u_4^B) = \tau^2(\frac{1}{2} + 2\theta_3^B \theta_4^B + 2\phi_3^B \phi_4^B);$$

$$\text{cov}(u_3^B, v_4^B) = \tau^2(2\theta_3^B \phi_4^B - 2\theta_4^B \phi_3^B);$$

$$\text{cov}(u_4^B, v_3^B) = -\tau^2(2\theta_3^B \phi_4^B - 2\theta_4^B \phi_3^B) = -\text{cov}(u_3^B, v_4^B).$$

Important point: We see that even though the original landmarks are independent, the shape variables become correlated when registering on a common edge. Hence, care should be taken with interpretation of PCs obtained from Bookstein coordinates, as discussed by Kent (1994).

It is immediately clear that the above covariances are invariant under similarity transformations of the mean configuration, that is transformations of the form

$$a(\mu_j + iv_j) + b, \quad a, b \in \mathbb{C}, \quad j = 1, \dots, k,$$

as the terms in a and b cancel out in Equation (11.21).

Although we have given the covariance structure for $k = 4$, for larger k the covariances between any pair of landmarks j and l will be exactly the same as in Corollary 11.11 but with a suitable change of subscripts. So, given a mean configuration $\mu_j + iv_j$, $j = 1, \dots, k$, we can write down the approximate covariances for Bookstein's shape variables for any particular choice of baseline using the above corollaries.

11.1.5   Exact moments

Result 11.5   *If using Kendall or Bookstein coordinates, then only the first moment exists. In particular, using Bookstein coordinates* (Mardia and Dryden 1994)

$$E(U^B) = \Theta^B[1 - \exp\{-1/(4\tau^2)\}]. \tag{11.22}$$

See Mardia and Dryden (1994) for a proof of the result.

We see from Equation (11.22) that the expected value of U^B is 'pulled in' towards the origin by the multiplicative factor $1 - \exp\{-1/(4\tau^2)\}$.

Computing the exact expectation of the shape variables involves finding the expectation of a linear term in a random vector X over a quadratic form in X, that is $a^T X / X^T B X$. From Jones (1987) we have $E[(l^T X / X^T B X)^n] < \infty$ if $n < \text{rank}(B)$. Here $\text{rank}(B) = 2$ and hence the first moment is the only finite positive integer moment.

11.1.6   Isotropy

We now show that a general isotropic distribution for landmarks about a mean shape in two dimensions gives rise to an isotropic distribution in the tangent space to the mean so that standard PCA is valid in the tangent space, following Mardia (1995). This result is of great significance and it underlies the development of the Procrustes tangent space by Kent (1994), discussed in Section 7.7.

Result 11.6   *If the original points have an isotropic covariance structure, then the partial tangent coordinates v are also isotropic.*

Proof: Consider the original landmarks z^o to be isotropic and take the Helmertized landmarks $z = Hz^o$ which will also be isotropic. The Helmertized landmarks z can be modelled by $z = v + \delta$, $\delta \in \mathbb{C}^{k-1}$, where δ is isotropic and $E(\delta) = 0$, so that the p.d.f. of δ is given by:

$$f(\delta) = h(\delta^* \delta), \quad \delta \in \mathbb{C}^{k-1}. \tag{11.23}$$

The partial Procrustes tangent projection is:

$$v = e^{-i\theta}(I - vv^*)z/\|z\|, e^{i\theta} = v^* z/\{\|v^* z\|\}.$$

Without any loss of generality, let us take $v = (1, 0, \ldots, 0)^T$ so that,

$$v^T = e^{-i\theta}(0, \Delta^T)/\{1 + \delta_1^* \delta_1 + 2\text{Re}(\delta_1) + \lambda^2\}^{1/2},$$

where $\delta = (\delta_1, \Delta^T)^T$ and $\lambda = \|\Delta\|$. Now, $\Delta|\{\delta_1 = c\}$ is isotropic since from Equation (11.23)

$$f(\Delta|\delta_1 = c) = \frac{h(c^* c + \lambda^2)}{\int h(c^* c + \lambda^2) d\Delta} \tag{11.24}$$

which is a function of λ^2 only. Also, since a new function of Δ depending on λ^2 (and λ is the radial component of Δ) is also isotropic given δ_1, it follows from Equation (11.24) that

$$u = \Delta\{[1 + \delta_1^*\delta_1 + 2\text{Re}(\delta_1)] + \lambda^2\}^{-1/2}$$

conditional on δ_1 is also isotropic and $(v^T|\delta_1) = (0, e^{-i\theta}u^T|\delta_1), \theta = \text{Arg}(1 + \delta_1)$ and $E(u^T|\delta_1) = 0$. Hence,

$$E[v|\delta_1] = 0. \tag{11.25}$$

Furthermore, we can write

$$E[vv^*|\delta_1] = E\left[\begin{pmatrix} 0 \\ u \end{pmatrix}(0, u^*)|\delta_1\right] = g_1(\delta_1^*\delta_1)\text{block-diag}(0, I_{k-2}). \tag{11.26}$$

Hence

$$E[vv^*] = E\{E[vv^*|\delta_1]\} \propto \text{block-diag}(0, I_{k-2}),$$

so that Equation (11.25) and Equation (11.26) yield:

$$\text{cov}(v) \propto \text{block-diag}(0, I_{k-2}). \qquad \square$$

The same argument applies for the PCA in real coordinates in the tangent space.

11.2   Offset normal shape distributions with general covariances

Shape distributions for the general multivariate normal model of Equation (11.1) with a general covariance structure in $m = 2$ dimensions have been studied by Dryden and Mardia (1991b). We write $X \sim N_{2k}(\mu, \Omega)$ in this general case, where $\mu = (\mu_1, \ldots, \mu_k, \nu_1, \ldots, \nu_k)^T$ and Ω is a $2k \times 2k$ symmetric positive definite matrix. The offset normal shape density in the most general case involves finite sums of generalized Laguerre polynomials and is very complicated. The general shape distribution depends on $2k - 4$ mean shape parameters Θ (the shape of μ) and a $2k - 2 \times 2k - 2$ matrix of covariance parameters $\Sigma = L\Omega L^T$, where $L = I_2 \otimes H$ and H is the Helmert submatrix. We write the p.d.f. of the Kendall shape variables in the general case as $f(u; \Theta, \Sigma)$, and the form of the density is given by Dryden and Mardia (1991b). Simplification of the density does occur when there is complex normal structure in the covariance matrix, which we now discuss.

11.2.1   The complex normal case

Let $x = \text{vec}(X)$ have a complex multivariate normal distribution with covariance matrix

$$\Omega = \frac{1}{2}\begin{bmatrix} \Omega_1 & \Omega_2 \\ -\Omega_2 & \Omega_1 \end{bmatrix}, \quad \Omega_2^{\mathrm{T}} = -\Omega_2,$$

where Ω_1 is a $k \times k$ positive definite symmetric matrix and Ω_2 is a $k \times k$ skew symmetric matrix [see Equation (10.3)]. The covariance matrix of the Helmertized landmarks $\text{vec}(HX)$ is:

$$\Sigma = L\Omega_0 L^T = \frac{1}{2}\begin{bmatrix} H\Omega_1 H^{\mathrm{T}} & H\Omega_2 H^{\mathrm{T}} \\ -H\Omega_2 H^{\mathrm{T}} & H\Omega_1 H^{\mathrm{T}} \end{bmatrix} = \frac{1}{2}\begin{bmatrix} \Sigma_1 & \Sigma_2 \\ -\Sigma_2 & \Sigma_1 \end{bmatrix} \quad (11.27)$$

say, where $\Sigma_2^{\mathrm{T}} = -\Sigma_2$, $\mu_H = L\mu$, and $L = I_2 \otimes H$.

Result 11.7   *In the complex normal case the shape density is given by:*

$$\frac{(k-2)!\exp\{-\kappa(1-\cos 2r)\}}{\pi^{k-2}((u^+)^{\mathrm{T}}\Sigma^{-1}u^+)^{k-1}|\Sigma|^{1/2}}\,{}_1F_1\{2-k; 1; -\kappa(1+\cos 2r)\}, \quad (11.28)$$

where r is defined as a generalized Procrustes distance given by:

$$\cos^2 r = \{(\mu_H^T\Sigma^{-1}u^+)^2 + (\mu_H^T\Sigma^{-1}v^+)^2\}/\{(u^+)^{\mathrm{T}}\Sigma^{-1}u^+\,\mu_H^T\Sigma^{-1}\mu_H\}, \quad (11.29)$$

and $\kappa = \mu_H^T\Sigma^{-1}\mu_H/4$.

The proof of this result was given by Dryden and Mardia (1991b). Le (1991b) has verified the complex normal shape distribution using stochastic calculus. The complex normal shape density is very similar in structure to the isotropic normal case, except that the Euclidean norm is replaced by a Mahalanobis norm. There are many useful models that have a complex normal covariance structure; such as the isotropic model of Section 11.1.2 and the cyclic Markov model (Dryden and Mardia 1991b).

11.2.2   General covariances: Small variations

Let us assume that not all the landmark means are equal, and consider the case where variations are small compared with the mean length of the baseline. We write the population shape as $\Theta = (\theta_3, \ldots, \theta_k, \phi_3, \ldots, \phi_k)^{\mathrm{T}}$ (i.e. the Kendall coordinates of $\{(\mu_j, v_j)^{\mathrm{T}}, j = 1, \ldots, k\}$), and without loss of generality we translate, rotate and rescale the population so that the vectorized Helmertized coordinates $X_H = LX$ have mean $\mu_H = (1, \theta_3, \ldots, \theta_k, 0, \phi_3, \ldots, \phi_k)^{\mathrm{T}}$, and general positive definite covariance matrix $\Sigma = 2\tau^2 L\Omega L^{\mathrm{T}}$, where $\tau = \sigma/[(\mu_2 - \mu_1)^2 + (v_2 - v_1)^2]^{1/2}$.

Result 11.8   *As $\tau \to 0+$,*

$$\frac{(U - \Theta)}{\tau} \xrightarrow{D} N_{2k-4}\{0, 2\tau^2 FL\Omega L^T F^T\}, \tag{11.30}$$

where

$$F = \left[-\Theta, \begin{array}{c} I_{k-2} \\ 0_{k-2} \end{array}, -\Phi, \begin{array}{c} 0_{k-2} \\ I_{k-2} \end{array}\right], \Phi = (-\phi_3, \dots, -\phi_k, \theta_3, \dots, \theta_k)^T,$$

and $I_j, 0_j$ are the $j \times j$ identity and zero matrices.

Proof: After transforming to the Helmertized coordinates as in Equation (11.12), we write $X_H = \mu_H + \epsilon^+$ where

$$\epsilon^+ = (\epsilon_1, \epsilon^T, \eta_1, \eta^T)^T \sim N_{2k-2}\{0, 2\tau^2 L\Omega L^T\}, \tag{11.31}$$

and ϵ and η are $(k-2) \times 1$ column vectors. Note that, if τ is small, then each element of ϵ^+ will be small. Transforming to Kendall coordinates, as in Equation (11.13), we have

$$U = [(1 + \epsilon_1)\{\Theta + (\epsilon^T, \eta^T)^T\} - \eta_1\{\Phi + (-\eta^T, \epsilon^T)^T\}]/[(1 + \epsilon_1)^2 + (\eta_1)^2]$$

$$= [(1 + \epsilon_1)\{\Theta + (\epsilon^T, \eta^T)^T\} - \eta_1\Phi]/\{1 + 2\epsilon_1\} + \sum_{j=1}^{2k-2} \sum_{l=1}^{2k-2} O[(\epsilon^+)_j(\epsilon^+)_l].$$

Thus,

$$U = \Theta - \epsilon_1\Theta - \eta_1\Phi + (\epsilon^T, \eta^T)^T + \sum_j \sum_l O[(\epsilon^+)_j(\epsilon^+)_l]$$

$$= \Theta + F\epsilon^+ + \sum_j \sum_l O[(\epsilon^+)_j(\epsilon^+)_l].$$

Thus, using Equation (11.31) and considering up to first-order terms [since we assume τ and hence $(\epsilon^+)_j$ are small], we have the result. $\square$

The covariance matrix of the normal approximation of Equation (11.30) depends on the mean Θ. For the isotropic case, with $\Omega = I_{2k}$, the covariance matrix of the normal approximation is given by:

$$2\tau^2 FF^T = 2\tau^2\{I_{2k-4} + \Theta\Theta^T + \Phi\Phi^T\},$$

as seen in Equation (11.18). In the complex normal case the normal approximation is also of complex normal form.

11.3   Inference for offset normal distributions

11.3.1   General MLE

Consider using the general offset normal shape distribution to model population shapes. We could have a random sample of n independent observation shapes $u_1, \dots, u_n$ from the most general offset normal shape distribution for $k \geq 3$, with population shape parameters Θ, and covariance parameters Σ. We have $2k^2 - k - 3$ parameters to estimate in total, although we expect only $2k^2 - 5k + 2$ to be identifiable, as there are $2k - 4$ mean parameters and $(2k - 4)(2k - 3)/2$ parameters in the covariance matrix of a general $2k - 4$ dimensional multivariate normal distribution. If we write the exact shape density as $f(u; \Theta, \Sigma)$, then the likelihood of the data is:

$$L(u_1, \dots, u_n; \Theta, \Sigma) = \prod_{i=1}^{n} f(u_i; \Theta, \Sigma)$$

and the log-likelihood is given by:

$$\log L(u_1, \dots, u_n; \Theta, \Sigma) = \sum_{i=1}^{n} \log f(u_i; \Theta, \Sigma). \tag{11.32}$$

The MLE for shape and covariance is given by:

$$(\hat{\Theta}, \hat{\Sigma}) = \arg \sup_{\Theta, \Sigma} \log L(u_1, \dots, u_n; \Theta, \Sigma).$$

We could proceed to maximize the likelihood with a numerical routine. For example, we could choose Newton–Raphson, Powell's method, simulated annealing or a large variety of other techniques. Estimation can be problematic for completely general covariance structures (because singular matrices can lead to non-degenerate shape distributions), and so we proceed by considering simple covariance structures with a few parameters. For many simple practical cases there are no estimation problems. We now deal with such an application, where the covariance structure is assumed to be isotropic.

11.3.2   Isotropic case

Consider a random sample of n independent observations $u_1, \dots, u_n$ from the isotropic offset normal shape distribution of Equation (11.11). We work with parameters $(\Theta^T, \tau)^T$, where Θ is the mean shape and τ is the coefficient of variation. From Equation (11.16), the exact log-likelihood of $(\Theta^T, \tau)^T$, ignoring constant terms, is:

$$\log L(\Theta, \tau) = \sum_{i=1}^{n} \log {}_1F_1[2 - k; 1; -(1 + \Theta^T\Theta)(1 + \cos 2\rho_i)/(4\tau^2)]$$

$$- \sum_{i=1}^{n} (1 + \Theta^T\Theta)(1 - \cos 2\rho_i)/(4\tau^2), \tag{11.33}$$

where $\rho_i = \rho(u_i, \Theta)$ is the Riemannian distance between u_i and the population mean shape Θ, given in Equation (11.17). Using the result (e.g. Abramowitz and Stegun 1970) that

$$\frac{d\{_1F_1(a; b; x)\}}{dx} = \frac{a}{b}{}_1F_1(a + 1; b + 1; x),$$

we can find the likelihood equations and solve using a numerical procedure. The root of the likelihood equations that provides a global maximum of Equation (11.33) is denoted by $(\hat{\Theta}^T, \hat{\tau})^T$, the exact MLEs of $(\Theta^T, \tau)^T$, with $\tau > 0$.

We can now write down likelihood ratio tests for various problems, for example testing for differences in shape between two independent populations or for changes in variation parameter. Large sample standard likelihood ratio tests can be carried out in the usual way, using Wilks' theorem (Wilks 1962).

Consider, for example, testing whether $H_0 : \Theta \in \mathcal{A}_0$, versus $H_1 : \Theta \in \mathcal{A}_1$, where $\mathcal{A}_0 \subset \mathcal{A}_1 \subseteq \mathbb{R}^{2k-4}$, with $\dim(\mathcal{A}_0) = p < 2k - 4$ and $\dim(\mathcal{A}_1) = q \leq 2k - 4$. Let

$$-2 \log \Lambda = 2 \sup_{H_1} \log L(\Theta, \tau) - 2 \sup_{H_0} \log L(\Theta, \tau)$$

then according to Wilks' theorem $-2 \log \Lambda \approx \chi^2_{q-p}$ under H_0, for large samples (under certain regularity conditions).

If we had chosen different shape coordinates for the mean shape, then inference would be identical as there is a one to one linear correspondence between Kendall and Bookstein coordinates as seen in Equation (2.12). Also, if a different baseline was chosen, then there is a one to one correspondence between Kendall's coordinates and the alternative shape parameters. Therefore, any inference will not be dependent upon baseline or coordinate choice with this exact likelihood approach.

Example 11.1   Consider the schizophrenia data described in Section 1.4.5. The isotropic model may not be reasonable here but we will proceed with an illustrative example of the technique. So, we consider the shape variables in group g to be random samples from the isotropic offset normal distribution, with likelihood given by Equation (11.33) with $2k - 3 = 23$ parameters in each group Θ_g, τ_g, where $g = 1$ (control group) or $g = 2$ (schizophrenia group). A hypothesis test for isotropy was considered in Example 9.7.

The isotropic offset normal MLEs were obtained using the R routine `frechet` which uses `nlm` for the maximization (see Section 11.3.3). Plots of the mean shapes in Bookstein coordinates using baseline *genu* (landmark 2) and *splenium* (landmark 1) are given in Figure 11.3. The exact isotropic MLE of shape is very close to the full Procrustes mean shape; in particular the Riemannian distance ρ from the exact isotropic MLE to the full Procrustes mean shapes in both groups are both extremely small. The MLE of the variation parameters are $\hat{\tau}_1 = 0.0344$ and $\hat{\tau}_2 = 0.0365$.

Testing for equality in mean shape in each group we have $-2 \log \Lambda$, distributed as χ^2_{22} under H_0, as 43.269. Since the p-value for the test is $P(\chi^2_{22} \geq 43.269) = 0.005$

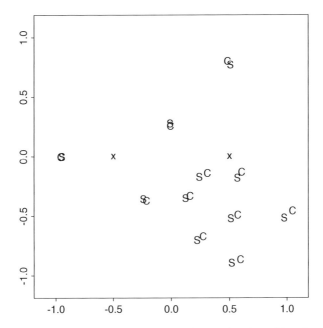

Figure 11.3   The exact isotropic MLE mean shapes for the schizophrenia patients (S) and control group (C), pictured in Bookstein coordinates with baseline 2, 1 (x).

we have strong evidence that the groups are different in mean shape, as is also seen in Example 9.7. Hence, there is evidence for a significant difference in mean shape between the controls and the schizophrenia patients. Note the large shape change in the triangle formed by the *splenium* and the two landmarks just below it. If we restrict the groups to have the same τ as well as the same mean shape in H_0, then $-2 \log \Lambda = 44.9$ and the p-value for the test is $P(\chi^2_{23} \geq 44.9) = 0.004$.

Of course, we may have some concern that our sample sizes may not be large enough for the theory to hold and alternative F and Monte Carlo permutation tests (with similar findings) are given in Example 9.7.   □

11.3.3   Exact isotropic MLE in R

The R code for the above example is as follows:

```
> anss<-frechet(schizophrenia$x[,,schizophrenia $group=="scz"],
      mean="mle")
> ansc<-frechet(schizophrenia$x[,,schizophrenia$group=="con"],
      mean="mle")
> anss$loglike
[1] 775.8128
> ansc$loglike
[1] 798.1256
```

```
> anss$tau
[1] 0.036485
> ansc$tau
[1] 0.03441964
> anss$kappa
[1] 1022.851
> ansc$kappa
[1] 1183.186
#pooled groups
> anspool<-frechet(schizophrenia$x,mean="mle")
> anspool$loglike
[1] 1551.484
# Wilks' statistic
> 2*(ansc$loglike+anss$loglike-anspool$loglike)
[1] 44.90941
```

Hence, we have verified the estimated variance parameters, and the final likelihood ratio statistic.

11.3.4   EM algorithm and extensions

Kume and Welling (2010) consider a different approach for maximum likelihood estimation for the offset-normal shape distribution using an EM algorithm. The approach involves working with the original distribution of the landmark coordinates but treating the rotation and scale as missing/hidden variables. The approach is also useful for size-and-shape estimation. The methodology has been demonstrated to be effective in a variety of examples and it has also been extended to 3D shape and size-and-shape analysis (Kume *et al.* 2015).

A further extension of the method has been carried out by Huang *et al.* (2015) who use a mixture of offset-normal distributions for clustering with mixing proportions from a regression model, with application to corpus callosum shapes. Burl and Perona (1996) used the offset normal distribution for recognizing planar object classes, in particular for detecting faces from clutter via likelihood ratios.

11.4   Practical inference

In order to carry out practical inference one needs to decide upon a suitable procedure. The offset normal shape models are over-parameterized and rather awkward to work with. The isotropic offset normal model is reasonably straightforward to use, although the complex Watson or complex Bingham distribution are much easier to use for practical inference because they are members of the exponential family. So, if the isotropic model is appropriate, then in practice one would use the complex Watson distribution. However, the equal independent isotropic covariance structure is too simplistic for most datasets and so a next stage is to consider the complex covariance structure. The complex Bingham distribution is much easier to use than the offset complex normal distribution and so would be preferred.

There is a close analogy here with inference in directional data analysis. The offset normal shape distribution is analogous to the offset normal distribution in directional statistics, yet in practice one uses the von Mises–Fisher distribution for practical analysis. As a next stage in modelling one considers the real Bingham or Fisher–Bingham family of distributions, rather than more complicated offset normal distributions.

Important point: Perhaps the most straightforward and preferred way to proceed with inference for planar shape is to use the complex Bingham distribution, which gives the full Procrustes mean as the MLE of mean shape. In order to assess the structure of variability it is best to proceed with a multivariate analysis approach in the real tangent space, as seen in Chapter 9 and Section 12.4.

There are various extensions of offset normal shape distributions that can be considered, for example Alshabani *et al.* (2007a) consider partial shape distributions, where the landmark data are in three dimensions but the rotation invariance is only in one plane. These distributions are closely related to the planar shape distributions, and can be considered the joint shape distribution with marks at each landmark. The development of this work was motivated by the Bayesian analysis of human movement data (Alshabani *et al.* 2007b) from Section 1.4.13.

11.5   Offset normal size-and-shape distributions

Under the general normal model of Equation (11.1) we wish to investigate the joint distribution of size-and-shape – the offset normal size-and-shape distribution. These distributions have been developed by Dryden and Mardia (1992), and in general the results are complicated. Bookstein (2013b) states that in many applications the data really do live in Euclidean space but many models are in non-Euclidean shape spaces. One could argue that models should also be constructed in Euclidean spaces, where covariances between atoms for example make sense, and scale is not removed.

11.5.1   The isotropic case

11.5.1.1   The density

In the isotropic case we have $\Omega = \sigma^2 I_{2k}$ and so $\Sigma = \sigma^2 I_{2k-2}$. It will be convenient to consider the centroid size S to measure size here. This model was used for first-order size-and-shape analysis by Bookstein (1986).

Corollary 11.4   *The size-and-shape density of (S, U), with respect to the Lebesgue measure $dSdU$, for the isotropic case is:*

$$f(s, u) = \frac{2s^{2k-3}e^{-(s^2+\zeta^2)/(2\sigma^2)}}{(2\sigma^2)^{k-1}(k-2)!} I_0(s\zeta \cos \rho/\sigma^2) f_\infty(u),$$
$$s > 0, -\infty < u_3, \ldots, u_k, v_3, \ldots, v_k < \infty, \tag{11.34}$$

where $\zeta = \|\mu_H\|$ is the population centroid size and ρ is the Riemannian distance between the population and observed shapes given by:

$$\cos^2 \rho = \frac{(\mu_H^T u^+)^2 + (\mu_H^T v^+)^2}{\|\mu_H\|^2 \|u^+\|^2},$$

and $f_\infty(u)$ is the uniform shape density in Kendall coordinates, from Equation (10.2).

A proof was given by Dryden and Mardia (1992) which involves a transformation and integrating out a rotation variable. The result was also obtained by Goodall and Mardia (1991). Let us write Θ for the Kendall coordinates obtained from μ_H. There is a total of $2k - 2$ parameters here, namely the population size ζ, the population shape Θ ($2k - 4$ parameters) and the variation σ^2 at landmarks.

11.5.1.2   Equal means

If all the means $(\mu_j, v_j)^T, j = 1, \dots, k$, are equal under the isotropic model, that is the population configuration is coincident ($\zeta = 0$), then the density of (S, U), with respect to the Lebesgue measure, is given by:

$$f(s, u) = \frac{2s^{2k-3}e^{-s^2/(2\sigma^2)}}{(2\sigma^2)^{k-1}(k-2)!}f_\infty(u), \quad s > 0, -\infty < u_3, \dots, u_k, v_3, \dots, v_k < \infty.$$

Proof: Let $\zeta \to 0$ in the density of Equation (11.34). $\qquad\qquad\qquad\square$

Note that size (a scaled χ_{2k-2} random variable) and shape are independent in this limiting case, as $\zeta \to 0$.

11.5.1.3   Small variations

Bookstein (1986) has shown that the centroid size and shape variables have approximately zero covariance for small variations under the isotropic normal model. We can also derive this result directly from the p.d.f. Let

$$\Theta = (\theta_3, \dots, \theta_k, \phi_3, \dots, \phi_k)^T$$

be the shape of the population configuration,

$$\Phi = (-\phi_3, \dots, -\phi_k, \theta_3, \dots, \theta_k)^T$$

and define

$$B(\Theta) = I_{2k-4} + \Theta\Theta^T + \Phi\Phi^T,$$

with

$$B^{-1}(\Theta) = I_{2k-4} - (\Theta\Theta^{\mathrm{T}} + \Phi\Phi^{\mathrm{T}})/(1 + \Theta^{\mathrm{T}}\Theta).$$

Result 11.9 *If (σ/ζ) is small, then the approximate isotropic size-and-shape distribution is:*

$$\begin{bmatrix} S \\ U \end{bmatrix} \simeq N_{2k-3} \left\{ \begin{bmatrix} \zeta \\ \Theta \end{bmatrix}, \begin{bmatrix} \sigma^2 & 0^{\mathrm{T}} \\ 0 & \sigma^2(1 + \Theta^{\mathrm{T}}\Theta)B(\Theta)/\zeta^2 \end{bmatrix} \right\}. \qquad (11.35)$$

Proof: Transform to $T = (S - \zeta)/\sigma$ and $W = \zeta(U - \Theta)/\sigma$, and write $\gamma = \sigma/\zeta$, where $\zeta > 0$. The density of (T, W) is:

$$f(t, w) = \frac{(t\gamma + 1)^{2k-3}\exp\{-(\gamma^{-2} + t\gamma^{-1} + \frac{1}{2}t^2)\}}{(2\pi)^{k-2}\gamma\{1 + (\gamma w + \Theta)^{\mathrm{T}}(\gamma w + \Theta)\}^{k-1}} I_0\{(t\gamma + 1)\cos\rho/\gamma^2\}.$$

Using the facts that, for large x,

$$I_0(x) = (2\pi x)^{-1/2}e^x\{1 + O(x)\}$$

and that, for small γ,

$$\cos\rho = 1 - \gamma^2 w^{\mathrm{T}}B^{-1}(\Theta)w/\{2(1 + \Theta^{\mathrm{T}}\Theta)\} + O(\gamma^4),$$

we see that

$$f(t, w) = \frac{e^{-\frac{1}{2}t^2}}{(2\pi)^{(2k-3)/2}(1 + \Theta^{\mathrm{T}}\Theta)^{k-1}}\exp\left\{-\frac{w^{\mathrm{T}}B^{-1}(\Theta)w}{2(1 + \Theta^{\mathrm{T}}\Theta)}\right\} + O(\gamma).$$

Thus, since $|B(\Theta)|^{1/2} = (1 + \Theta^{\mathrm{T}}\Theta)$, we deduce that, as $\gamma \to 0+$,

$$\begin{bmatrix} T \\ W \end{bmatrix} \sim N_{2k-3} \left\{ \begin{bmatrix} 0 \\ 0 \end{bmatrix}, \begin{bmatrix} 1 & 0^{\mathrm{T}} \\ 0 & (1 + \Theta^{\mathrm{T}}\Theta)B(\Theta) \end{bmatrix} \right\}.$$

The result follows. □

This result could also be obtained using Taylor series expansions for S and U, and Bookstein (1986) used such a procedure without reference to the offset normal size-and-shape distribution.

11.5.1.4 Conditional distributions

Using the standard result for the distribution of S under the isotropic model (e.g. Miller 1964, p. 28), it follows that the conditional density of U given $S = s$ is:

$$f_\infty(u)\{s\zeta/(2\sigma^2)\}^{k-2} I_0(s\zeta \cos \rho/\sigma^2)/I_{k-2}(s\zeta/\sigma^2) ,$$

where $f_\infty(u)$ is the uniform shape density of Equation (10.2). The conditional density of S given U involves the ratio of a modified Bessel function and a confluent hypergeometric function. Using Weber's first exponential integral (Watson 1944, p. 393), it can be seen that

$$E(S^2|U = u) = \int_0^\infty s^2 f(s|u)\mathrm{d}s = 2\sigma^2(k-1)\frac{{}_1F_1\{1-k;1;-\zeta^2 \cos^2 \rho/(2\sigma^2)\}}{{}_1F_1\{2-k;1;-\zeta^2 \cos^2 \rho/(2\sigma^2)\}}.$$

If $\gamma = \sigma/\zeta$ is small, then, since for large $x > 0$ (Abramowitz and Stegun 1970, p. 504),

$$_1F_1(a;b;-x) = \Gamma(b)/\Gamma(b-a)(x)^{-a}\{1 + O(|x|^{-1})\},$$

and since $\cos^2 \rho = 1 - O(\gamma^2)$, it follows that

$$E(S^2|U = u) = \zeta^2\{1 + O(\gamma^2)\}.$$

This verifies the result of Bookstein (1986) that S^2 and U are approximately uncorrelated for small γ, which can also be seen from Equation (11.35).

11.5.2   Inference using the offset normal size-and-shape model

The size-and-shape distribution can be used for likelihood-based inference. Inference for models with general covariance structures will be very complicated. Placing some restrictions on the covariance matrix is necessary for the same identifiability reasons as in the shape case, discussed in Section 11.3.

Let us assume that we have a random sample (s_i, u_i), $i = 1, \ldots, n$, from the isotropic size-and-shape distribution of Equation (11.34). We have the $2k - 2$ parameters to estimate: population mean size ζ, population mean shape Θ (a $2k - 4$-vector), and population variance σ^2. The log likelihood is given, up to a constant with respect to the parameters, by:

$$\log L(\zeta, \Theta, \sigma^2) = -(k-1)n \log(\sigma^2) - \frac{n\zeta^2}{2\sigma^2} - \sum_{i=1}^{n} \frac{s_i^2}{2\sigma^2} + \sum_{i=1}^{n} \log I_0\left(\frac{s_i\zeta \cos \rho_i}{\sigma^2}\right),$$

$$(11.36)$$

where ρ_i is the Riemannian distance between the shapes u_i and Θ given by Equation (4.21). We denote the MLEs by $(\hat{\zeta}, \hat{\Theta}, \hat{\sigma}^2)$.

The likelihood equations must be solved numerically, although there is the following relationship between the MLEs of σ^2 and ζ:

$$\hat{\sigma}^2 = \frac{1}{n(2k-2)} \sum_{i=1}^{n}(s_i^2 - \hat{\zeta}^2).$$

The numerical maximization of Equation (11.36) requires the accurate evaluation of $\log I_0(\kappa)$. This can be performed using a polynomial approximation such as mentioned by Abramowitz and Stegun (1970, p. 378).

Consider, for example, testing for symmetry in a sample of triangles. One aspect of the shape that could be investigated is asymmetry about the axis from the midpoint of the baseline to landmark 3.

Example 11.2   Consider a random sample of 30 T1 mouse vertebrae with $k = 3$ landmarks on each bone. Each triangle is formed by landmarks 1, 2 and 3 as in Figure 11.4.

Let us propose that the data form a random sample from the isotropic size-and-shape distribution. As a model check, since variations are small we would expect the Kendall coordinates and centroid size (U, V, S) to be approximately normal and uncorrelated from Equation (11.35). There is no reason to doubt normality using Shapiro–Wilk W tests, and the sample correlations for $(U, V), (U, S)$ and (V, S) are only $-0.07, -0.06$ and -0.07, respectively. In addition, the standard deviations of U and V should be approximately equal and here the sample standard deviations of U and V are 0.023 and 0.034, respectively. Thus, we conclude that the model is fairly reasonable and there appears to be no linear relationship between size and shape in this particular dataset.

The MLEs are found by maximizing Equation (11.36). The values of the MLEs are $\hat{\theta} = 0.0127, \hat{\phi} = 0.4694, \hat{\zeta} = 173.2$ and $\hat{\sigma} = 4.147$.

It is suggested that the population mean triangle could be isosceles with axis of symmetry $\theta = 0$. We can investigate such asymmetry by testing $H_0 : \Theta = (0, \phi)^T$ versus $H_1 : \Theta = (\theta, \phi)^T, \Theta$ unrestricted, where $\phi, \zeta > 0$ and $\sigma^2 > 0$ are not restricted

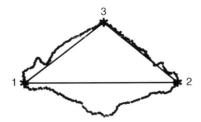

Figure 11.4   The three labelled landmarks taken on each T1 mouse vertebra. The baseline was taken as 1–2 in the calculations described in the text. Source: Dryden & Mardia 1992. Reproduced with permission of Oxford University Press.

under either hypothesis. Standard likelihood ratio tests can be performed in the usual way using Wilks' theorem for large samples. Let

$$-2\log \Lambda = -2 \left\{ \sup_{H_0} \log L(\zeta, \Theta, \sigma^2) - \sup_{H_1} \log L(\zeta, \Theta, \sigma^2) \right\}.$$

Then according to Wilks' theorem $-2\log \Lambda \simeq \chi_1^2$ under H_0, for large samples. This test could of course be carried out using shape information alone, ignoring size. For this T1 mouse vertebrae example the statistic is $-2\log \Lambda = 5.52$ with approximate p-value 0.02 calculated from the χ_1^2 approximation. Thus we have evidence to reject symmetry about $\theta = 0$.   □

11.6   Distributions for higher dimensions

11.6.1   Introduction

For practical data analysis in three and higher dimensions we recommend the use of Procrustes analysis and tangent space inference (Chapters 7 and 9). The study of suitable models for higher dimensional shape is much more difficult than in the planar case. We outline some distributional results for normally distributed configurations, although the work is complicated.

11.6.2   QR decomposition

Goodall and Mardia (1991, 1992, 1993) have considered higher dimensional offset normal size-and-shape and shape distributions by using the QR decomposition. In particular, the decomposition of Bartlett (1933) described below leads to the main results.

Although we have concentrated on $k \geq m + 1$ so far, in this section on higher dimensional distributions we also allow $k < m + 1$. Invariance with respect to rotation if $k < m + 1$ is the same as invariance with respect to the set of $k - 1$ orthonormal frames in m dimensions, which is the Stiefel manifold $V_{k-1,m}$ (e.g. see Jupp and Mardia 1989). The Stiefel manifold $V_{n,m}$ of orthonormal n frames in m dimensions can be regarded as the set of $(n \times m)$ matrices A satisfying $AA^T = I_n$. When $k > m$ then $V_{n,m} = V_{m,m} = O(m)$.

First of all we consider the general QR decomposition described in Section 5.6.2 in more detail, for general values of k and m. Consider the QR decomposition of the Helmertized coordinates $Y = HX$, namely

$$Y = T\Gamma, \quad \Gamma \in V_{n,m}, \tag{11.37}$$

where $n = \min(k - 1, m)$, $V_{n,m}$ is the Stiefel manifold and $T((k - 1) \times n)$ is lower triangular with non-negative diagonal elements $T_{ii} \geq 0, i = 1, \dots, n$. When $k > m$ we have $\Gamma \in O(m)$ and $\det(\Gamma) = \pm 1$, and hence the QR decomposition removes orientation *and reflection* in this case.

Before in Section 5.6.2 we just considered $k > m$ but restricted Γ so that reflections were not allowed. In that case we removed orientation only for $k > m$ and so we had $\Gamma \in SO(m), \det(\Gamma) = +1$ and T unrestricted. However, when $k \leq m$ there is no distinction to make between allowing reflections or not – the shape is invariant under reflection in this case.

For $k > m$, $T, \{T_{ij} : 1 \leq i \leq j \leq m\}$ is the size-and-shape if $\Gamma \in SO(m)$ and the reflection size-and-shape if $\Gamma \in O(m)$. If $k \leq m$, then the distinction is irrelevant and we just call T the size-and-shape.

To remove scale we divide by the Euclidean norm of T,

$$W = T/\|T\|, \quad \|T\| > 0.$$

For $k > m$, W is the shape of our configuration if $\Gamma \in SO(m)$ and reflection shape if $\Gamma \in O(m)$. If $k \leq m$ the distinction is again irrelevant and we call W the shape. If $\|T\| = 0$, then the shape is not defined.

11.6.3   Size-and-shape distributions

11.6.3.1   Coincident means

If X has the multivariate isotropic normal model

$$\text{vec}(X) \sim N_{km}(\mu^*, \sigma^2 I_{km}), \tag{11.38}$$

then $YY^T = HXX^T H^T = T\Gamma\Gamma^T T^T = TT^T$ has a central Wishart distribution if $\text{rank}(\mu) = 0$, where $\mu = E[Y] = H\text{vec}_m^{-1}(\mu^*)$, and $m > k - 1$. Bartlett's decomposition gives the distribution of the lower triangular matrix T, where $YY^T = TT^T$. Hence T is the reflection size-and-shape in the QR decomposition of Y in Equation (11.37).

Result 11.10   *(Bartlett decomposition). Under the normal model of Equation (11.38) with $\mu = 0$ the distribution of $(T)_{ij}$ is:*

$$T_{ll}^2 \sim \sigma^2 \chi_{m-l+1}^2 \;, \quad l = 1, \ldots, N, \quad T_{ij} \sim N(0, \sigma^2) \tag{11.39}$$

for $i > j, i = 1, \ldots, k - 1, \; j = 1, \ldots, N$, and all these variables are mutually independent $\{N = \min(k - 1, m)\}$.

The proof involves random rotation matrices. For example, for $k - 1 < m$ see Kshirsagar (1963). Goodall and Mardia (1991) have shown that this result holds also for $m < k$ (even when the Wishart matrix is singular) and thus Equation (11.39) is the reflection size-and-shape distribution for the coincident means ($\mu = 0$) case. If the size-and-shape distribution without reflections is required, then the result of Equation (11.39) holds, except that $T_{mm} \sim N(0, \sigma^2)$ for $k > m$.

Goodall and Mardia (1991) derive further results for planar and higher rank means. The densities are complicated in general. One simple case is that of triangles in higher dimensions.

If $k = 3$ and $m \geq 3$, then we can consider Kendall's spherical coordinates on the hemisphere $S^2_+(\frac{1}{2})$. In this case we have $W_{11} = (1 + \sin \theta \cos \phi)^{1/2}/\sqrt{2}$, $W_{22} = \cos \theta (1 + \sin \theta \cos \phi)^{-1/2}/\sqrt{2}$ and $J(u) = \sin \theta (1 + \sin \theta \cos \phi)^{-1/2}/(2\sqrt{2})$. Hence the shape density on $S^2_+(\frac{1}{2})$ is given by,

$$(m - 1) \cos^{m-2} \theta \sin \theta /(2\pi). \tag{11.40}$$

The uniform measure on the hemisphere is $\sin \theta /(2\pi) d\theta d\phi$ and $\theta/2$ is the great circle distance to the north pole (the equilateral triangle shape). This result was derived by Kendall (1988) and indicates that the higher the dimension m, the greater the strength of the mode at the equilateral triangle.

If $m = 2$ then we divide Equation (11.40) by 2 to obtain the shape density without reflection on the sphere $S^2(1/2)$,

$$\sin \theta /(4\pi),$$

which is the uniform density on $S^2(\frac{1}{2})$.

Diffusions on planar shape spaces were studied by Ball et al. (2008) and the explicit expressions for Brownian motion were given. Also Ornstein–Uhlenbeck processes in planar shape space were defined using Goodall–Mardia coordinates. A drift term was added in the Ornstein–Uhlenbeck process along the geodesic towards the mean shape. Inference for these models was given by Ball et al. (2006) who discussed exact retrospective sampling of Ornstein–Uhlenbeck processes, with a particular application to modelling cell shape.

Other types of diffusions include Radon shape diffusions, where the shape of random projections of shapes into subspaces of lower dimension is considered (Panaretos 2006, 2008). In related work Panaretos (2009) investigates the recovery of a density function from its Radon transform at random and unknown projection angles, and a practical application is a random tomography technique for obtaining sparse reconstructions of proteins from noisy random projections using the LASSO (Panaretos and Konis 2011).

11.6.4   Multivariate approach

An alternative way to proceed is to consider an approach using results from multivariate analysis. The QR decomposition is again considered $Y = T\Gamma$, but one transforms to the joint distribution of (T, Γ) and then integrates out Γ over the Stiefel manifold to obtain the size-and-shape distribution. Integrating out size then gives the marginal shape distribution.

The key result for integrating out the rotation (and reflections with $k \geq m + 1$) is that, from James (1964):

$$\int_{O(m)} \exp\{\text{trace}(\mu^T T\Gamma)\}[\Gamma d\Gamma^T] = \frac{2^N \pi^{Nm/2}}{\Gamma_N(m/2)} {}_0F_1\{m/2; \mu^T TT^T \mu/(4\sigma^2)\},$$

where $[\Gamma d\Gamma^T]$ is the invariant unnormalized measure on $O(m)$ and ${}_0F_1(\cdot; \cdot)$ is a hypergeometric function with matrix argument, which can be written as a sum of zonal polynomials. If $\text{rank}(\mu) < m$, then it follows that

$$\int_{O(m)} \exp\{\text{trace}(\mu^T T\Gamma)\}[\Gamma d\Gamma^T] = \frac{1}{2} \int_{SO(m)} \exp\{\text{trace}(\mu^T T\Gamma)\}[\Gamma d\Gamma^T]$$

and so the density without reflection invariance is fairly straightforward to find in this case. If $\text{rank}(\mu) = m$, for $m \geq 3$ then the density without reflection invariance cannot be computed in this way.

Goodall and Mardia (1992) discuss in detail the multivariate approach for computing shape densities from normal distributions. The case where the covariance matrix is $I_m \otimes \Sigma$ is dealt with and size-and-shape and shape distributions are given for all cases except $\text{rank}(\mu) = m$, for $m \geq 3$.

In conclusion, marginal shape distributions from normal models are very complicated in higher than two dimensions, but for low rank means (rank ≤ 2) the distributions are tractable.

11.6.5 Approximations

When variations are small Goodall and Mardia (1993) have given some normal approximations to the size-and-shape and shape distributions.

Result 11.11 *Let T^P be the Procrustes size-and-shape coordinates after rotating the size-and-shape T to the mean μ (diagonal and of rank m) given in Equation (7.14). For small σ, it follows that $(T^P)_{ij}$ follow independent normal distributions with means μ_{ij} and variances σ^2, for $i = j$ or $i > m$, and $\sigma^2/(1 + \mu_{ii}^2/\mu_{jj}^2)$ for $j < i \leq m$.*

Important point: The normal approximation is much simpler to use for inference than the offset normal distribution, and Goodall and Mardia (1993) consider an $m = 3$ dimensional application of size-and-shape analysis in the manufacture of chairs. A decomposition of the approximate distribution for size and shape is also given, and it can be shown for general m that size and shape are asymptotically independent (as $\sigma \to 0$) under isotropy. The 2D ($m = 2$) case was seen earlier in Result 11.9. Thus the result gives a method for small σ for shape analysis. Inference will be similar to using the tangent approximation inference of Section 9.1.

12

Deformations for size and shape change

12.1 Deformations

12.1.1 Introduction

In biology and other disciplines one is frequently interested in describing the difference in shape or size-and-shape between two or more objects. A measure of distance, such as the Riemannian distance, gives us a numerical measure of size and shape comparison, but this global measure does not indicate locally where the objects differ and the manner of the difference. The biological goal in many studies is to depict or describe the morphological changes in a study.

First of all we consider the case when only two objects are available and we wish to describe the size and shape difference between them. Sometimes there will be a natural ordering of the two objects, for example a juvenile growing into an adult, and sometimes not, for example a male and a female of a species. In order to describe the difference in size and shape we compute the transformation of the space in which the first object lies into the space of the second object. The transformation will give us information about the local and global shape differences.

Definition 12.1 *By **global** differences we mean large scale changes (with small amounts of bending per unit Procrustes distance), such as an overall affine transformation.* **Local** *differences are on a smaller scale (with larger amounts of bending per unit Procrustes distance), for example highlighting changes in a few nearby landmarks. Global differences are smooth changes between the figures, whereas local changes are the remainder of the components of a deformation and are less smooth.*

Statistical Shape Analysis, with Applications in R, Second Edition. Ian L. Dryden and Kanti V. Mardia.
© 2016 John Wiley & Sons, Ltd. Published 2016 by John Wiley & Sons, Ltd.

The main method for describing deformations will be the thin-plate spline transformation. Following Bookstein (1989, 1991) we shall decompose a thin-plate spline deformation (defined below) into a global affine transformation and a set of local deformations which highlight changes at progressively smaller scales. This decomposition is in close analogy with a Fourier series decomposition, with the constant term in the Fourier series being the global parameter and the coefficients of the trigonometric terms being local parameters at successively smaller scales.

12.1.2   Definition and desirable properties

We consider the case where we have two objects in $m \geq 2$ dimensions.

Definition 12.2   *Consider two k landmark configuration matrices in $\mathbb{R}^m$, $T = (t_1, \ldots, t_k)^{\mathrm{T}}$ and $Y = (y_1, \ldots, y_k)^{\mathrm{T}}$ both $k \times m$ matrices, and we wish to deform T into Y, where $t_j, y_j \in \mathbb{R}^m$. We use the notation that the m-vector t_j is written as $t_j = (t_j[1], \ldots, t_j[m])^{\mathrm{T}}$. A* **deformation** *is a mapping from $t \in \mathbb{R}^m$ to $y \in \mathbb{R}^m$ defined by the transformation*

$$y = \Phi(t) = (\Phi_1(t), \Phi_2(t), \ldots, \Phi_m(t))^{\mathrm{T}}.$$

Here T is the **source** and Y is the **target**. The multivariate function $\Phi(t)$ should have certain desirable properties. In particular, we would want as many of the following properties to hold as possible for the deformation:

1. continuous;

2. smooth (which implies 1. of course);

3. bijective (one to one and onto);

4. not prone to gross distortions (e.g. not folding, which will be guaranteed if 3. holds);

5. equivariant under relative location, scale and rotation of the objects, that is the transformation is not affected by a common translation, rotation and scaling of the figures;

6. an interpolant, that is $y_j = \Phi(t_j)$ for $j = 1, \ldots, k$.

If the interpolation property is not satisfied then we call the deformation a smoother. Note that the deformation is from the whole space $\mathbb{R}^m$ to $\mathbb{R}^m$, rather than just from a set of landmarks to another or an outline to another.

Applications of deforming images will be considered in Section 17.4.1.

12.1.3   D'Arcy Thompson's transformation grids

D'Arcy Thompson (1917) considered deformations from one species to another in order to explain size and shape differences. A regular square grid pattern was drawn

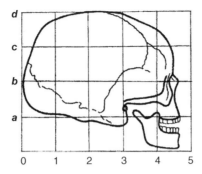

Figure 12.1 An initial square grid placed on a human skull. Source: Thompson 1917. Reproduced with permission of Cambridge University Press.

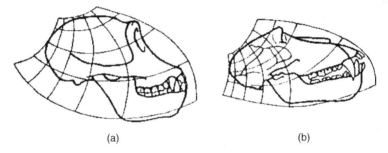

(a)   (b)

Figure 12.2 Cartesian transformation grids from the human skull in Figure 12.1 to a chimpanzee (a) and a baboon (b). Source: Thompson 1917. Reproduced with permission of Cambridge University Press.

on one object and the grid was deformed to lie on the second object, with corresponding biological parts located in the corresponding grid blocks. In Figure 2.1, Figure 12.1, Figure 12.2 and Figure 12.3 we see some examples. These grids are known as Cartesian transformation grids. The transformation grids enable a biologist to describe the size and shape change between the two species, albeit in a rather

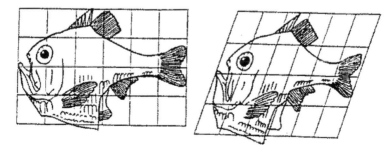

Figure 12.3 Cartesian transformation grids from one species of fish to another. The transformation is an affine deformation. Source: Thompson 1917. Reproduced with permission of Cambridge University Press.

subjective way. D'Arcy Thompson's figures were drawn by hand and were not always very accurate, for example in Figure 2.1 the correspondence in the lower and upper fins at $(4.5, b)$ and $(4.5, c)$ is not good (Bookstein, 1994b). There have been many attempts since 1917 to recreate these figures more objectively, as will be discussed in Section 12.4.1.

12.2   Affine transformations

12.2.1   Exact match

The simplest possible size and shape change is that of an affine transformation, as seen in Figure 12.3. In this case the square grid placed on the first fish is deformed uniformly and affinely into a parallelogram grid on the second fish.

Definition 12.3   *The **affine** transformation from a point $t \in \mathbb{R}^m$ to a point $y \in \mathbb{R}^m$ is:*

$$y = At + c, \tag{12.1}$$

*where A is $m \times m$, c is $m \times 1$ and A is non-singular. A **linear** transformation is the affine transformation of Equation (12.1) but with $c = 0$.*

Consider selecting a subset of $m + 1$ landmarks on each figure, and denote the subsets by T_{sub} and Y_{sub}, both $(m + 1) \times m$ matrices. Writing $B = [c, A]^T$ then, since we match the $m + 1$ points exactly, the unknown parameters B are found from the solution to:

$$Y_{sub} = [1_{m+1}, T_{sub}]B. \tag{12.2}$$

By selecting $m + 1$ landmarks we obtain a unique solution in general, because there are $(m + 1)m$ unknown parameters in B and $(m + 1)m$ constraints in Equation (12.2). Write $X = [1_{m+1}, T_{sub}]$ for the 'design matrix'. Solving Equation (12.2) for B we have:

$$Y_{sub} = XB$$
$$\Leftrightarrow X^T Y_{sub} = X^T XB$$
$$\Leftrightarrow \hat{B} = (X^T X)^{-1} X^T Y_{sub}$$

assuming $(X^T X)^{-1}$ exists, which will be true if X is of full rank $m + 1$. This solution is the same as the least squares estimator when we wish to estimate a transformation from $k = m + 1$ points, and here the fit is exact and the residuals are all zero.

12.2.2   Least squares matching: Two objects

For $k > m + 1$ the affine match is not exact in general and so we estimate the affine transformation by minimizing some criterion. Write $X_D = [1_k, T]$ for the $k \times (m + 1)$ design matrix. A general regression model is given later in Equation (18.1). Taking $h(T, B) = X_D B$, where B is an $(m + 1) \times m$ matrix, gives an affine match of T to Y. We exclude configurations that do not have full rank, so $\text{rank}(X_D) = m + 1 = \text{rank}(Y)$. Equation (18.1) becomes:

$$
\begin{aligned}
Y &= X_D B + E \\
&= TA + 1_k c^T + E,
\end{aligned}
$$

where A is an $m \times m$ matrix and c an m-vector for translation.

The general shape space (see Section 18.4) with affine invariance has been studied in detail by Ambartzumian (1982, 1990). Some distributional results for offset normal models were given by Goodall and Mardia (1993). Also Bhavnagri (1995b) and Sparr (1992) have considered the affine shape space which, if configurations of less than full rank are ignored, is the Grassmann manifold of m-planes in $\mathbb{R}^{m(k-1)}$ (e.g. Harris, 1992; Patrangenaru and Mardia, 2003). A suitable distance in the affine shape space between T and Y is given by:

$$
d_A^2(Y, T) = \inf_{A, c} \| Y(Y^T Y)^{-1/2} - \left(TA + 1_k c^T \right) \|^2. \tag{12.3}
$$

The post-multiplication of Y by $(Y^T Y)^{-1/2}$ prevents degenerate solutions and ensures that $d_A(Y, T) = d_A(T, Y)$.

It is sometimes convenient to use vector notation: writing $y = \text{vec}(Y)$, $\beta = \text{vec}(B)$, and $\epsilon = \text{vec}(E)$, the linear model is

$$
y = X_d \beta + \epsilon,
$$

where ϵ is taken to be a zero mean random vector and $X_d = I_m \otimes X_D$ is a $km \times m(m + 1)$ block diagonal design matrix. The match is obtained by estimating β with $\hat{\beta}$, where

$$
\hat{\beta} = \text{argmin } s^2[\text{vec}^{-1}(\epsilon)] = \text{argmin } \epsilon^T \epsilon
$$

where $\text{vec}^{-1}(\epsilon) = E$ is the reverse of vectorizing (see Definition 4.5).

Using least squares matching, the solution $\hat{\beta}$ is given explicitly and is unique provided $X_D^T X_D$ is of full rank $(m + 1)$. The solution is obtained by differentiating

$$
s^2(E) = (y - X_d \beta)^T (y - X_d \beta)
$$

with respect to β (a discussion of matrix differentiation is given e.g. in Mardia *et al.* 1979, Appendix A.9). The resulting normal equations are:

$$-2X_d^T y + 2\beta X_d^T X_d = 0,$$

and hence the solution is:

$$\hat{\beta} = \left(X_d^T X_d\right)^{-1} X_d^T y,$$

where $(X_d^T X_d)^{-1}$ is assumed to exist. The matrix representation of the solution is:

$$\hat{B} = \left(X_D^T X_D\right)^{-1} X_D^T Y.$$

The second derivative matrix is $2X_d^T X_d$ which is strictly positive definite and hence the solution is a minimum.

The residual vector $r = y - X_d \hat{\beta}$ is given by:

$$r = \left[I_{km} - X_d \left(X_d^T X_d\right)^{-1} X_d \right] y = (I_{km} - H_{hat})y,$$

where H_{hat} is the usual 'hat' matrix. Note that $H_{hat} H_{hat} = H_{hat}$ and so H_{hat} is idempotent. Also, $(I_{km} - H_{hat})$ is idempotent. The minimized value of the objective function is

$$s^2[\text{vec}^{-1}(r)] = r^T r = y^T (I_{km} - H_{hat})y.$$

It is important to note that matching from T to Y is not the same as matching from Y to T in general. This property is generally true for most registration methods. One way to introduce symmetry into the problem is by obtaining an average object 'half-way' between T and Y, and then match T and Y separately into the average object.

The affine case where the residuals are zero (exact matches) was considered earlier in more detail in Section 12.2.1.

Example 12.1   Horgan *et al.* (1992) consider affine matches of two electrophoretic gels (gel A and gel B) described in Section 1.4.14. In Figure 1.25 we saw two configurations of $k = 10$ landmarks located on each gel which are taken from different proteins from malaria parasites. The landmarks are called invariant points in this application because they are common to all proteins in this study. The idea is to match the two configurations with an affine transformation. By matching the gels to each other or to a reference figure one can identify the malaria parasite by using the positions of the other (variant) spots in the image. In Figure 12.4 the fitted gel A image has been registered onto the gel B image. New grey levels x_{ij}^N have been calculated from the grey levels from the fitted gel A ($\hat{x}_{ij}^A$) and gel B (x_{ij}^B), using $x_{ij}^N = 2\hat{x}_{ij}^A - x_{ij}^B$. The

Figure 12.4   The superimposed gel A onto gel B using an affine match, using the 10 invariant spots chosen by an expert. The four grey levels indicate the correspondence of dark and light pixels in the registered images. The key used is: black, dark pixel in gel A and a light pixel in gel B; dark grey, matched dark pixels in gels A and B; light grey, matched light pixels in gels A and B; and white, dark pixel in gel B and a light pixel in gel A.

figures have been thresholded to four levels for display purposes, indicating which spots were in gel A, gel B or both, and which region was not dark in either gel. The affine fitted gel A image matches quite well with the gel B image, because there are many dark grey common spots in both images. The parameters of the fit are:

$$\hat{\beta} = (-36.08, 0.973, -0.023, 66.64, 0.039, 0.904)^{\mathrm{T}}$$

and so there is a shift in location to the left and upwards $(-36.08, 66.64)$ and approximately a 10% vertical shrinkage and a 3% horizontal shrinkage.   □

Mardia *et al.* (2012) have addressed a specific problem for matching gels under affine transformation in which each configuration contains a subset of points ('markers') whose labels correspond with high probability, with the remaining points having arbitrary labels ('non-markers').

Gordon (1995) provides a matching procedure based on different affine transformations in subsets of points formed by dividing regions of the plane, with an application to face matching.

12.2.3   Least squares matching: Multiple objects

Consider a random sample of configurations $T_1, \ldots, T_n$ from a population with mean μ. The design matrices are $X_{Di} = [1_k, T_i], i = 1, \ldots, n$. If least squares is chosen for both the estimation of the mean and the point configuration matching, then we wish to minimize

$$\sum_{i=1}^{n} \text{trace} \left\{ (\mu - X_{Di}B_i)^{\mathrm{T}}(\mu - X_{Di}B_i) \right\} = \sum_{i=1}^{n} \|\text{vec}(\mu) - X_{di}\beta_i\|^2,$$

where $\beta_i = \text{vec}(B_i)$ and $X_{di} = I_m \otimes X_{Di}, i = 1, \ldots, n$. It is clear that, given μ,

$$\hat{\beta}_i = \left(X_{di}^{\mathrm{T}} X_{di} \right)^{-1} X_{di}^{\mathrm{T}} \text{vec}(\mu).$$

Hence,

$$\text{vec}(R_i) = (I_{km} - H_i)\text{vec}(\mu),$$

where H_i is the 'hat' matrix for X_{di}, that is

$$H_i = X_{di} \left(X_{di}^{\mathrm{T}} X_{di} \right)^{-1} X_{di}, \quad i = 1, \ldots, n.$$

Therefore,

$$\text{vec}(\hat{\mu}) = \text{argmin} \ \text{vec}(\mu)^{\mathrm{T}} A_H \ \text{vec}(\mu)$$

subject to the constraints $\mu^{\mathrm{T}}\mu = I_m$, where

$$A_H = nI - \sum_{i=1}^{n} H_i. \tag{12.4}$$

Let $\check{A}$ be the $k \times k$ matrix such that $A_H = I_m \otimes \check{A}$. Therefore the solution is given by the m eigenvectors of $\check{A}$ with smallest eigenvalues. This is a well-known result and can be simply seen by repeatedly applying the standard result that the minimum value of

$$\frac{x^{\mathrm{T}} A_H x}{x^{\mathrm{T}} x}$$

is given by the smallest eigenvalue of A_H [e.g. Mardia *et al.* 1979, Equation (A.9.11)].

Even with a general choice of $s(E)$ if $\phi(E) = E^T E$ in Equation (18.2), then to obtain the new $\hat{\mu}$ from the $\hat{B}_i$ we have:

$$\hat{\mu} = \frac{1}{n} \sum_{i=1}^{n} X_i \hat{B}_i,$$

which can lead to a simple implementation.

The above generalized affine matching result was termed the linear spectral theorem by Goodall (1991, p. 335) and was also presented by Rohlf and Slice (1990) and Hastie and Kishon (1991) and for planar shape analysis using complex notation by Kent (1994).

It is clear that affine matching and statistical inference can be tackled (with care) by existing regression and multivariate analysis procedures.

12.2.4   The triangle case: Bookstein's hyperbolic shape space

In two dimensions the most general transformation between two triangles is an affine transformation and we study this case in more detail.

Consider two (different) triangles in a plane $T = (t_1, t_2, t_3)^T$ and $Y = (y_1, y_2, y_3)^T$, both (3×2) matrices, and the ratio of baseline lengths is denoted as $\gamma = \|y_2 - y_1\| / \|t_2 - t_1\|$ and Bookstein's shape coordinates for each triangle are $(\theta_T, \phi_T)^T$ and $(\theta_Y, \phi_Y)^T$. Let us exclude all collinear triplets of points (flat triangles) and all shape changes from a triangle with an apex above the baseline to a triangle with an apex below the baseline (we are excluding reflections). Consider the change in shape from triangle T to triangle Y. First of all the triangles are translated, rotated and rescaled to $T^* = (t_1^*, t_2^*, t_3^*)^T$ and $Y^* = (y_1^*, y_2^*, y_3^*)^T$ so that they have common baselines $(-\frac{1}{2}, 0)^T$ to $(\frac{1}{2}, 0)^T$. The transformation between the shapes of the two triangles is described by the affine transformation

$$y_j^* = A t_j^* + c, \ j = 1, 2, 3, \tag{12.5}$$

where A is a 2×2 matrix, c is a 2×1 vector and

$$t_1^* = y_1^* = \left(-\frac{1}{2}, 0\right)^T, \ t_2^* = y_2^* = \left(\frac{1}{2}, 0\right)^T, t_3^* = (\theta_T, \phi_T)^T, \ y_3^* = (\theta_Y, \phi_Y)^T.$$

Solving the six equations in Equation (12.5) we find that

$$(A)_{11} - 1, \ (A)_{21} = 0, \ (A)_{12} = (\theta_Y - \theta_T)/\phi_T, \ A_{22} = \phi_Y/\phi_T, \ c = 0. \tag{12.6}$$

If we inscribe a circle of unit diameter in triangle T, this is linearly transformed into an ellipse in triangle Y, and the major and minor axes of the ellipse have lengths

$\delta_1 \geq \delta_2 > 0$ and are called the **principal strains**. Calculation of δ_1 and δ_2 is very simple using a singular value decomposition of A

$$A = U\Lambda V^{\mathrm{T}}, \quad \text{where} \quad \Lambda = \mathrm{diag}(\delta_1, \delta_2),$$

where the columns u_1 and u_2 (both 2×1) of U are the eigenvectors of AA^{T} and the columns v_1 and v_2 of V are the eigenvectors of $A^{\mathrm{T}}A$. Hence, δ_1 and δ_2 are the square roots of the eigenvalues of both AA^{T} and $A^{\mathrm{T}}A$. The **principal axes** in triangle Y are u_1 and u_2, corresponding to axes v_1 and v_2 in T. Also,

$$Av_1 = \delta_1 u_1, \quad Av_2 = \delta_2 u_2$$

and so we can see that the principal strains have lengths $\delta_1 \geq \delta_2 > 0$ after the deformation where they were of unit length before.

The direction of the principal axes and the ratio δ_1/δ_2 are invariant under similarity transformations of the original coordinates and are suitable shape change descriptors (provided there is no reflection).

The quantity δ_1/δ_2 is the ratio of the largest stretch and the greatest compression. Taking the distance between two triangle shapes as $\log(\delta_1/\delta_2)$ we see that this shape change space for triangles is different from the shape space with the Riemannian metric. The distance $\log(\delta_1/\delta_2)$ is approximately δ/h, where δ is the Euclidean distance between the Bookstein coordinates and h is the V^B coordinate of the first (or second) triangle, see Figure 12.5.

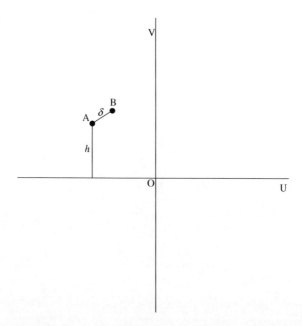

Figure 12.5   Bookstein's hyperbolic shape space considers the distance between two shapes of triangles A and B as approximately δ/h for small shape differences.

In fact Bookstein's shape space for triangles with distance $\log(\delta_1/\delta_2)$ is a space of constant negative curvature and is known in geometry as the hyperbolic space or the Poincaré half plane. Further details have been given by Small (1988), Bookstein (1991, Appendix A2), Small and Lewis (1995) and Kendall (1998). Kendall (1998) gives some diffusion results for triangles in this space. The hyperbolic space is somewhat restrictive due to its applicability only to triangles when we have 2D data. Comparisons of other configurations of more than three points could proceed by defining a triangulation of the points and working separately on each triangle (i.e. consider a union of hyperbolic spaces). Such triangulations and finite elements will be considered in Section 12.4.2. Small and Lewis (1995) describe the hyperbolic geometry for comparing the shapes of $m + 1$ points in $m \geq 2$ dimensions, with tetrahedra in $m = 3$ dimensions being the most useful generalization.

12.3 Pairs of thin-plate splines

In practice when there are more than $m + 1$ landmarks in m dimensions an affine transformation in general will not fit exactly and so a non-affine shape deformation is required. Various approaches have been suggested, although not all the desirable properties listed in Section 12.1.2 are always satisfied.

We shall mainly concentrate on the important $m = 2$ dimensional case, with deformations given by the bivariate function

$$y = \Phi(t) = (\Phi_1(t), \Phi_2(t))^{\mathrm{T}}.$$

Bookstein (1989) has developed a highly successful approach for deformations using a pair of thin-plate splines for the functions $\Phi_1(t)$ and $\Phi_2(t)$. The thin-plate spline is a sensible choice since it minimizes the bending required to take the first form into the second, and it does not suffer from problems with very large bending towards the periphery.

12.3.1 Thin-plate splines

The thin-plate spline is a natural interpolating function for data in two dimensions and plays a similar rôle in $m = 2$ dimensions to the natural cubic spline for interpolation in the 1D case. The natural cubic spline in one dimension is the unique interpolant $g(x)$ which minimizes the roughness penalty

$$\int \left| \frac{\partial^2 g}{\partial x^2} \right|^2 \mathrm{d}x,$$

subject to interpolation at the knots (landmarks). However, there are important differences in our application to shape analysis. The cubic spline need not be monotone and is not usually required to be, whereas the thin-plate spline is often desired to be bijective. A monotone cubic spline is analogous to a bijective thin-plate spline.

We shall see that the thin-plate spline is a natural interpolant in two dimensions because it minimizes the amount of bending in transforming between two configurations, which can also be considered a roughness penalty. For an introduction to natural cubic splines and thin-plate splines see Wahba (1990) and Green and Silverman (1994).

We concentrate on the important $m = 2$ dimensional case, the theory of which was developed by Duchon (1976) and Meinguet (1979). Consider the (2×1) landmarks $t_j, j = 1, \ldots, k$, on the first figure mapped exactly into $y_i, i = 1, \ldots, k$, on the second figure, that is there are $2k$ interpolation constraints,

$$(y_j)_r = \Phi_r(t_j), \quad r = 1, 2, \quad j = 1, \ldots, k, \tag{12.7}$$

and we write $\Phi(t_j) = (\Phi_1(t_j), \Phi_2(t_j))^T$, $j = 1, \ldots, k$, for the 2D deformation. Let

$$T = [\, t_1 \quad t_2 \quad \ldots \quad t_k \,]^T, \; Y = [\, y_1 \quad y_2 \quad \ldots \quad y_k \,]^T$$

so that T and Y are both $(k \times 2)$ matrices.

Definition 12.4   *A **pair of thin-plate splines** (PTPS) is given by the bivariate function*

$$\begin{aligned}
\Phi(t) &= (\Phi_1(t), \Phi_2(t))^T \\
&= c + At + W^T s(t),
\end{aligned} \tag{12.8}$$

where t is (2×1), $s(t) = (\sigma(t - t_1), \ldots, \sigma(t - t_k))^T, (k \times 1)$ and

$$\sigma(h) = \begin{cases} \|h\|^2 \log(\|h\|), & \|h\| > 0, \\ 0, & \|h\| = 0. \end{cases} \tag{12.9}$$

The $2k + 6$ parameters of the mapping are $c(2 \times 1), A(2 \times 2)$ and $W(k \times 2)$. There are $2k$ interpolation constraints in Equation (12.7), and we introduce six more constraints in order for the bending energy in Equation (12.14) to be defined:

$$1_k^T W = 0, \quad T^T W = 0. \tag{12.10}$$

The pair of thin-plate splines which satisfy the constraints of Equation (12.10) are called **natural thin-plate splines**. Equation (12.7) and Equation (12.10) can be re-written in matrix form

$$\begin{bmatrix} S & 1_k & T \\ 1_k^T & 0 & 0 \\ T^T & 0 & 0 \end{bmatrix} \begin{bmatrix} W \\ c^T \\ A^T \end{bmatrix} = \begin{bmatrix} Y \\ 0 \\ 0 \end{bmatrix}, \tag{12.11}$$

where $(S)_{ij} = \sigma(t_i - t_j)$ and 1_k is the k-vector of ones. The matrix

$$\Gamma = \begin{bmatrix} S & 1_k & T \\ 1_k^T & 0 & 0 \\ T^T & 0 & 0 \end{bmatrix}$$

is non-singular provided S in non-singular. Hence,

$$\begin{bmatrix} W \\ c^T \\ A^T \end{bmatrix} = \begin{bmatrix} S & 1_k & T \\ 1_k^T & 0 & 0 \\ T^T & 0 & 0 \end{bmatrix}^{-1} \begin{bmatrix} Y \\ 0 \\ 0 \end{bmatrix} = \Gamma^{-1} \begin{bmatrix} Y \\ 0 \\ 0 \end{bmatrix},$$

say. Writing the partition of Γ^{-1} as

$$\Gamma^{-1} = \begin{bmatrix} \Gamma^{11} & \Gamma^{12} \\ \Gamma^{21} & \Gamma^{22} \end{bmatrix},$$

where Γ^{11} is $k \times k$, it follows that

$$W = \Gamma^{11} Y$$
$$\begin{bmatrix} c^T \\ A^T \end{bmatrix} = [\hat{\beta}_1, \hat{\beta}_2] = \Gamma^{21} Y, \qquad (12.12)$$

giving the parameter values for the mapping. If S^{-1} exists, then we have:

$$\Gamma^{11} = S^{-1} - S^{-1} Q (Q^T S^{-1} Q)^{-1} Q^T S^{-1},$$
$$\Gamma^{21} = (Q^T S^{-1} Q)^{-1} Q^T S^{-1} = (\Gamma^{12})^T, \qquad (12.13)$$
$$\Gamma^{22} = -(Q^T S^{-1} Q)^{-1},$$

where $Q = [1_k, T]$, using for example Rao (1973, p. 39).

Using Equation (12.12) and Equation (12.13) we see that $\hat{\beta}_1$ and $\hat{\beta}_2$ are generalized least squares estimators, and

$$\text{cov}((\hat{\beta}_1, \hat{\beta}_2)^T) = -\Gamma^{22}.$$

Mardia *et al.* (1991) gave the expressions for the case when S is singular.

Definition 12.5   *The $k \times k$ matrix B_e is called the **bending energy matrix** where*

$$B_e = \Gamma^{11}. \qquad (12.14)$$

There are three constraints on the bending energy matrix

$$1_k^T B_e = 0, \quad T^T B_e = 0$$

and so the rank of the bending energy matrix is $k - 3$.

It can be proved that the transformation of Equation (12.8) minimizes the total bending energy of all possible interpolating functions mapping from T to Y, where the total bending energy is given by:

$$J(\Phi) = \sum_{j=1}^{2} \int \int_{\mathbb{R}^2} \left(\frac{\partial^2 \Phi_j}{\partial x^2} \right)^2 + 2 \left(\frac{\partial^2 \Phi_j}{\partial x \partial y} \right)^2 + \left(\frac{\partial^2 \Phi_j}{\partial y^2} \right)^2 dxdy. \quad (12.15)$$

A simple proof is given by Kent and Mardia (1994a). The minimized total bending energy is given by:

$$J(\Phi) = \mathrm{trace}(W^T S W) = \mathrm{trace}(Y^T \Gamma^{11} Y). \quad (12.16)$$

For the deformation grid we want to have as little bending as possible to provide a simple interpretation, both locally near data points and further away from the data. Hence the minimum bending property of the PTPS is highly suitable for many applications. The thin-plate spline has also proved popular because of the simple analytical solution. The deformation can be generalized into m dimensions when a multiple of m thin-plate splines (MTPS) will be required, and this is studied in Section 12.5.

12.3.2   Transformation grids

Following from the original ideas of Thompson (1917) we can produce similar transformation grids, using a pair of thin-plate splines for the deformation from configuration matrices T to Y. A regular square grid is drawn over the first figure and at each point where two lines on the grid meet t_i the corresponding position in the second figure is calculated using a PTPS transformation $y_i = \Phi(t_i), i = 1, \ldots, n_g$, where n_g is the number of junctions or crossing points on the grid. The junction points are joined with lines in the same order as in the first figure, to give a deformed grid over the second figure. The PTPS can be used to produce a transformation grid, say from a regular square grid on the first figure to a deformed grid on the second figure. The resulting interpolant produces transformation grids that 'bend' as little as possible. We can think of each square in the deformation as being deformed into a quadrilateral (with four shape parameters). The PTPS minimizes the local variation of these small quadrilaterals with respect to their neighbours.

Example 12.2   Consider describing the square to kite transformation which was considered by Bookstein (1989) and Mardia and Goodall (1993). Given $k = 4$ points in $m = 2$ dimensions the matrices T and Y are given by:

$$T = \begin{bmatrix} 0 & 1 \\ -1 & 0 \\ 0 & -1 \\ 1 & 0 \end{bmatrix}, \quad Y = \begin{bmatrix} 0 & 0.75 \\ -1 & 0.25 \\ 0 & -1.25 \\ 1 & 0.25 \end{bmatrix}.$$

We have here

$$S = \begin{bmatrix} 0 & a & b & a \\ a & 0 & a & b \\ b & a & 0 & a \\ a & b & a & 0 \end{bmatrix},$$

where $a = \sigma(\sqrt{2}) = 0.6931$ and $b = \sigma(2) = 2.7726$. In this case, the bending energy matrix is:

$$B_e = \Gamma^{11} = 0.1803 \begin{bmatrix} 1 & -1 & 1 & -1 \\ -1 & 1 & -1 & 1 \\ 1 & -1 & 1 & -1 \\ -1 & 1 & -1 & 1 \end{bmatrix}.$$

It is found that

$$W^{T} = \begin{bmatrix} 0 & 0 & 0 & 0 \\ -0.1803 & 0.1803 & -0.1803 & 0.1803 \end{bmatrix}, \quad c = 0, \ A = I_2 ,$$

and so the PTPS is given by $\Phi(t) = (\Phi_1(t), \Phi_2(t))^{T}$, where

$$\Phi_1(t) = t[1], \tag{12.17}$$

$$\Phi_2(t) = t[2] + 0.1803 \sum_{j=1}^{4} (-1)^j \sigma(\|t - t_j\|).$$

Note that in Equation (12.17) there is no change in the $t[1]$ direction. The affine part of the deformation is the identity transformation.

    We consider Thompson-like grids for this example, displayed in Figure 12.6. A regular square grid is placed on the first figure and deformed into the curved grid on the kite figure. We see that the top and bottom most points are moved downwards with respect to the other two points. If the regular grid is drawn on the first figure at a different orientation, then the deformed grid does appear to be different, even though the transformation is the same. This effect is seen in Figure 12.6 where both figures have been rotated clockwise by 45° in the second row.　□

Example 12.3　In Figure 12.7a we see a square grid drawn on the estimate of mean shape for the Control group in the schizophrenia study of Section 1.4.5. Here there are $n_g = 40 \times 39 = 1560$ junctions and there are $k = 13$ landmarks. In Figure 12.7b we see the schizophrenia mean shape estimate and the grid of new points obtained from the PTPS transformation. It is quite clear that there is a shape change in the centre of the brain, around landmarks 1, 6 and 13.　□

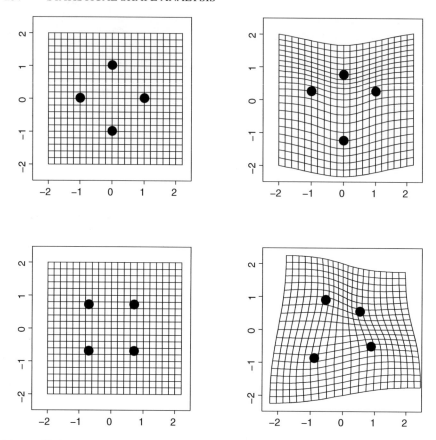

Figure 12.6   Transformation grids for the square (left column) to kite (right column). In the second row the same figures as in the first row have been rotated by 45° and the deformed grid does look different, even though the transformation is the same. Source: Adapted from Bookstein 1989.

12.3.3   Thin-plate splines in R

Transformation grids can be drawn using the `shapes` package in R with the command `tpsgrid`. So, for example we can produce the grid of Figure 12.7 using the commands:

```
data(schizophrenia)
control<-procGPA(schizophrenia$x[,,1:14])$mshape
patient<-procGPA(schizophrenia$x[,,15:28])$mshape
patient<-procOPA(control,patient)$Bhat
              #align patient mean to control
tpsgrid(control,patient,opt=2,ngrid=40,mag=3)
```

Also, we can retain the scale by considering the transformation between the size-and-shape means, by using the option `scale=FALSE` in `procGPA`. Note that the options here are `opt=2` which plots the grids on both figures (whereas `opt=1` just

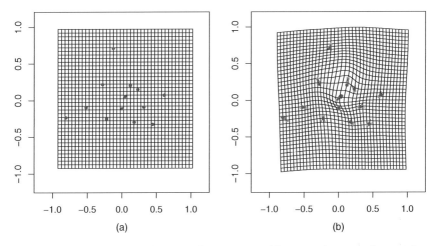

Figure 12.7   A thin-plate spline transformation grid between the control mean shape estimate and the schizophrenia mean shape estimate, from the schizophrenia study. The square grid is placed on the estimated mean control shape (a) and the curved grid is pictured on the estimated mean shape of the schizophrenic patients (b), magnified three times, with arrows drawn from the control mean (red) to patient mean (green). For a colour version of this figure, see the colour plate section.

plots the deformed grid on the second figure), `ngrid=40` gives the dimension of the grid, and `mag=3` exaggerates the deformation by a magnification factor of 3 here.

Example 12.4   Consider again the schizophrenia data from Section 1.4.5 but now the transformation from the patient mean shape to the control mean shape is displayed in Figure 12.8. It is clear that a feature of the grid is that a derivative is approximately zero at the point of major shape change. This feature is known as a **crease**, see Bookstein (2000, 2014, p. 444). Crease analysis is described in detail by Bookstein (2000) and the method involves finding a magnification, orientation and scaling for a more precise location of the crease. Creases have been hypothesized to be important stable features during evolution (Bookstein, 2002). In Figure 12.8(b) the crease is visually obvious, and a specific test for group differences at the crease is suggested (Bookstein, 2000).   □

Example 12.5   Consider describing the growth of the sooty mangabey skull from the dataset in Section 1.4.12, and in particular in relation to 12 landmarks in the midline section of the skull. Five juveniles are taken at increasing dental age (an approximate age of the animal based on the progress in growth of the teeth). A PTPS is used to deform a square grid from age stages 1 to 2, from age stages 1 to 3, from age stages 1 to 4 and from age stages 1 to 5 (Figure 12.9). The series of grids is a useful tool for describing perceived growth. In particular there is a clear extension to the lower left of the face as the age of the monkeys increase. In this case the monkeys are all different and can be assumed to be independent, within the class of sooty mangabeys. In the example the PTPS transformation is useful for describing size-and-shape change.   □

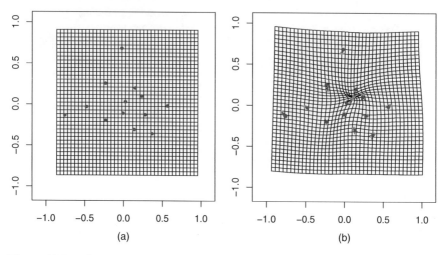

Figure 12.8   The thin-plate spline transformation grid from the patient mean (a) to the control mean (b), with arrows drawn from the patient mean (red) to control mean (green). The shape change has been magnified three times. For a colour version of this figure, see the colour plate section.

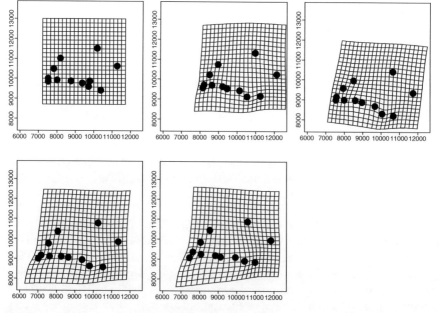

Figure 12.9   A series of grids showing the shape changes in the skull of some sooty mangabey monkeys (read across the rows from left to right): (first row) age stage 1, age stages 1 to 2, 1 to 3; (second row) age stages 1 to 4, 1 to 5.

Definition 12.6   *Transforming from one object to another is called* **warping,** *particularly when the transformation is from one image to another image.*

Further properties

The (single) thin-plate spline can be viewed as a particular case of kriging – a technique for spatial prediction developed in geostatistics; for example, see Cressie (1993) and Kent and Mardia (1994a). Discussion of this link is found in Section 12.5.

The thin-plate spline consists of an affine and a non-affine part. Only the non-affine part contributes to the bending energy and hence two configurations which are affine transformations of each other have zero total bending energy. If $k = 3$ points are available, then there is only the affine part.

The thin-plate spline is so named because it solves the mechanical problem of determining the minimum energy shape of an infinitely thin stiff metal plate under certain constraints. The PTPS is usually bijective (one to one) for practical situations where the shape change is not far from affine, but it is not guaranteed to be bijective. There is no mathematical restriction to prevent folding, and if the shape change is highly non-affine, then folding may occur.

A further desirable property is that the PTPS is equivariant under location, rotation and scale (but not a general affine deformation). In addition, as t becomes farther away from the first object the thin-plate spline deformation becomes approximately affine.

Various modifications could be made to the deformation. In particular, one may wish to consider the domain to be finite, D say, rather than $\mathbb{R}^2$. In this case the thin-plate spline can be modified and details can be found in Stone (1988) and Green and Silverman (1994). The finite domain interpolant has important differences from the infinite domain thin-plate spline. In particular, there are usually differences around the edges, with reduced variability in these regions.

Also, instead of second derivatives in the energy integral in Equation (12.15) one could consider rth ($r > 2$) derivatives, leading to a smoother interpolant (see Section 12.5.3). Such a spline will have $2r - 3$ continuous derivatives everywhere but its $2r - 2$th derivative has a singularity at each data point. Details of higher order splines can be found in Duchon (1976); Meinguet (1979); Green and Silverman (1994); Kent and Mardia (1994a); and Mardia *et al.* (1996e) for example. For further details of thin-plate splines refer to Wahba (1990); and Green and Silverman (1994); Kent and Mardia (1994a) and the references therein.

Smoothing splines

Rather than interpolating landmarks exactly one could consider a pair of smoothing thin-plate splines. The pair of smoothing thin-plate splines consists of the bivariate function $\Phi(t)$ which minimizes

$$\sum_{j=1}^{k} \operatorname{trace}\{[y_j - \Phi(t_j)]^{\mathrm{T}}[y_j - \Phi(t_j)]\} + \lambda J(\Phi),$$

where $\lambda > 0$ is a smoothing parameter, and $J(\Phi)$ is defined in Equation (12.15). The term $\lambda J(\Phi)$ can be thought of as a roughness penalty. Using Duchon (1976) (for a single thin-plate spline) we see that the solution has exactly the same form as in Equation (12.8) except that the unknown coefficients are obtained from the matrix equation:

$$
\begin{bmatrix} S + \lambda I_k & 1_k & T \\ 1_k^T & 0 & 0 \\ T^T & 0 & 0 \end{bmatrix} \begin{bmatrix} W \\ c^T \\ A^T \end{bmatrix} = \begin{bmatrix} Y \\ 0 \\ 0 \end{bmatrix}.
\tag{12.18}
$$

Hence, dealing with smoothing thin-plate splines involves an almost identical algorithm as that required for interpolating splines, except that S is replaced by $S + \lambda I_k$. The smoothing parameter controls how smooth the spline is and the value of λ must be pre-specified. If λ is very large, then the resulting spline resembles an affine function (which is completely smooth) and if λ is very small the smoothing spline is close to the interpolating spline. The smoothing spline is not equivariant under scaling of the configurations.

An alternative to pre-specifying λ is to use cross-validation to estimate λ. Consider the 2D observation y_j corresponding to t_j as a new observation by omitting it from the dataset. Denote the smoothing thin-plate spline from the rest of the data points as $\Phi^{-j}(t_j, \lambda)$. We can judge how well the smoother predicts the deleted observation by examining the magnitude of

$$
\|y_j - \Phi^{-j}(t_j, \lambda)\|^2.
$$

The cross-validation score is an overall measure of the prediction squared error

$$
CV(\lambda) = \frac{1}{k} \sum_{j=1}^{k} \|y_j - \Phi^{-j}(t_j, \lambda)\|^2
$$

and so an automatic choice for λ can be obtained by minimizing the cross-validation score $CV(\lambda)$ over λ. The minimization is often not straightforward, and a simple grid search is often the best approach (Green and Silverman, 1994, p. 30).

12.3.4   Principal and partial warp decompositions

Bookstein (1989, 1991) introduced principal and partial warps to help explain the components of a thin-plate spline deformation. The principal and partial warps are useful for decomposing the thin-plate spline transformations into a series of large scale and small scale components.

Definition 12.7   *Consider the PTPS transformation from $t \in \mathbb{R}^2$ to $y \in \mathbb{R}^2$, which interpolates the k points T to Y ($k \times 2$) matrices. An eigen-decomposition of the $k \times k$ bending energy matrix B_e of Equation (12.14) has non-zero eigenvalues $\lambda_1 \leq$*

$\lambda_2 \leq \ldots \leq \lambda_{k-3}$ *with corresponding eigenvectors* $\gamma_1, \gamma_2, \ldots, \gamma_{k-3}$. *The eigenvectors* $\gamma_1, \gamma_2, \ldots, \gamma_{k-3}$ *are called the* **principal warp eigenvectors** *and the eigenvalues are called the* **bending energies**. *The functions,*

$$P_j(t) = \gamma_j^{\mathrm{T}} s(t) \ , \ j = 1, \ldots, k-3, \qquad (12.19)$$

are the **principal warps**, *where* $s(t) = (\sigma(t - t_1), \ldots, \sigma(t - t_k))^{\mathrm{T}}$.

Here we have labelled the eigenvalues and eigenvectors in this order (with λ_1 as the smallest eigenvalue corresponding to the first principal warp) to follow Bookstein's labelling of the order of the warps (Bookstein, 1996b). The principal warps do not depend on the second figure Y. The principal warps will be used to construct an orthogonal basis for re-expressing the thin-plate spline transformations. The principal warp deformations are univariate functions of 2D t, and so could be displayed as surfaces above the plane or as contour maps. Alternatively one could plot the transformation grids from t to $y = t + (c_1 P_j(t), c_2 P_j(t))^{\mathrm{T}}$ for each j, for particular values of c_1 and c_2. Note that the principal warp eigenvectors are orthonormal.

Definition 12.8   *The* **partial warps** *are defined as the set of* $k - 3$ *bivariate functions* $R_j(t), j = 1, \ldots, k - 3$, *where*

$$R_j(t) = Y^{\mathrm{T}} \lambda_j \gamma_j P_j(t) = Y^{\mathrm{T}} \lambda_j \gamma_j \gamma_j^{\mathrm{T}} s(t). \qquad (12.20)$$

The jth **partial warp scores** *for Y (from T) are defined as:*

$$(p_{j1}, p_{j2})^{\mathrm{T}} = Y^{\mathrm{T}} \gamma_j \ , \ j = 1, \ldots, k - 3, \qquad (12.21)$$

and so there are two scores for each partial warp.

Bookstein (2015a) also considers deflated partial warp scores, each multiplied by the square root of λ_1 / λ_j for 2D and multiplied by λ_1 / λ_j for 3D. Since

$$W^{\mathrm{T}} s(t) = \sum_{j=1}^{k-3} R_j(t),$$

we see that the non-affine part of the PTPS transformation can be decomposed into the sum of the partial warps. The jth partial warp corresponds largely to the movement of the landmarks which are the most highly weighted in the jth principal warp. The jth partial warp scores indicate the contribution of the jth principal warp to the deformation from the source T to the target Y, in each of the Cartesian axes.

The partial warps are bivariate functions and can be viewed in a plot as deformations of a square grid placed over T. In particular if t is a point in the plane of the original configuration, then we obtain the grid for the jth partial warp by deforming to $y = t + R_j(t)$. Hence, interpretation of the partial warps can lead to an increased understanding of the deformation. It is also often helpful to display the affine transformation as a deformed grid, giving a complete decomposition of the thin-plate spline deformation.

Let the figure T be scaled to unit size, centred with coordinates $(\mu_1, v_1), \ldots, (\mu_k, v_k)$, and rotated so that $\sum_{j=1}^{k} \mu_j v_j = 0$. Let $\alpha = \sum \mu_j^2, \beta = \sum v_j^2$ with $\alpha + \beta = 1$, and denote

$$u_1 = (\alpha v_1, \alpha v_2, \ldots, \alpha v_k, \beta \mu_1, \beta \mu_2, \ldots, \beta \mu_k)^{\mathrm{T}} / (\alpha \beta)^{1/2}, \qquad (12.22)$$

$$u_2 = (-\beta \mu_1, -\beta \mu_2, \ldots, -\beta \mu_k, \alpha v_1, \alpha v_2, \ldots, \alpha v_k)^{\mathrm{T}} / (\alpha \beta)^{1/2}. \qquad (12.23)$$

Note that u_1 and u_2 are orthogonal to the block diagonal matrix $\mathrm{diag}(B_e, B_e)$ (and to T). The affine component is often called the zeroth-order warp. The affine scores (zeroth partial warp scores) are given by:

$$(a_1, a_2)^{\mathrm{T}} = (u_1, u_2)^{\mathrm{T}} \mathrm{vec}(Y).$$

Note that

$$\left\{ \left(\gamma_i^{\mathrm{T}}, 0_k^{\mathrm{T}}\right)^{\mathrm{T}}, \left(0_k^{\mathrm{T}}, \gamma_j^{\mathrm{T}}\right)^{\mathrm{T}}, : i, j = 1, \ldots, k-3 \right\}$$

and u_1 and u_2 together form an orthonormal basis of tangent space.

The partial warps represent non-affine aspects of the deformation ordered according to the amount of bending that is required to move points in the first figure. In practice this results in the partial warps representing the deformation at various scales – the larger values of the eigenvalues (i.e. larger j) represent small scale local deformation and the smaller eigenvalues (smaller j) represent large scale global deformation. Hence, the first principal and partial warps will be associated with an overall large scale bending of the figure and the last principal and partial warps will usually be associated with the landmarks that are closest together, at the smallest scale.

An analogy is with a Fourier series representation of a function. An orthonormal trigonometric basis is chosen which represents features at a variety of successively finer scales, for example $\cos x, \sin x, \cos 2x, \sin 2x, \cos 3x, \sin 3x, \ldots$.

We take the inverse bending energy matrix as the Moore–Penrose inverse, given in Definition 9.1:

$$B_e^- = \sum_{j=1}^{k-3} \lambda_j^{-1} \gamma_j \gamma_j^{\mathrm{T}}. \qquad (12.24)$$

A summary of the different types of warps is given in Table 12.1. The principal warps are detemined by one configuration only, the partial warps from a transformation between two configurations, and the relative warps from a sample of n configurations (described in Section 12.3.6).

Table 12.1   Types of warps and notation.

Type	Notation	Subscripts	Equation
Principal warp	$P_j(t)$	$j = 1, \ldots, k-3$	(12.19)
Partial warp	$R_j(t)$	$j = 1, \ldots, k-3$	(12.20)
Partial warp score	(p_{j1}, p_{j2})	$j = 1, \ldots, k-3$	(12.21)
Relative warp	f_j	$j = 1, \ldots, 2k-6$	(12.25)
Relative warp score	a_{ij}	$i = 1, \ldots, n; j = 1, \ldots, 2k-6$	(12.25)

Example 12.6   Bookstein (1991, Section 7.5.2) considers a non-symmetric T configuration ($k = 5$ points) deformed into a second figure (Figure 12.10a,b). It is found that the two eigenvectors of the bending energy matrix are:

$$\gamma_1^T = (-0.494, -0.242, -0.337, 0.470, 0.603),$$
$$\gamma_2^T = (0.215, -0.327, 0.135, -0.655, 0.632),$$

with corresponding eigenvalues $\lambda_1 = 0.296$ and $\lambda_2 = 0.568$. Note that eigenvalues of Bookstein (1991) differ by a factor of 2 because he uses $\sigma(\|h\|) = \|h\|^2 \log(\|h\|^2) = 2\|h\|^2 \log(\|h\|)$ for the covariance function. We can inspect the loadings of each eigenvector to see which coefficients are large, and hence interpret the effect. Alternatively we can plot the principal warps as surfaces and in Figure 12.10c,d we see the two principal warps. The first principal warp corresponds to a relatively large feature – the difference in landmarks 4 and 5 relative to 1, 2 and 3. The second (and last) principal warp reflects the relatively small feature differences in the displacements of the two nearest points, landmarks 4 and 5. In Figure 12.11 we see the deformation grids for the partial warps and the affine part. The first (larger scale change) involves landmarks 4 and 5 moving down relative to points 1, 2 and 3. The second partial warp is largely composed of the relative movement of landmarks 4 and 5 together (smaller scale change). Also in Figure 12.11 we see the affine component, which is a shear of the left region upwards and the right region downwards.   □

Example 12.7   Consider describing the shape change from an estimate of the mean size-and-shape female gorilla skull midline to an estimate of the mean size-and-shape of the male gorilla skull midline, for the data described in Section 1.4.8. There are $k = 8$ landmarks in the plane. The thin-plate spline deformation is given in Figure 12.12. Although interpretation is far from straightforward, a possible explanation is that the main size-and-shape difference is a greater protrusion in the male face compared with the female face – there is less size-and-shape difference at the back of the skull.

There are five principal warps from the pooled Procrustes mean, shown as surfaces in Figure 12.13. It is clear that the first few principal warps are smoother and spread over a larger scale than the less smooth, smaller scale, last few principal warps. The principal warps are also displayed as deformation grids in Figure 12.14. We can

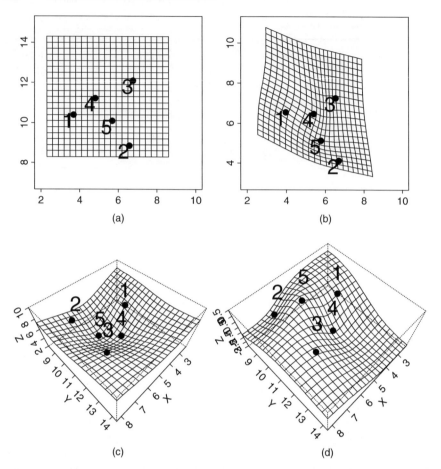

Figure 12.10   The pair of thin-plate splines deformation from a non-symmetric T configuration of k = 5 landmarks (a) deformed into a second figure (b), and the principal warps for the non-symmetric T example drawn as surfaces above the Cartesian plane: (c) the first principal warp; and (d) the second principal warp. Source: Adapted from Bookstein 1989.

see the larger scale deformation of the first principal warp which includes the face (left half) and braincase (right half) moving downwards relative to the middle of the skull. The face is also stretched outwards. The second principal warp includes the relative shrinking in height of the braincase, and the bottom left three landmarks moving away from the rest of the skull. The principal warps are concerned with successively smaller scale deformations, and at the smaller scales the fourth principal warp involves the movement in *ba* and *o* and the final principal warp includes a contrast between leftmost landmarks *pr* and *na* and the rest of the skull. The landmark labels were given in Figure 1.18. If using either of the individual male or female group means rather than the pooled mean, then the plots appear virtually identical to the eye.

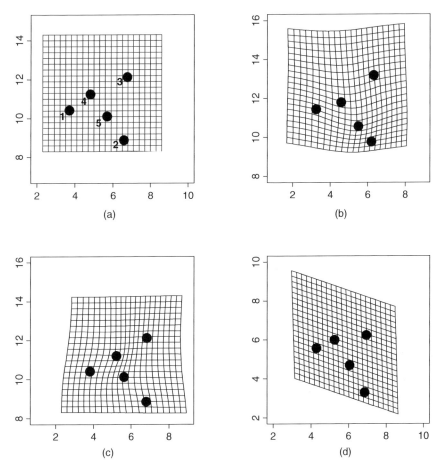

Figure 12.11   (a) The initial grid on the non-symmetric T, (b) the first partial warp, (c) the second partial warp, and (d) the affine component. Source: Adapted from Bookstein 1989.

The five partial warp deformation grids for the female mean shape to the male mean shape are shown in Figure 12.15 together with the affine grid (all magnified three times). We see that the face region is pulled up and outwards in partial warp 1, and the face is taller in height in partial warp 2. There is very little contribution in partial warp 3. In partial warp 4 landmarks *ba* and *o* become closer together and in partial warp 5 landmarks *pr* and *na* become closer together.

The affine and partial warp scores for the gorilla skull data are displayed in Figure 12.16. For each partial warp there are two scores (one for each Cartesian axis *x*, *y*). For each warp we plot the pairs of scores for the *y*-axis against the *x*-axis, labelling the males and females with 'm' and 'f', respectively. We can see that there are fairly good separations of the males and females in the affine scores and the partial warp scores, except for partial warp 3. The percentage of pooled variability explained by the affine component is 26.1%, and for the partial warps 1 to 5 the percentages explained

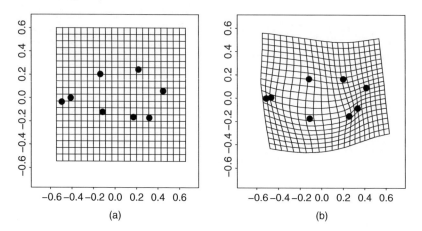

Figure 12.12   A thin-plate spline transformation grid between a female and a male gorilla skull midline. (a) The square grid is placed on the estimated mean female skull and (b) the curved grid is pictured on the estimated mean male skull. The deformation is magnified three times here, for ease of interpretation.

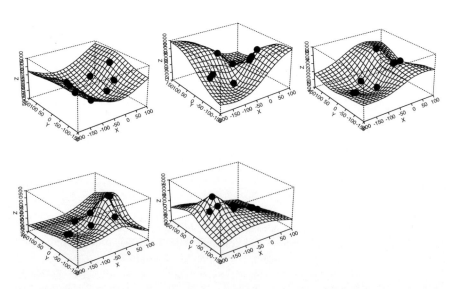

Figure 12.13   The five principal warps for the the pooled mean shape of the gorillas. The top row shows the first three principal warps (first on the left to third on the right) and the bottom row shows the last two principal warps (fourth on the left, fifth on the right).

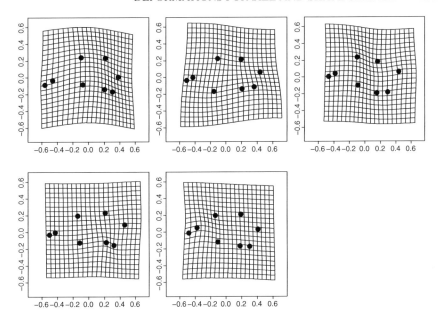

Figure 12.14   The five principal warps for the pooled mean shape of the gorillas. The top row shows the first three principal warps (first on the left to third on the right) and the bottom row shows the last two principal warps (fourth on the left, fifth on the right). The principal warps are pictured by deforming the full Procrustes mean shape for the pooled dataset of all 59 gorillas to figures which have a partial warp score of 0.15 along each warp.

are 32.7, 17.8, 12.3, 6.5 and 4.4%, respectively. Note that the variance of the partial warps drops with warp number as in Bookstein (2015a). The shape variability is not spread evenly over the warps. So, the variability in the scores is not spherical – there is much more shape variability in the first few partial warp scores and the smaller scale warps summarize only a small fraction of the shape variability.   □

Note that the principal warps and partial warp scores can be obtained in R using procGPA with the option alpha=1. In particular, the output is in $principalwarps, $partialwarpscores and the percentage contributions are in $partialwarp.percent. For example, for the pooled gorilla data:

```
ans<-procGPA( abind(gorf.dat,gorm.dat), alpha=1)
> ans$partialwarps.percent
[1] 32.719875 17.787477 12.304224 6.506736 4.443423
```

The affine components, scores and percentage contributions are obtained using option affine=TRUE and are given in $pcar, $scores, $percent respectively. For the pooled gorilla data we have:

```
ans<-procGPA( abind(gorf.dat,gorm.dat), alpha=1, affine=TRUE)
> ans$percent
[1] 17.529125 8.565122
```

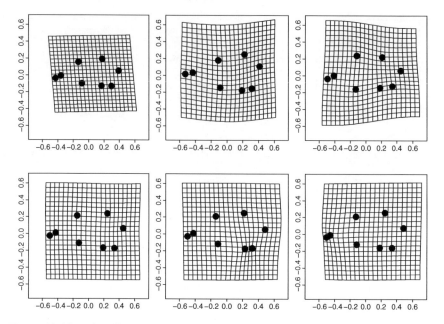

Figure 12.15   The affine component and the partial warps for the deformation from the female to the male gorilla (all magnified three times). The top row shows the affine component and the first two partial warps (affine on the left, first in the middle and second on the right) and the bottom row shows the last three partial warps (third on the left, fourth in the middle and fifth on the right).

12.3.5   PCA with non-Euclidean metrics

Principal component analysis of a random sample of shapes can be carried out with respect to the bending energy matrix or its generalized inverse, emphasizing large or small scale variations.

Definition 12.9   *Define the* **pseudo-metric space** *($\mathbb{R}^p, d_A$) as the real p-vectors with* **pseudo-metric** *given by:*

$$d_A(x_1, x_2) = \sqrt{(x_1 - x_2)^{\mathrm{T}} A^-(x_1 - x_2)},$$

where x_1 and x_2 are p-vectors, and A^- is a generalized inverse (or inverse if it exists) of the positive semi-definite matrix A.

The Moore–Penrose inverse is a suitable choice of generalized inverse (see Definition 9.1). If A is the population covariance matrix of x_1 and x_2, then $d_A(x_1, x_2)$ is the Mahalanobis distance. The norm of a vector x in the metric space is

$$\|x\|_A = d_A(x, 0) = (x^{\mathrm{T}} A^- x)^{1/2}.$$

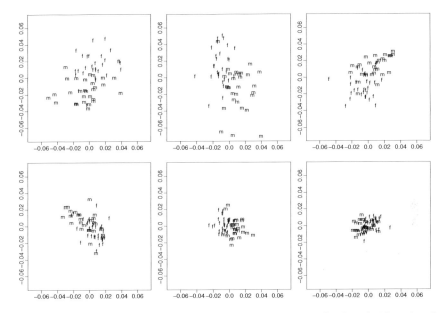

Figure 12.16   The affine scores and the partial warp scores for female (f) and male (m) gorilla skulls. The top row shows the affine scores and the first two partial warp scores (affine on the left, first in the middle and second on the right) and the bottom row shows the last three partial warp scores (third on the left, fourth in the middle and fifth on the right).

We could carry out statistical inference in the metric space rather than in the usual Euclidean ($A = I$) space (Mardia 1977, 1995; Bookstein 1995, 1996b; Kent, 1995, personal communication). A simple way to proceed is to transform from $x \in (\mathbb{R}^p, d_A)$ to $y \in (A^-)^{1/2}x$ in Euclidean space. For example, consider PCA of n centred p-vectors $x_1, \ldots, x_n$ in the metric space. Transforming to $y_i = (A^-)^{1/2}x_i$ the PC loadings are the eigenvectors of

$$S_y = \frac{1}{n} \sum_{i=1}^{n} y_i y_i^{\mathrm{T}}.$$

Denote the eigenvectors (p-vectors) of S_y as $\gamma_{yj}, j = 1, \ldots, p$ (assuming $p \leq n - 1$), with corresponding eigenvalues $\lambda_{yj}, j = 1, \ldots, p$. The **PC scores** for the jth PC on the ith individual are:

$$r_{ij} = \gamma_{yj}^{\mathrm{T}} y_i = \left(\gamma_{yj}^{\mathrm{T}} (A^-)^{1/2} \right) x_i, \quad i = 1, \ldots, p.$$

So, the (unnormalized) PC loadings on the original data are $(A^-)^{1/2}\gamma_{yj}$ which are the eigenvectors of $(A^-)^{1/2}S_x(A^-)^{1/2}$, where $S_x = \frac{1}{n} \sum_{i=1}^{n} x_i x_i^{\mathrm{T}}$, using standard linear algebra, see Mardia *et al.* (1979, Appendix). The first few PCs in the metric space (with loadings given by the eigenvectors of $(A^-)^{1/2}S_x(A^-)^{1/2}$) can be useful for interpretation, emphasizing a different aspect of the sample variability than the usual PCA in Euclidean space. If our analysis is carried out in the pseudo-metric space, then we say the our analysis has been carried out *with respect to A*.

For example, in canonical variate analysis in multivariate analysis observations from several groups are available. An eigendecomposition of the between group covariance matrix $S_{between}$ is carried out with respect to S_{within}, the within group covariance matrix (e.g. see Mardia *et al.* 1979, p. 338). Another form of weighted PCA was carried out by Polly *et al.* (2013) who consider PCA weighted by phylogenetic distance.

Potentially very useful metrics for shape analysis are the bending energy and inverse bending energy metrics in the tangent space. All metrics here depend on quadratic forms, and the procedures use Mahalanobis distances. A very different approach could be obtained by using, say, the L_1 norm

$$\|x\|_1 = \sum_{i=1}^{p} |x_i|,$$

instead of the Euclidean L_2 norm

$$\|x\| = \|x\|_2 = \sqrt{\sum_{i=1}^{p} x_i^2}.$$

12.3.6   Relative warps

If a random sample of shapes is available, then one may wish to examine the structure of the within group variability in the tangent space to shape space. This problem was approached in Section 7.7 using PCA with respect to the Euclidean metric, but an alternative is the method of relative warps. Relative warps are PCs with respect to the bending energy or inverse bending energy metrics in the shape tangent space. Bookstein (2015a) also considered deflated relative instrinsic warps which involves deflating partial warp scores.

Consider a random sample of n shapes represented by Procrustes tangent coordinates $v_1, \ldots, v_n$ (each is a $2k - 2$-vector), where the pole μ is chosen to be an average pre-shape such as from the full Procrustes mean, see Section 8.3. The sample covariance matrix in the tangent plane is denoted by S_v and the sample covariance matrix of the centred tangent coordinates $x_i = (I_2 \otimes H^T)v_i$, $i = 1, \ldots, n$ is denoted by S_c ($2k \times 2k$). In our examples we have used the covariance matrix of the Procrustes fit coordinates of Section 7.7.4. The bending energy matrix is calculated for the average shape B_e and then the tensor product is taken to give $B_2 = I_2 \otimes B_e$, which is a $2k \times 2k$ matrix of rank $2k - 6$. We write B_2^- for a generalized inverse of B_2 (e.g. the Moore–Penrose generalized inverse of Definition 9.1).

We consider PCA in the tangent space with respect to a power of the bending energy matrix, in particular with respect to B_e^α.

Definition 12.10 *Let the non-zero eigenvalues of $(B_2^-)^{\alpha/2}S_c(B_2^-)^{\alpha/2}$ be $l_1, \dots, l_{2k-6}$ with corresponding eigenvectors $f_1, \dots, f_{2k-6}$ and*

$$(B_2^-)^{\alpha/2} = \sum_{r=1}^{2k-6} \lambda_r^{-\alpha/2} \gamma_r \gamma_r^{\mathrm{T}},$$

with $\lambda_1, \dots, \lambda_{2k-6}$ the eigenvalues of B_2 with corresponding eigenvectors $\gamma_1, \dots, \gamma_{2k-6}$. The eigenvectors $f_1, \dots, f_{2k-6}$ are called the **relative warps**. *The* **relative warp scores** *are:*

$$a_{ij} = (f_j)^{\mathrm{T}}(B_2^-)^{\alpha/2}x_i, \quad j = 1, \dots, 2k - 6, \quad i = 1, \dots, n. \tag{12.25}$$

Important point: The relative warps and the relative warp scores are useful tools for describing the non-affine shape variation in a dataset. In particular the effect of the jth relative warp can be viewed by plotting

$$H^{\mathrm{T}}\mu \pm cB_2^{\alpha/2} f_j l_j^{1/2},$$

for various values of c, where

$$B_2^{\alpha/2} = \sum_{r=1}^{2k-6} \lambda_r^{\alpha/2} \gamma_r \gamma_r^{\mathrm{T}}.$$

The procedure for PCA with respect to the bending energy requires $\alpha = +1$ and emphasizes **large scale variability**. Principal component analysis with respect to the inverse bending energy requires $\alpha = -1$ and emphasizes **small scale variability**. If $\alpha = 0$, then we take $B_2^0 = I_{2k}$ as the $2k \times 2k$ identity matrix and the procedure is exactly the same as PCA of the Procrustes tangent coordinates; see Section 7.7. Bookstein (1996b) has called the $\alpha = 0$ case PCA with respect to the Procrustes metric. Bookstein (2014) argues that isotropic covariance structure is unrealistic for most applications, and that covariance matrices based on bending energy are often preferable.

Affine variation can also be considered. Let the Procrustes average shape be scaled to unit size and rotated with the principal axes horizontal and vertical, and let u_1 and u_2 be constructed from the mean shape as in Equation (12.22) and Equation (12.23). The vectors u_1 and u_2 are orthogonal to the bending energy matrix. Write $U = (u_1, u_2)$ for the $2k \times 2$ matrix. In order to view the affine contribution to variability we plot

$$H^{\mathrm{T}}\mu \pm cU, \quad i = 1, 2,$$

for various values of c.

Example 12.8 We consider a relative warp analysis of the male and female gorilla data, continuing from Example 12.7. We saw that the affine component explained

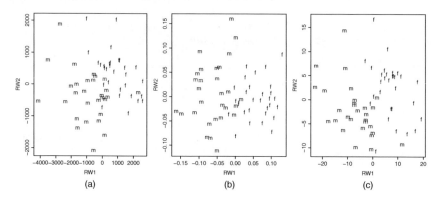

Figure 12.17   A plot of the first two relative warp scores with respect to (a) bending energy (α = 1), (b) inverse bending energy (α = − 1) and (c) Procrustes metric (α = 0) for the female (f) and male (m) gorilla skulls.

26.1% of the total pooled variability. We calculate B_e using the overall full Procrustes mean from the pooled data. Carrying out a relative warp analysis with respect to B_e (α = 1) we have the ratio of eigenvalues of the first two relative warps to the sum of the eigenvalues for all the relative warps as 63.4 and 19.6%, respectively. A plot of the first two relative warp scores is given in Figure 12.17a, and there is good separation between the males and females. The effect of the first two relative warps is demonstrated in Figure 12.18 by placing a grid on the pooled mean shape and deforming to a shape along each of the relative warps. We see that the relative warps match quite

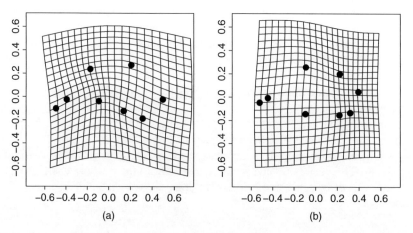

Figure 12.18   Deformation grids for the first two relative warps from the gorilla data with respect to bending energy (α = 1), emphasizing large scale variation. A square grid is drawn onto the pooled mean shape (not shown) and then deformed (by PTPS) to a shape at +6 standard deviations along the first relative warp (a) and −6 standard deviations along the second relative warp (b).

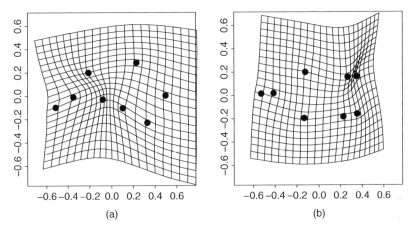

Figure 12.19   Deformation grids for the first two relative warps with respect to the inverse bending energy (α = − 1), emphasizing the small scale variation. A square grid is drawn onto the pooled mean shape (not shown) and then deformed (by PTPS) to shapes at 6 standard deviations along the relative warps [first in (a), second in (b)].

closely to the first two principal warps (seen in Figure 12.13), emphasizing the large scale variability in the data.

Carrying out a relative warp analysis with respect to B_e^- ($α = -1$) we have the ratio of eigenvalues of the first two relative warps to the sum of the eigenvalues for all the relative warps as 34.8, and 17.4%, respectively. A plot of the first two relative warp scores is given in Figure 12.17b, and again there is good separation between the males and females. The effect of the first two relative warps is demonstrated in Figure 12.19 and we see again that the first relative warp matches quite closely to the first principal warp, although some small scale movement of landmarks *na* and *pr* and movement of landmarks *ba* and *o* is emphasized. The second relative warp has some similarities with the second principal warp although there is a different emphasis in the braincase region (*l* is pushed farther leftwards).

Finally we carry out a relative warp analysis with respect to the identity matrix ($α = 0$), which is the same as PCA of the full Procrustes tangent coordinates of the pooled data. We have the percentages of shape variability explained by the first two relative warps (PCs) as 45.9 and 16.9%, respectively. A plot of the first two relative warp scores is given in Figure 12.17c, and again there is good separation between the males and females. The effect of the first two relative warps is demonstrated in Figure 12.20 and we see again that the relative warps match quite closely to the first and second principal warps.   □

In R we can carry out relative warp analysis using the command `procGPA` with option `alpha=1` for relative warp scores with respect to bending energy, with option

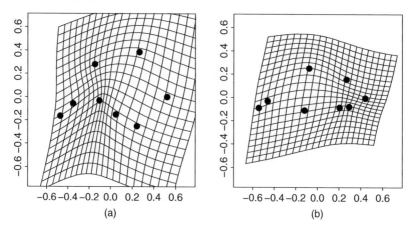

Figure 12.20 *Deformation grids for the first two relative warp scores with respect to the identity matrix ($\alpha = 0$), for the gorilla data. A square grid is drawn onto the pooled mean shape (not shown) and then deformed (by PTPS) to a shape at +6 standard deviations along the first relative warp (a) and −6 standard deviations along the second relative warp (b).*

alpha=-1 with respect to inverse bending energy and with option alpha=0 for the Procrustes tangent space PCA:

```
s1<-procGPA( apes$x[,,1:59] , alpha=1)$scores
s2<-procGPA( apes$x[,,1:59] , alpha=-1)$scores
s3<-procGPA( apes$x[,,1:59] , alpha=0)$scores
par(mfrow=c(1,3))
#relative warps with respect to Be
plot(s1[,1],s1[,2],type="n",xlab="RW1",ylab="RW2")
text(s1[,1],s1[,2],c(rep("f",times=30),rep("m",times=29)))
title(sub="(a)")
#relative warps with respect to inverse Be
plot(s2[,1],s2[,2],type="n",xlab="RW1",ylab="RW2")
text(s2[,1],s2[,2],c(rep("f",times=30),rep("m",times=29)))
title(sub="(b)")
#relative warps with respect to inverse Be
plot(-s3[,1],-s3[,2],type="n",xlab="RW1",ylab="RW2")
text(-s3[,1],-s3[,2],c(rep("f",times=30),rep("m",times=29)))
title(sub="(c)")
```

Important point: We may ask: which of the choices of α leads to the most informative analysis? The answer depends on the particular application. If large scale effects are the most important, then $\alpha = 1$ would be a good choice. If small scale effects are the most important, then $\alpha = -1$ is a good choice. Otherwise, $\alpha = 0$ would be an appropriate choice.

In the example of the gorilla data, large scale variation seems more important so it could be argued that $\alpha = 1$ and $\alpha = 0$ are more useful than $\alpha = -1$.

12.4   Alternative approaches and history

12.4.1   Early transformation grids

Similar grids to the Cartesian transformation grids of D'Arcy Thompson were popular with Renaissance artists in the 16th century. For example, grids were drawn on a figure such as a skull and then deformed into a new skull by painting corresponding parts in the appropriate part of the deformed grid. A famous example is in the painting 'The Ambassadors' by Hans Holbein the Younger (1533, National Gallery, London). Such a deformation is known as 'anamorphosis', and in 'The Ambassadors' a distorted skull lies diagonally across the bottom of the picture. Other examples included the drawing of grids on different objects to compare them and particularly good examples were given by Dürer (1528) (Figure 12.21) for comparing human form (Bookstein, 1996b).

Medawar (1944) approached the problem mathematically after D'Arcy Thompson's original ideas. Using the motivation of modelling the shape of humans, Medawar (1944) modelled the dataset known as 'Stratz's plane figures' which have been commented on by Jackson (1915) (Figure 12.22). The data are the frontal elevations of the male human being, at six stages through life. Medawar used quadratic functions for the deformation function $\Phi_2(t)$ which only depends on height, and coefficients dependent on the age of the individual. The deformation function $\Phi_1(t)$ is not modelled explicitly. Here $t = (t[1], t[2])^{\mathrm{T}}$ is again a point on the first figure.

Medawar (1945) also considered explicit analytic modelling of both $\Phi_1(t)$ and $\Phi_2(t)$ in transforming an outline of a distorted spleen culture.

Sneath (1967) fitted cubic functions for $\Phi_1(t)$ and $\Phi_2(t)$ after initially centring, rotating and rescaling the two figures by a procedure which is identical to ordinary Procrustes analysis (see Section 7.2). One of the problems with this technique is that it allows for considerable bending away from the centre of the forms and results in gross distortions of the grid towards the periphery. See Figure 12.23 for an example, where undesirable folding of the grid can be seen.

Cheverud and Richtsmeier (1986) also obtain transformation grids but use the finite element scaling method. Their work is in three dimensions but includes sections of grids in two dimensions. Again the bending of the grids is very large in localized regions, see Figure 12.24 for an example.

12.4.2   Finite element analysis

If two figures are available (e.g. the average shape in each of two groups), then another possible method of shape change description is to consider a finite element description of the size and shape difference. Each figure is represented by a collection of elements or building blocks, for example triangles and quadrilaterals in two

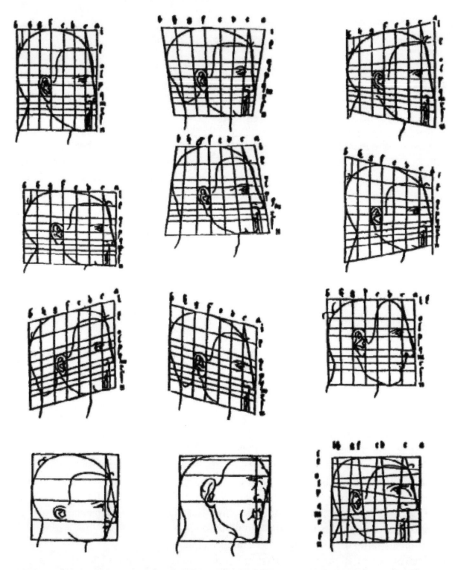

Figure 12.21   Early transformation grids of human profiles. Source: Bookstein 1996b. Reproduced with permission from Springer Science+Business Media.

dimensions, or tetrahedrons and cuboids in three dimensions. A separate transformation is considered in each element and the approach usually leads to interpolants which are continuous but not smooth across element boundaries. The essence of the finite element approach is that a shape difference can be described in terms of the directions and magnitudes of the principal strains (see Section 12.2.4 for triangles in two dimensions) in the transformation of elements in one form to another.

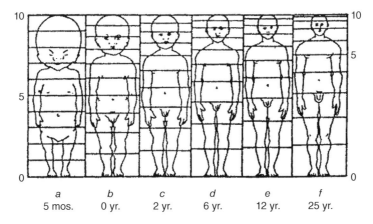

Figure 12.22   Early transformation grids modelling six stages through life. Source: Medawar 1944.

Homogeneous finite element methods (e.g. Bookstein, 1978; Moss *et al.* 1987) involve the assumption that shape differences (strains) are uniformly distributed throughout each element. Elements may contain different numbers of landmarks but the simplest possible are triangles in 2D figures with apices as homologous landmarks between two forms. In this case shape differences are necessarily affine between two triangles, as described in Section 12.2.4. Other types of finite element methods for shape analysis use non-homogeneous finite elements. Note that with all finite element analyses interpretation of shape differences will depend on element design (Zienkiewicz, 1971). Finite element scaling analysis (e.g. Lew and Lewis, 1977) is non-homogeneous and has been applied in studies of craniofacial growth and sexual dimorphism in three dimensions (Cheverud *et al.* 1983; Richstmeier, 1986; Cheverud and Richtsmeier, 1986; Richstmeier, 1989). An alternative is to use the design of elements that arise from a Delaunay triangulation (O'Higgins and Dryden, 1993; Kent and Mardia, 1994b; Okabe *et al.* 2000).

Recent work has provided a bridge between geometric morphometrics and finite element analysis, via the use of large scale relative warps (Bookstein, 2013a).

Example 12.9   Consider the size-and-shape difference in the midline between female and male gorilla crania, described in Section 1.4.8. First of all the average female and the average male are obtained. We saw in Section 9.4 that the mean shape difference is statistically significant.

In Figure 12.25 we see the average female drawn with finite elements indicating how the triangles change size and shape into the average male form. The crosses on the diagram are the principal strains (see Section 12.2.4) indicating how the unit length axes of a circle change into the major and minor axes of the deformed circle (ellipse).

There is not much shape difference in the back of the skull (b, ba, l) with smaller sum of strains here. However, there is more of a change in the face region

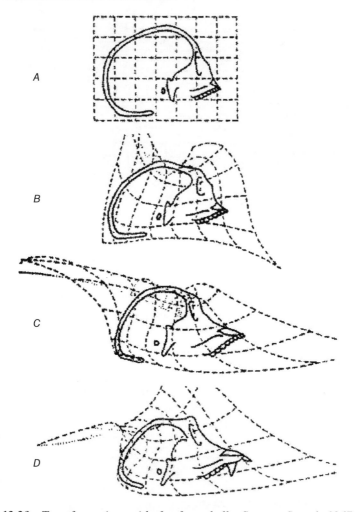

Figure 12.23   Transformation grids for four skulls. Source: Sneath 1967. Repro-
duced with permission from John Wiley & Sons.

(pr, na, st, n, ba) where the front of the face is more protruded in the male than in the
female. The PTPS deformation for the same example is given in Figure 12.12.   □

12.4.3   Biorthogonal grids

Biorthogonal grids were introduced by Bookstein (1978) and give numerical information as to the directions and magnitudes of shape change. These grids give a map
of infinitesimal strains at each point in the interior of the first figure, as it is deformed
into the second. The biorthogonal grid shows the maximum and minimum amounts
of the size-and-shape difference and can be viewed as the first derivative of the

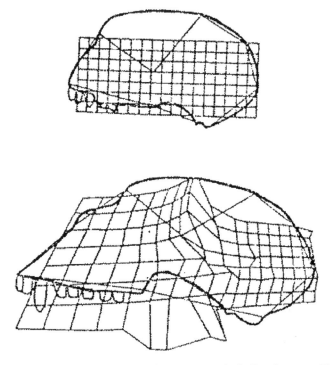

*Figure 12.24   Transformation grids for a pair of skulls. Source: Cheverud &
Richtsmeier 1986. Reproduced with permission of Oxford University Press.*

interpolation. The choice of interpolant will be crucial to our perception of shape
change. At each point in the interior of the figure there are two differentials which
are perpendicular before and after the transformation (principal strains). The curves
following the directions of the differentials form a grid whose intersections are at
90° in both figures, and these grids are called biorthogonal grids. Applications of
the biorthogonal grid have been made to a wide variety of situations (e.g. Bookstein,
1978, 1986), including describing the shape difference in the faces of children with
and without fetal alcohol syndrome (Bookstein and Sampson, 1990). Sampson *et al.*
(1991) have also used the biorthogonal grid but in a different setting in environmental
statistics. They use the biorthogonal grid as an explanatory tool in the description of
spatial covariance structures and their routine for displaying the grids uses the thickness of the lines to indicate the strength of the principal strains.

12.5   Kriging

12.5.1   Universal kriging

To obtain deformations we could consider kriging, a commonly used method of
prediction in spatial statistics. We initially consider prediction from a univariate

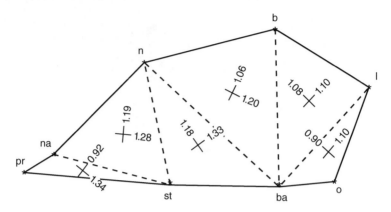

Figure 12.25   One finite element description for the shape change from the average female (shown) to the average male gorilla. The crosses represent the principal axes of the deformation in each triangle – the magnitudes of the largest and least amount of stretching are given, together with the directions of the strains. Source: O'Higgins & Dryden 1993. Reproduced with permission from Elsevier.

random process $Y(t) \in \mathbb{R}$ at an m-dimensional site t, given k univariate observations $\check{y} = (y_1, \dots, y_k)^T$ taken at sites $t_1, \dots, t_k \in \mathbb{R}^m$. Kriging involves the construction of the unbiased linear predictor $\hat{Y}(t) = \gamma^T \check{y}$ which minimizes the mean square prediction error

$$E[(Y(t) - \hat{Y}(t))^2].$$

Let $g(t) = (g_1(t), \dots, g_p(t))^T$ be a vector of known functions. Consider the general linear model

$$Y(t) = \beta^T g(t) + \epsilon(t),$$

where $\beta = (\beta_1, \dots, \beta_p)^T$ and $\epsilon(t)$ is a random field. If $\epsilon(t)$ is a zero mean stationary random field, with positive definite covariance function

$$\text{cov}(\epsilon(s), \epsilon(t)) = \text{cov}(Y(s), Y(t)) = \sigma(s - t)$$

with $\sigma(0) < \infty$, then the prediction is called **universal kriging**. If $\beta^T g(t) = \mu$ is constant, then the prediction is called **ordinary kriging**. We consider universal kriging and more complete details can be found, for example, in Cressie (1993, p. 151). We follow the treatment of Mardia and Marshall (1984) and Mardia and Little (1994).

The term $\beta^T g(t)$ is called the drift of the underlying trend. Writing $Y_{site} = (Y(t_1), \dots, Y(t_k))^T$ as the random vector of the process at the k sites and $\check{y}$ as a realization of Y_{site}, then the regression predictor is the conditional mean of $Y(t)$ given Y_{site}, namely

$$E[Y(t)|Y_{site}] = E[Y(t)] + \text{cov}(Y(t), Y_{site})\{\text{cov}(Y_{site})\}^{-1}\{\breve{y} - E[Y_{site}]\} \quad (12.26)$$
$$= g(t)^T\beta + s(t)^T S^{-1}(\breve{y} - D\beta), \quad (12.27)$$

where $(S)_{ij} = \sigma(t_i - t_j)$, $s(t) = (\sigma(t - t_1), \dots, \sigma(t - t_k))^T$ and the jth row of D is $g(t_j)^T$. Under the assumption of a Gaussian random field Equation (12.26) is well known (e.g. Mardia *et al.* 1979, p. 63). The universal kriging predictor (Mardia and Marshall, 1984) is equivalent to replacing β in Equation (12.27) with the generalized least squares predictor

$$\hat{\beta} = (D^T S^{-1} D)^{-1} D^T S^{-1} \breve{y}.$$

Definition 12.11 *The **universal kriging predictor** at site t is:*

$$\hat{Y}(t) = \gamma^T \breve{y},$$

where

$$\gamma^T = g(t)^T (D^T S^{-1} D)^{-1} D^T S^{-1} + s(t)^T S^{-1}\{I_k - D(D^T S^{-1} D)^{-1} D^T S^{-1}\}.$$

There is an alternative representation where the predictor is regarded as a (non-linear) function of t, that is

$$\hat{Y}(t) = a^T g(t) + w^T s(t),$$

where

$$a = (D^T S^{-1} D)^{-1} D^T S^{-1} \breve{y}, \quad w = S^{-1}\{I_k - D(D^T S^{-1} D)^{-1} D^T S^{-1}\}\breve{y}. \quad (12.28)$$

In matrix form we can write Equation (12.28) as:

$$\begin{bmatrix} S & D \\ D^T & 0 \end{bmatrix} \begin{bmatrix} w \\ a \end{bmatrix} = \begin{bmatrix} \breve{y} \\ 0 \end{bmatrix}, \quad (12.29)$$

and these equations are known as the dual-kriging equations. When written in this alternative form it is quite straightforward to see the link between splines and kriging predictors. We make the full link in the discussion of generalized kriging in Section 12.5.3. There is also a close connection with the radial basis functions (Arad *et al.* 1994; Mardia *et al.* 1996e). Radial basis functions in the numerical analysis literature (Powell, 1987) are approriate to use as isotropic covariance functions.

12.5.2   Deformations

In our applications we are interested in deformations from $\mathbb{R}^m \to \mathbb{R}^m$. Hence we are interested in predicting the multivariate process $Y(t) = (Y(t)[1], \dots, Y(t)[m])^{\mathrm{T}}$ using a collection of m independent kriging predictors. So, if the k observations in the m dimensions at the sites $T = (t_1, \dots, t_k)^{\mathrm{T}}$ are written as a $k \times m$ matrix Y_{obs}, then it follows that the universal kriging predictors can be written as:

$$\hat{Y}(t) = Cg(t) + W^{\mathrm{T}}s(t), \tag{12.30}$$

where C is $m \times p$ and W is $k \times m$, satisfying

$$\begin{bmatrix} S & D \\ D^{\mathrm{T}} & 0 \end{bmatrix} \begin{bmatrix} W \\ C^{\mathrm{T}} \end{bmatrix} = \begin{bmatrix} Y_{obs} \\ 0 \end{bmatrix}. \tag{12.31}$$

Therefore, kriging can be used for deformations for a given positive definite covariance function $\sigma(t)$. The same deformations using Equation (12.30) are given in the generalized kriging framework which follows, where the class of functions for $\sigma(t)$ is widened. The kriging deformations were first introduced in Mardia *et al.* (1991).

12.5.3   Intrinsic kriging

In Section 12.5.1 the positive definite covariance function must exist. However, the technique of intrinsic or generalized kriging involves prediction from the model $Y(t) = \beta^{\mathrm{T}}g(t) + \epsilon(t)$, where $\epsilon(t)$ is an intrinsic random field, which we now define. Intrinsic processes were first developed by Matheron (1973).

Definition 12.12   *Let G be a space of known functions. Given sites* $t_1, \dots, t_k$, *we call*

$$(a_1, t_1, a_2, t_2, \dots, a_k, t_k)$$

an **increment** *with respect to G if*

$$\sum_{i=1}^{k} a_i h(t_i) = 0 \quad \textit{for all } h \in G.$$

An **intrinsic random field** *is a stochastic process Y(t) which has, at increments,* $\sum a_i Y(t_i)$ *distributed with zero mean and variance*

$$\sum a_i a_j \sigma^*(t_i, t_j) \geq 0.$$

The function $\sigma^*(t_i, t_j)$ *is called a* **conditional positive definite (c.p.d.) covariance function** *with respect to G.*

In time series analysis taking differences annihilates a mean trend. Likewise taking increments annihilates trends or drifts expressed as functions in G. So an intrinsic random field corresponds to an equivalence class of stochastic processes which will have different drifts expressed as functions in G.

The linear predictor that minimizes mean square prediction at site t is called the intrinsic or generalized kriging predictor, and is also given by Equation (12.30), but with $\sigma(t)$ a c.p.d. covariance function and a basis for G given by $g(t) = (g_1(t), \ldots, g_p(t))^\mathrm{T}$. Technical details can be found in Mardia *et al.* (1996e).

Consider the important class of intrinsic random processes with c.p.d. covariance function $\sigma^*(t, t + h) = \sigma_\alpha(h)$ indexed by $\alpha > 0$,

$$\sigma_\alpha(h) = \begin{cases} (-1)^{[\alpha]+1} \|h\|^{2\alpha}, & \alpha \text{ not an integer,} \\ (-1)^{\alpha+1} \|h\|^{2\alpha} \log \|h\|, & \alpha \text{ an integer,} \end{cases} \quad (12.32)$$

for $h \in \mathbb{R}^m$. These c.p.d. covariance functions are of interest because they are self-similar, that is $\sigma_\alpha(ch) \equiv c^{2\alpha} \sigma_\alpha(h)$ for $c > 0$ where '$\equiv$' means equal up to an even polynomial in h of degree $2[\alpha]$, where $[\cdot]$ is the 'integer part' function. Thus, $\sigma_\alpha(h)$ and $\sigma_\alpha(ch)$ yield the same predictions. Let the space of functions G_r be the space of polynomials in h of degree $\leq r$. It is well known that $\sigma_\alpha(h)$ defines a c.p.d. function with respect to G_r provided $r \geq [\alpha]$ (e.g. Kent and Mardia, 1994a). The thin-plate spline for $h \in \mathbb{R}^2$ is exactly the same as the intrinsic kriging predictor using the self-similar random field with $\alpha = 1$, $\sigma(0) = 0$ and c.p.d. covariance function $\sigma_1(h)$ taken with respect to G_1, the linear functions. Full details are given in Kent and Mardia (1994a). For $h \in \mathbb{R}^3$ we will use $\alpha = 1/2$ and $\sigma_{1/2}(h) = -\|h\|$ taken with respect to G_1.

Bookstein (2015a) explores self-similarity in the analysis of morphometric data, including an application to the rat growth data of Section 1.4.11.

The link between kriging and splines has been noted by many authors and discussed in some detail initially by Kimeldorf and Wahba (1971) (e.g. see also Matheron, 1981, p. 180, Cressie 1993; Kent and Mardia 1994a). However, the aims of the two approaches are quite different: for kriging the aim is to provide a linear predictor of $Y(t)$ using the available observations $y_1, \ldots, y_k$, whereas for splines the aim is to predict $Y(t)$ with a smooth function of the site location t.

Comparing Equation (12.31) and Equation (12.18) we see that the intrinsic kriging predictor is the same as a smoothing spline with a c.p.d. covariance function equal to $\sigma(t)$, with smoothing parameter

$$\lambda = \sigma(0) - \lim_{|h| \to 0} \sigma(h).$$

If $\lambda = 0$, then the kriging predictor is an interpolant [i.e. $\hat{Y}(t_j) = y_j$, $j = 1, \ldots, k$] and this corresponds to an interpolating spline.

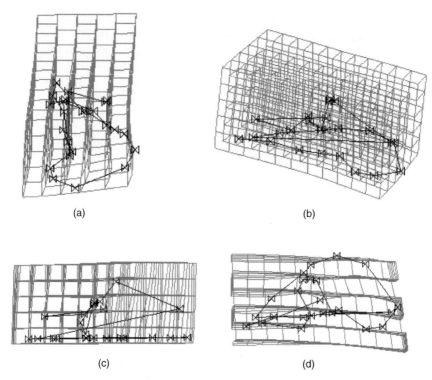

(a)   (b)

(c)   (d)

Figure 12.26   The deformed cuboid grid from the full Procrustes mean male to an icon which is three standard deviations along the first PC of shape variability. We see four different views of the icon in 3D space. In particular, view (c) is the top view of the skull and view (d) is the side view. We can see that there is more deformation in the grid at the back and top of the skull, showing the region of greatest shape variability. The deformed grid has been calculated using intrinsic kriging.

Example 12.10   Consider describing the main aspect of shape variability in the macaque data of Section 1.4.3, with $k = 24$ landmarks. We consider a kriging transformation from the full Procrustes mean male macaque to an icon which is three standard deviations away from the mean along the first PC. The transformation will demonstrate which parts of the skull are varying the most. The data are $m = 3$ dimensional and the c.p.d. covariance function is taken as $\sigma_{1/2}(h) = -\|h\|$. A regular cuboid grid is evaluated on the mean male and then deformed using the kriging transformation of the mean male to the icon. In Figure 12.26 we see four different views of the grids on the deformed male at three standard deviations along the first PC. The 3D representation is best viewed with an interactive spinning plot. However, we see from the static plots that the main shape variability in this PC is in the top of the braincase moving relative to the bottom of the braincase.   □

12.5.4   Kriging with derivative constraints

Bookstein and Green (1993) introduced the term 'edgel' to mean the first derivative information at a landmark. They extended the interpolating thin-plate spline for 2D deformations to include derivative constraints. As well as matching landmarks from one figure to the second, derivative information such as the tangent directions to the outline at a landmark were included. This approach allows higher order information to be included in the deformation. Mardia and Little (1994) and Mardia *et al.* (1996e) consider kriging with derivative constraints in a more general framework. The method can be extended to kriging with up to pth-order derivative constraints, provided the $2p$th-order derivatives of the c.p.d. covariance function exist. Details are given in Mardia *et al.* (1996e). In particular, one may be interested in a deformation where there are landmark, tangent direction and curvature constraints. Note that away from the objects some undesirable bending or folding can occur. Mardia *et al.* (2004) have given a model-based approach for an edgel at a triangle shape whereas Mardia (2013a) has given a characterization for two edgels at two landmarks in three dimensions.

12.5.5   Smoothed matching

In some applications the data may be recorded imprecisely, and there may be quite a large amount of noise. For example, Hastie and Kishon (1991) considered affine matching of handwritten signatures where there was a great deal of added measurement noise. It can be desirable to impose some smoothing into the analysis, in order to obtain smooth estimates of shape and shape variability.

A suitable approach to the problem is to introduce a roughness penalty into the objective function, in the manner of Rice and Silverman (1991) and Green and Silverman (1994). Consider the regression equation

$$Y = h(T, B) + E.$$

For matching two configurations one could minimize

$$s^2[Y - h(T, B)] + \lambda J[Y - h(T, B)],$$

where λ is a smoothing parameter, $s^2(\cdot)$ is an objective function for measuring shape discrepancy and $J(\cdot) \geq 0$ is a roughness penalty. For example, in two dimensions consider smooth shape matching of the complex k-vector configurations T to Y by minimizing

$$\|Y - \beta T \mathrm{e}^{i\theta} - 1_k c^{\mathrm{T}}\|^2 + \lambda J(Y - \beta T \mathrm{e}^{i\theta} - 1_k c^{\mathrm{T}}),$$

where β is a scale, θ is a rotation and $c \in \mathbb{C}$ is a translation.

If we consider m-dimensional affine matching from T to Y (the $k \times m$ matrices of coordinates from two configurations), then we would minimize

$$\|Y - [1_k, T]B\|^2 + \lambda J(Y - [1_k, T]B),$$

where B is an $(m + 1) \times m$ matrix.

One possible penalty function is the quadratic penalty $J(E) = \text{trace}(E^*\Omega E)$, where Ω is the $k \times k$ integrated square derivative matrix corresponding to a smoothing spline, as suggested by Hastie and Kishon (1991) for smoothed affine matching. The bending energy of Equation (12.16) is another possible choice, and the solution is the smoothing thin-plate spline of Section 12.3.3.

In the case of $J(E) = \text{trace}(E^T\Omega E)$ and $s^2(E) = \text{trace}(E^T E)$ a very simple adaptation of the Procrustes algorithm provides a solution. In the affine least squares case the solution is simply given by the m eigenvectors of $\check{A} + \lambda\Omega$ with smallest eigenvalues [$\check{A}$ is the matrix defined after Equation (12.4)].

12.6   Diffeomorphic transformations

There are many other types of deformations that could be considered. A rather different approach using techniques of fluid mechanics was considered by Christensen *et al.* (1996). A smooth deformation is obtained by applying a vector field transformation to the starting figure T. Although large scale smoothness is guaranteed, large magnitude deformations on a local scale are not heavily penalized. The result is that dramatic diffeomorphic deformations can be achieved, for example from an image of a small wedge-shaped patch into a letter 'C'. A diffeomorphism is a one to one and onto transformation, that is a diffeomorphic transformation. The class of diffeomorphic transformations is very appealing, as the transformation is invertible.

Younes (2010) gives a comprehensive account of diffeomorphic transformations in shape analysis, including using Euler–Lagrange equations and flow-based matching. Although the treatment is far more generally applied to curves and surfaces, the methodology is also appropriate for comparing point configurations as a special case. Grenander and Miller (2007) also provide a comprehensive treatment of the deformation approach to shape analysis.

Beg *et al.* (2005) introduce the large deformation diffeomorphic metric mapping (LDDMM) methodology, which is a method for computing diffeomorphisms, usually between medical images in the field of computational anatomy. In particular the estimated diffeorphic mapping ϕ_1 from image I_0 to image I_1 is the end point of a flow of a time-dependent velocity field v_t specified by the ordinary differential equation (ODE) $\dot{\phi}_t = v_t(\phi_t)$. Here the time index t indicates a series of transformations, where ϕ_0 is the identity transformation, and the end point ϕ_1 is the transformation of interest. There are many such paths, and the LDDMM method involves obtaining the smoothest path according to some criterion. In particular the solution is obtained from a variational problem by minimizing over the space of velocity fields the expression

$$\int_0^1 \|v_t\|_V^2 + \frac{1}{\sigma^2} \left\| I_0 \circ \phi_1^{-1} - I_1 \right\|^2 ,$$

where $\|v_t\|_V$ is a norm on a space of smooth velocity fields. Beg *et al.* (2005) derive the Euler–Lagrange equations and provide a gradient algorithm for numerical computation. Miller *et al.* (2002) make connections with Euler–Poincaré equations to describe geodesic motions on the space of diffeomorphism groups known as EPDiff (Mumford and Michor, 2013).

Miller *et al.* (2006) use the concept of geodesic shooting for computing the diffeomorphism. Geodesic shooting involves estimating an inital velocity field of the transformation, and then evolving the deformorphic transformation which is determined by the initial velocities. A measurement of error or goodness of fit is chosen to represent how well the diffeomorphism represents the transformation, and the initial velocities are then adjusted iteratively to minimize the goodness of fit measure. LDDMM is usually applied to medical images, although it can also be applied to landmark configurations (Joshi and Miller, 2000). Further related work is given by Younes *et al.* (2008b) and Bauer *et al.* (2014).

Arsigny *et al.* (2006) introduce an efficient computational framework for diffeomorphisms which defines an exponential mapping from the vector space of smooth stationary velocity fields to diffeomorphisms. The exponential exp(*u*) of a velocity field is given by the flow at a time of a stationary ODE. In this log-Euclidean framework the vector field is stationary, and one only needs the flow at a single time to compute the diffeomorphism. Hence the method is very efficient numerically. Vercauteren *et al.* (2009) describe the method of diffeomorphic demons, which is an efficient tool for image registation following from Arsigny *et al.* (2006). Ashburner (2007) gives an introduction to deformations in general, making the distinction clear between small and large deformation frameworks. Ashburner (2007)'s method DARTEL has similarities with the log-Euclidean diffeomorphism methods. Other techniques of interest include Glaunès *et al.* (2004) for landmark matching via large deformation diffeomorphisms on the sphere; Durrleman *et al.* (2008, 2009) for comparing brain surfaces based on currents; and Sommer *et al.* (2013) who consider LDDMM and a kernel bundle variational formulation, including deformations at various scales and sparsity. Some practical comparisons of various deformation methods for tensor-based morphometry are given by Villalon *et al.* (2011). In particular they compare surface constrained deformations, diffeomorphic demons and fluid based registration. A general survey of defomable medical image registration methods is given by Sotiras *et al.* (2013).

13

Non-parametric inference and regression

13.1 Consistency

The consistency of mean shape and size-and-shape estimators has been of long-standing interest. Consistency is a desirable property for an estimator, and informally a consistent estimator for a population quantity has the property that with more and more independent observations the sample estimator should become closer and closer to the true population quantity. Consider a random sample of n configurations given by $X_1, \ldots, X_n$ from a distribution with population mean shape $[\mu]$. Let a sample estimator of $[\mu]$ be obtained from $X_1, \ldots, X_n$ and denoted by $[\hat{\mu}]$. We say that the estimator is consistent if for any $\varepsilon > 0$,

$$\lim_{n \to \infty} P\big(d([\hat{\mu}], [\mu]) \geq \varepsilon\big) = 0, \tag{13.1}$$

where $d(,)$ is a choice of shape distance. We write

$$[\hat{\mu}] \overset{P}{\to} [\mu] \quad \text{as} \quad n \to \infty,$$

and say that $[\hat{\mu}]$ converges in probability to $[\mu]$.

Similarly a mean size-and-shape estimator is consistent if

$$[\hat{\mu}]_S \overset{P}{\to} [\mu]_S \quad \text{as} \quad n \to \infty.$$

It was shown by Lele (1993) that Procrustes mean estimators might not be consistent for the shape of a mean configuration under various Gaussian perturbation

Statistical Shape Analysis, with Applications in R, Second Edition. Ian L. Dryden and Kanti V. Mardia.
© 2016 John Wiley & Sons, Ltd. Published 2016 by John Wiley & Sons, Ltd.

models (Lele and McCulloch 2002). However, Procrustes estimators are often consistent for a population Fréchet mean shape using the appropriate Procrustes distance. We shall investigate this issue in more detail below. For some models the shape of the means and population Fréchet mean shape coincide, and in this case the Procrustes estimators are consistent for the shape of the means, and in particular the full and partial Procrustes mean are consistent for the shape of the means for 2D isotropic distributions (Kent and Mardia 1997).

An early paper which gives a strong law of large numbers in separable quasi-metric spaces is by Ziezold (1977). In the paper, Ziezold (1977) defines the sample Fréchet mean and then shows that as $n \to \infty$ the sample Fréchet mean tends to an element of the equivalence class of the population Fréchet mean. This classic paper is one of the earliest in size-and-shape analysis, and in particular introduces partial Procrustes analysis and gives the 2D case in some detail using complex arithmetic. Further papers expanding on the topic of planar size-and-shape and shape that also introduce further statistical inference procedures are also by Ziezold (1989, 1994).

Ziezold (1977)'s results are for intrinsic means. An alternative is to consider an embedding and use the corresponding limit theorems in the embedded space, and these measures of location are extrinsic means (see Section 6.1). Early results include Hendriks and Landsman (1996a,b) studying the asymptotic properties of the mean location on manifolds, with a focus on the the sample mean direction on spheres in Hendriks *et al.* (1996). Further inference, including large sample hypothesis tests and confidence regions on manifolds, is given by Hendriks and Landsman (1998, 2007).

Two important papers in non-parametric shape analysis are by Bhattacharya and Patrangenaru (2003, 2005) who lay out the framework for consistency for intrinsic and extrinsic mean estimators. Also, they provide central limit theorems for both intrinsic and extrinsic means. Results for extrinsic means are studied in particular detail. The two high-profile papers provide a thorough, careful treatment of non-parametric estimation and inference on Riemannian manifolds. Several examples using landmark shape data on Kendall's shape space are given, particularly emphasizing the extrinsic means. As well as being more straightforward to deal with than the intrinsic mean, the extrinsic mean in the embedded space is unique, whereas conditions for uniqueness of the intrinsic mean are more restrictive and require concentrated data.

13.2   Uniqueness of intrinsic means

Intrinsic means are unique provided the distribution is sufficiently concentrated, (Karcher 1977; Kendall 1990b; Le 1995; Afsari 2011). A history of intrinsic means and their uniqueness has been given by Afsari (2011), and the notion of Riemannian centres of mass was introduced by Grove and Karcher (1973) and developed further in Grove *et al.* (1974, 1975); Karcher (1977); and Buser and Karcher (1981).

Definition 13.1   *A Riemannian L^p centre of mass in a Riemannian manifold M is defined as a minimizer of*

$$f_p(\mu) = \frac{1}{p}\int d(x,\mu)^p\, dF \;=\; \frac{1}{p}E_X[d_X(X,\mu)^p], \tag{13.2}$$

where dF is the probability measure on the manifold and $d(x,\mu)$ is the Riemannian distance between $x,\mu \in M$.

The global and local minimum of f_2 are often called the 'Fréchet mean' and 'Karcher mean', respectively (see Section 6.1). Likewise we call the global and local minimum of f_1 the 'Fréchet median' and 'Karcher median', respectively.

In order to discuss uniqueness of the means we require the definition of a regular geodesic ball, from Kendall (1990a).

Definition 13.2   *For a complete Riemannian manifold the geodesic ball $B_r(p)$ of radius r centred at p is regular if the supremum of sectional curvatures Δ is less than $\{\pi/(2r)\}^2$ and the cut locus of p does not meet the ball $B_r(p)$.*

Note that the cut locus at a point p comprises points which do not have unique minimum geodesics to p, for example the cut locus of a sphere at the north pole is the south pole. Kendall (1990b) showed that any two points in $B_r(p)$ are joined by one and only one minimal geodesic, and also proved the following result:

Result 13.1   (Kendall 1990b) *For a probability distribution with support in $B_r(p)$, there is one and only one Karcher mean in $B_r(p)$.*

Le (1991b) demonstrated global uniqueness by extending a result of Karcher (1977).

Result 13.2   (Le 1991b) *For a probability distribution with support in the regular ball $B_{r/2}(p)$ the global minimizer of f_2 is unique, that is the Fréchet mean is unique.*

Clearly the results hold for either population or sample means, where for the sample mean from a random sample of size n the probability measure is replaced with probability $1/n$ at each data point.

An example where we can check uniqueness is for size-and-shape, using Le and Kendall (1993). Let X be a $k - 1 \times m$ matrix representing a Helmertized configuration of $k > m$ points in m dimensions. Write

$$X^T = U\mathrm{diag}(\tilde{\Lambda},0)V,$$

with $U \in SO(m), V \in SO(k-1)$ and $\tilde{\Lambda} = \mathrm{diag}(\tilde{\lambda}_1, \ldots, \tilde{\lambda}_m)$, then the maximum sectional curvature is the maximum value out of all of the following terms for any (i,j) of:

$$3\frac{\tilde{\lambda}_i^2 + \tilde{\lambda}_j^2 - \tilde{\lambda}_r^2}{\left(\tilde{\lambda}_i^2 + \tilde{\lambda}_j^2\right)^2}, \quad (r = i \text{ or } r = j) \tag{13.3}$$

or

$$3\frac{\tilde{\lambda}_s^2\left(\tilde{\lambda}_i^2 + \tilde{\lambda}_j^2 - \tilde{\lambda}_r^2\right)}{(\tilde{\lambda}_i^2 + \tilde{\lambda}_j^2)(\tilde{\lambda}_i^2 + \tilde{\lambda}_s^2)(\tilde{\lambda}_j^2 + \tilde{\lambda}_s^2)}, \quad (r = i \text{ or } r = j; s \neq i, j). \tag{13.4}$$

Note that this uniqueness result was used by Mitchelson (2013) who extended Procrustes estimation for the mean size-and-shape for temporal human movement data to the case where some of the landmarks are not visible at times. The resulting MOSHFIT (motion shape fitting) algorithm defines a visibility graph, and provided this is connected the mean size-and-shape can be computed and checked for uniqueness.

Further discussion of the cut locus of a Fréchet mean is given by Le and Barden (2014). A detailed relevant discussion is found in Afsari *et al.* (2013) who describe the best possible gradient descent algorithm for computing the mean, and provide an in-depth review.

Example 13.1   Here we consider the example of estimation of the mean size-and-shape of the male macaque data of Section 1.4.3. The space of interest is the size-and-shape space and the sectional curvatures are obtained using Equation (13.3) and Equation (13.4). The sectional curvatures are computed, the maximum taken, and then compared with the maximum radius of the data using the Riemannian distance:

```
library(shapes)
x<-macm.dat
H<-defh(dim(x)[1]-1)
n<-dim(x)[3]
m<-dim(x)[2]
maxsc<-0
for (ii in 1:n){
y<-H%*%x[,,ii]
lam<-svd(y)$d**2
for (i in 1:m){
for (j in 1:m){
for (r in c(i,j)){
sc<-3*(lam[i]+lam[j]-lam[r])/(lam[i]+lam[j])**2
if (sc>maxsc){
maxsc<-sc
}
```

```
for (s in 1:m){
if ((s != i)&&(s!=j)){
sc<-3*(lam[s]*(lam[i]+lam[j]-lam[r])/
  ((lam[i]+lam[j])*(lam[i]+lam[s])*(lam[j]+lam[s])) )
if (sc>maxsc){
maxsc<-sc
}}}}}}

msh<-procGPA(x,scale=FALSE)$mshape
maxdist<-0
for (i in 1:(n)){
rho<-ssriemdist(x[,,i],msh)
if (rho>maxdist){
maxdist<-rho
}}
print(maxsc)
print(maxdist)
rball<-maxdist
print((pi/(4*rball))**2)
check<- (((pi/(4*rball))**2 - maxsc)>0)
print(check)
```

The output from running the above method is:

```
> print(maxsc)
[1]  0.0006405375
> print(maxdist)
[1]  12.71219
> rball<-maxdist
> print((pi/(4*rball))**2)
[1]  0.003817147
> check<- (((pi/(4*rball))**2 - maxsc)>0)
> print(check)
[1]  TRUE
```

Here the data do lie inside a regular ball, and hence the size-and-shape mean is unique.   □

Example 13.2   A simpler case where uniqueness is easily checked is for planar shape. In this case the maximal sectional curvature is 4 (Kendall 1984). Hence we need $4 < [\pi/(2r^*)]^2$, that is $r^* < \pi/4$. Therefore the data should lie within a ball of radius $\pi/8 \approx 0.3927$ to guarantee uniqueness. For the T2 mouse vertebrae all the observations lie within radius 0.1271 of the full Procrustes mean, and hence uniqueness is guaranteed for the intrinsic mean. However, for the digit 3 data the observations lie within 0.7061 of the full Procrustes mean, which does not guarantee uniqueness.   □

The calculations of the above example are as follows:

```
max(procGPA(qset2.dat)$rho)
[1] 0.1271043
max(procGPA(digit3.dat)$rho)
[1] 0.706119
```

13.3   Non-parametric inference

13.3.1   Central limit theorems and non-parametric tests

In Section 9.1.3 we introduced permutations tests and bootstrap tests based on tangent space approximations to the shape space, when the amount of variability in the data is small. However, the testing procedures can also be appropriate in situations where the data are less concentrated, provided the sample sizes are large and a central limit theorem is available.

Bhattacharya and Patrangenaru (2003, 2005) discussed large sample theory of intrinsic and extrinsic sample means on manifolds, and in particular provide central limit theorems for intrinsic and extrinsic means. The extrinsic mean central limit theorem is more straightforward and less restrictive, after projecting the data with a Euclidean embedding [also see Hendriks and Landsman (1998) for related work]. Amaral et al. (2007, 2010b) also considered central limit theorems for one and multi-sample shapes in two dimensions. Even if data are quite dispersed for large samples the means will have an approximate multivariate Gaussian distribution in a tangent space, and suitable pivotal statistics can be obtained where the asymptotic distribution does not depend on unknown parameters (e.g. chi-squared distributions). Further discussion of asymptotic distributions of sample means on Riemannian manifolds is given by Bhattacharya (2008) and Bhattacharya and Bhattacharya (2008, 2012).

Huckemann (2011a) explores the analysis of landmark shapes and size-and-shapes in three dimensions. In particular, he explores consistency and asymptotic normality of various types of mean shape and mean size-and-shape. Also, the degenerate rank cases are explored in detail, referring to motivating examples of diffusion tensors in medical image analysis. Discussion of the perturbation model and inconsistency is also carried out. Huckemann (2011a) extends the central limit theorem of Bhattacharya and Patrangenaru (2005) for extrinsic means (using the Veronese–Whitney embedding) to the full Procrustes sample mean for 3D data. From Dryden et al. (2014) the limiting distributions obtained using embeddings will generally depend on the chosen embedding. The intrinsic limiting distributions for intrinsic Fréchet means obtained in Kendall and Le (2011) eliminate this dependence and, at the same time, reveal explicitly the role played by the curvature of a Riemannian manifold M in the limiting behaviour of empirical intrinsic Fréchet means.

Amaral et al. (2007) considered pivotal bootstrap methods for k-sample problems for 2D shape analysis, and the method is available in the R routine `resampletest` (see Section 9.1.3 for some examples). In addition Amaral et al. (2010b) provide bootstrap confidence regions for the planar mean shape. In Section 9.1.3 we described both permutation tests and bootstrap tests on tangent spaces where the data are concentrated, but in cases where there is a central limit theorem available the tests are also often appropriate for large sample problems where the variability of the data may not necessarily be small. Some of the simulation studies of Amaral *et al.* (2007) demonstrate the utility of both permutation and bootstrap tests when the variability of the data is not small, but the sample sizes are large.

An early two sample non-parametric test was developed by Ziezold (1994) who considered a test statistic based on Mann–Whitney U statistics of partial Procrustes distances from each of the sample observations to the group means. Other non-parametric tests have been developed by Brombin and Salmaso (2009, 2013) who implement their multi-aspect permutation tests with small sample size to shape analysis applications; Brombin *et al.* (2011) who emphasize high-dimensional applications in shape analysis; and Brombin *et al.* (2015) who use non-parametric combination-based tests in dynamic shape analysis with applications to face gestures.

As well as non-parametric tests, the machinery that can be used for non-parametric Euclidean data analysis can be adapted to manifold valued data analysis. For example, kernel density estimation on Riemannian manifolds (Pelletier 2005); non-parametric Bayesian density estimation and consistency, with particular reference to planar shapes (Bhattacharya and Dunson 2010, 2012b); non-parametric regression estimation on Riemannian manifolds (Pelletier 2006); robust non-parametric regression (Henry and Rodriguez 2009); non-parametric Bayesian classification (Bhattacharya and Dunson 2012a); kernel based classification on Riemannian manifolds (Loubes and Pelletier 2008); k-means classification (Amaral *et al.* 2010a); and empirical likelihood for planar shapes (Amaral and Wood 2010).

13.3.2   M-estimators

The minimizers of the objective function (13.2) for sample data are types of M-estimators. Kent (1992) considered various estimators that minimize expressions of the form

$$\sum_{i=1}^{n} \phi^*(\rho(X_i, \mu)), \tag{13.5}$$

where $\rho(\cdot)$ is the Riemannian distance of Equation (4.12). These estimators can be considered M-estimators (maximum likelihood type estimators) for shape, and include the sample measure of location from (13.2). Although Kent (1992) utilized complex notation in the $m = 2$ case, such estimators are also valid for $m \geq 3$ dimensions. Estimators of this type were considered in Section 6.3.

When the objective function in (13.5) has $\phi^*(\rho) = \rho$ this is the same as $p = 1$ in the sample version of (13.2). Minimizing this objective function leads to the M-estimator being equal to the Fréchet/Karcher median which is discussed by Fletcher *et al.* (2009), with application to robust atlas estimation, and Koenker (2006).

13.4   Principal geodesics and shape curves

13.4.1   Tangent space methods and longitudinal data

Since the shape space or size-and-shape space is not a flat Euclidean space, we cannot simply apply the classical methods of linear regression and spline fitting directly to manifold valued data if the data are not concentrated. If the sample has little variability, the problem can be transferred to a tangent space (e.g. at the Procrustes mean of these shapes or size-and-shapes) and then Euclidean fitting procedures can be performed in this space. This is the approach of Morris *et al.* (1999) and Kent *et al.* (2001) who developed models where tangent space coordinates are modelled as polynomial functions of time for analysing growth in faces and growth in the rat data of Section 1.4.11. A pole is chosen in the middle of the dataset (e.g. the overall Procrustes mean) and the data are then projected into the tangent space at that point. Standard multivariate procedures can be carried out in the tangent plane, for example multivariate regression with covariates.

Let us consider the following multivariate regression growth curve model in a tangent space to shape or size-and-shape space, where there are n individuals available each observed at N_t time points. Denote the shape or size-and-shape tangent coordinates as v_{it} ($p \times 1$) vector for the ith individual at the tth time point ($i = 1, \ldots, n$, $t = 1, \ldots, N_t$). All individuals are assumed to be independent (although dependencies between individuals could be modelled if desired). A suitable regression model is:

$$v_{it} = X_f \alpha + X_r \beta_i + \epsilon_i \qquad (13.6)$$

for $i = 1, \ldots, n$, $t = 1, \ldots, N_t$, where X_f is the $p \times f$ design matrix for the f fixed effects α, and X_r is the $p \times r$ design matrix for the r random effects β_i. In addition, the β_i are taken to have a $N_r(0, \Delta)$ distribution and $\epsilon_i \sim N(0, \sigma^2)$ independently. The model can then be fitted by maximum likelihood or restricted maximum likelihood, using for example the R library `lme4` or `nlme` (Bates 2005). Any approach that is standard for multivariate longitudinal data can be implemented for shape or size-and-shape data in this manner provided the variability is small, although one should take care with the ranks of the tangent plane coordinates. A suitable approach could be to reduce the dimensionality of the dataset by choosing the first p PC scores which summarize most of the variability in the tangent plane data. Applications of such methodology include: Bock and Bowman (2006); Bowman and Bock (2006); Barry and Bowman (2008); and Bowman (2008), who study linear mixed models for longitudinal shape data with applications to facial modelling, particularly with regard to modelling cleft-lip changes after surgery.

Non-parametric regression models for shape using B-splines have been developed for modelling dynamic shape or size-and-shape measures over time, including modelling ratios of distances (Faraway 2004b) or specific local coordinate systems such as relative positions and angles of articulated limbs (Faraway 2004a) as functions of time. Faraway (2004b) modelled dynamic smile movement in facial shape and

Faraway (2004a) described various human movement studies. Alshabani *et al.* (2007b) applied Bayesian analysis of human movement shape curves to the application introduced in Section 1.4.13, where inference about the change-points for the start and end of the movements is particularly important. Indeed in all these applications where the shape or size-and-shape occurs at different rates for different individuals it is necessary to warp the curves to a common reference time before different individuals can be compared. The need for registration in time is a common issue in the analysis of functional data (e.g. see Kneip and Gasser 1992; Silverman 1995; Ramsay and Li 1998; Kneip *et al.* 2000; Claeskens *et al.* 2010; Srivastava *et al.* 2011b; Zhou *et al.* 2014; Cheng *et al.* 2016).

Time series models can be used where there are long temporal sequences of shapes or size-and-shapes. For example Dryden *et al.* (2002, 2010) investigate size-and-shape analysis of DNA molecular dynamics simulations, and in particular introduce the stationary Gaussian time-orthogonal PC model, where each PC score is independent, with different temporal autoregressive covariance structures. The model is applied to molecular dynamics simulations in pharmacy, where it is of interest to estimate the entropy of the molecule. A very small subset of the DNA data was described in Section 1.4.7, whereas in most practical studies there are often more than 10 000 up to millions of observations.

Das and Vaswani (2010) consider non-stationary models for filtering and smoothing of shape sequences, such as tracking human activity in video sequences. Vaswani *et al.* (2005) study shape activity with continuous state hidden Markov models for temporal sequences of shapes, with an application to detection of abnormal activity.

13.4.2   Growth curve models for triangle shapes

Goodall and Lange (1989) consider growth curve models for triangle shapes. This situation is particularly simple because the Kendall (1983) shape space is a sphere $S^2(\frac{1}{2})$. In particular Goodall and Lange (1989) give an algorithm for fitting a great circle growth curve shape model through the data. There are five stages in the algorithm:

1. Fit a great circle through the data using spherical regression (e.g. see Fisher *et al.* 1987, Section 8.3).

2. Rotate the fitted great circle to the equator.

3. Fit a growth curve model (possibly with covariates) where the angles at the equator (longitudes) depend on fixed and random effects. In the simple case of a fixed and random intercept/slope model we have:

$$\phi_i = \alpha_0 + \alpha_1 t_i + \beta_{0i} + \beta_{1i} t_i + \epsilon_i,$$

where $i = 1, \ldots, n$ and normal models are assumed $(\beta_{0i}, \beta_{1i})^{\mathrm{T}} \sim N_2(0, \Delta)$ and $\epsilon_i \sim N(0, \sigma^2)$ independently, and all individuals are assumed to be independent.

The parameters of the model are fit by maximum likelihood estimation or restricted maximum likelihood estimation.

4. The estimated mean growth curve $(\hat{\alpha}_0, \hat{\alpha}_1)$ is rotated back to the data giving the 'best' fitting great circle through the data.

5. Steps 1–4 are repeated until convergence.

More fixed and random effects could be included to help explain the change in shape over time.

13.4.3   Geodesic model

Le and Kume (2000a) introduced the idea of **testing for geodesics** to examine whether shape change over time follows a geodesic in shape space. The method does not rely on concentrated data. This type of shape change has also been called the geodesic hypothesis (Hotz *et al.* 2010). Le and Kume (2000a) applied the methodology to the rat skulls data of Section 1.4.11, and the method involves working with Riemannian distances in the shape space and then carrying out MDS. The resulting Euclidean embedding is 1D if and only if the data in the original manifold follow a geodesic path, and hence we can investigate whether classical MDS plots are 1D. When investigating growth of individuals we would like to examine if the objects grow along a geodesic in shape space. In further examples the geodesic model has been examined by Hotz *et al.* (2010) and Huckemann (2011b) in both tree ring and leaf shape applications.

Example 13.3   For the rat skulls data of Section 1.4.11 we calculate all pairwise Riemannian distances, and then carry out classical MDS using the command `cmdscale` in R:

```
library(shapes)
data(rats)
distmat<-matrix(0,144,144)
for (i in 1:144){
for (j in 1:144){
distmat[i,j]<-riemdist(rats$x[,,i],rats$x[,,j])
}
}
plot(cmdscale(distmat))
```

The resulting plot is given in Figure 13.1 and we see that there is clear non-linear structure, and hence the growth is not along a geodesic. The plot is similar to that in several earlier analyses of this dataset, e.g. Bookstein (2014, p. 415).   □

13.4.4   Principal geodesic analysis

Fletcher *et al.* (2004) developed principal geodesic analysis (PGA) as a method for obtaining the analogue of PCs on manifolds. The idea is to first find the intrinsic mean, then find the PC axes in the tangent space to the intrinsic mean, and then finally to

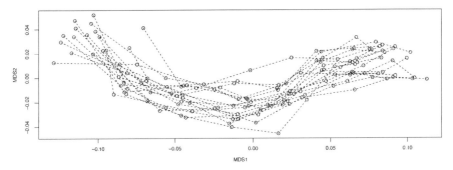

Figure 13.1 Multidimensional scaling plots of the rat data, using the pairwise Riemannian distances between all 144 rat skull shapes, with shapes for each individual rat joined by lines.

project back from the PC tangent vectors to the geodesic subspaces using the exponential map. Fletcher *et al.* (2004) applied the methodology to M-reps, which are medial axis representations used in medical imaging. A further example of PGA is given by Fotouhi and Golalizadeh (2012), in exploring the variability of DNA molecules. Fletcher (2013) provides a summary of geodesic regression and the theory of least squares on Riemannian manifolds.

Huckemann *et al.* (2010) have developed a comprehensive approach to modelling geodesics for shapes, and developing analogues of PCA for planar and 3D shapes using intrinsic methodology. Principal geodesics in triangular shape space were investigated initially by Huckemann and Ziezold (2006), who provided methodology for fitting a best geodesic through a set of triangle shapes (i.e. points on a sphere). In this work an alternative sample mean for the data is proposed which does not coincide with the usual Procrustes means, but rather the mean lies on the principal geodesic. This idea was then extended by Huckemann and Hotz (2009) to PC geodesics for planar shape spaces, and more generally in Huckemann *et al.* (2010) to the idea of geodesic PCA. A set of orthogonal geodesics are obtained in the shape space, and then PCA is carried out using projections onto the geodesic axes. The position of crossing of the geodesics is an alternative definition of mean shape of the data.

Related methodology is given in Kenobi *et al.* (2010) who also develop explicit methodology for fitting principal geodesics, and then proceed to define curvilinear paths through data (analogous to polynomials) by using principal geodesic axes. Further developments include applications of fitting geodesics in planar shape space and testing for common mean principal geodesics in different groups, applied to leaf shape growth using large sample asymptotics (Huckemann 2011b) and tree ring growth for planar outline shapes (Hotz *et al.* 2010). The applications considered by Kenobi *et al.* (2010) involve human movement data and lumbar spinal shape, where departures from the geodesic model are detected using likelihood ratio tests. Faraway and Trotman (2011) investigate shape change along geodesics applied to examples from cleft lip surgery. Further techniques of interest are found in Sozou *et al.* (1995) who use polynomial regression to account for non-linear variation in tangent space point distribution models; Hinkle *et al.* (2014) who introduce intrinsic polynomials for regression on Riemannian manifolds; and Piras *et al.* (2014) who develop a Procrustes motion analysis method based on a linear shift (parallel transport) in order to describe the 3D shape changes over time in the left ventricular heart cycle.

13.4.5   Principal nested spheres and shape spaces

Other related methodology which involves a very different backwards fitting approach is that of Jung *et al.* (2012) who consider fitting a sequence of principal nested spheres. The procedure involves fitting a sequence of successively lower dimensional spheres to the data. The penultimate nested sphere in the sequence is a best fitting 'small circle' curve in general, and a special case leads to a best fitting geodesic ('great circle' on the sphere). The final zero-dimensional 'sphere' (point) in the sequence is another alternative definition of a mean. Principal nested spheres builds on the methodology of Jung *et al.* (2011) on principal arc analysis on direct product manifolds, where applications to M-reps (Pizer *et al.* 2003) which are medial representations of medical images. In particular, Jung *et al.* (2011) use principal arcs to model the prostate in medical images. Also, Pizer *et al.* (2013) consider nested sphere statistics of skeletal models. Jung *et al.* (2012) also develop principal nested spheres for planar shape data, making use of the fact that the pre-shape sphere is a sphere, and at all stages the data are in optimal rotational alignment (Procrustes registered).

13.4.6   Unrolling and unwrapping

Consider the situation where we are given data consisting of n shapes or size-and-shapes observed at successive times, and we are interested in fitting a smooth curve to these shapes.

In situations where the shape or size-and-shape change is large, the tangent space approximation will not be appropriate, and so other methods are required. In the case of triangle shapes in the plane there is a method available for fitting smooth curves through shape data. Since Σ_2^3 is isometric with the 2D sphere of radius $1/2$, the problem for planar triangles can be solved by the novel method proposed by Jupp and Kent (1987), who introduced an algorithm for fitting spherical smoothing splines to spherical data based on the techniques of **unrolling** and **unwrapping** onto an appropriate tangent space. Le (2003) extended the results to unrolling in shape spaces and Kume *et al.* (2007) extended the fitting method of Jupp and Kent (1987) to data in Σ_2^k for the unrolling and unwrapping along a piecewise geodesic in Σ_2^k, using complex linear methods, and also a method for fitting splines to data in tangent spaces to Σ_2^k. The resulting paths are called shape-space smoothing splines. The shape-space smoothing splines are obtained by using the fact that Σ_2^k is a Riemannian quotient space of S_2^k, the pre-shape space of configurations obtained after removing the information on location and scale. This allows the identification of geodesics in Σ_2^k with particular geodesics in S_2^k and identification of tangent spaces to Σ_2^k at various shapes with particular subspaces of the appropriate tangent spaces to S_2^k.

Kume *et al.* (2007) describe in detail how to **unroll** a geodesic G_1 between z_0 and z_1 onto a tangent space $\mathcal{T}_0$ at a point z_0 (pole of the tangent space) on the geodesic. The unrolling is a straight line in the tangent space passing through 0 to the inverse exponential map projection of z_1, and the length of the geodesic from z_0 to z_1 is the same as the length of the line in the tangent space. To **unwrap** a point w with respect to the geodesic from z_0 to z_1 first one projects w on to the tangent space at z_1, denoted $\mathcal{T}_1$, using the inverse exponential map (so that angles with respect to the geodesic and lengths are preserved). Then one needs to see how the projected point in $\mathcal{T}_1$ maps to $\mathcal{T}_0$, when $\mathcal{T}_0$ rolls along the geodesic (using parallel transport) and touches the shape space at z_1. After unrolling and unwrapping we have the geodesic G_1 and the point w mapped to a line segment and a point, respectively, in the tangent space $\mathcal{T}_0$.

The procedure is then extended to piecewise geodesics, which are unrolled to give piecewise linear segments in a single base tangent space $\mathcal{T}_0$, and data points in the shape space can be unwrapped with respect to the piecewise geodesic path. Statistical fitting procedures such as smoothing splines, can be fitted in the base tangent space $\mathcal{T}_0$. The reverse geometrical transformations of rolling and wrapping are employed to map the base tangent points back to the shape space with respect to the piecewise geodesic path. An illustration of unrolling and unwrapping with respect to a piecewise geodesic path is given in Figure 13.2.

Kume *et al.* (2007) describe an iterative fitting algorithm whereby the piecewise geodesic path is updated after each spline fitting. The algorithm usually converges within a small number of iterations, and the smoothing parameters in the spline can be chosen using cross-validation. Since any continuous path can be approximated

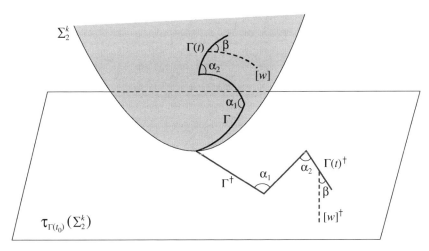

Figure 13.2   A diagrammatic view of unrolling and unwrapping with respect to a piecewise geodesic curve. Source: Kume et al. *2007. Reproduced with permission of Oxford University Press. For a colour version of this figure, see the colour plate section.*

by a continuous piecewise geodesic path, the algorithm gives discretized versions of shape-space smoothing splines.

Note that unrolling and unwrapping can be generalized to the shape spaces of configurations in $\mathbb{R}^m, m \geq 3$. Unrolling and unwrapping procedures for $m \geq 3$ are available and are given in Le (2003) using matrix representations. The procedures involve solutions of homogeneous first-order linear differential equations which must be solved numerically in general. For 2D data ($m = 2$) the use of complex arithmetic leads to explicit expressions for the unrolling and unwrapping of piecewise geodesics, which leads to a much more straightforward and transparent implementation for this important case. Also see Koenderink (1990) for discussion of discretized unrolling of geodesics.

Example 13.4   We consider the human movement data from Section 1.4.13 where $k = 4$ landmarks are recorded in $m = 2$ dimensions for five movements, each with 10 equally spaced time points. It is of interest to model the shape change over time, and an analysis was carried out by Kume *et al.* (2007) as in the following.

For each time observation we find the corresponding Procrustes mean shape of the five shapes observed at that time. We then take the corresponding fitted shape space smoothing spline to these 10 mean points. In order to obtain a sensible representation of our data we use the 'fitted mean path' to unwrap the observed data points at the tangent space of its starting point. The first two PC scores of the resulting data in this tangent space are plotted in Figure 13.3. Figure 13.3a shows fitted smoothing splines ($\lambda = 0.00013$), and Figure 13.3b shows fitted approximate geodesics ($\lambda = 60658.8$). The cubic splines were fitted with the R routine `smooth.spline()` with parameters `spar = 0.3`, `spar = 1.5`, respectively. The percentage of variability explained by the first two PCs is 95.5, 3.8% and 97.9, 1.9% for each plot, and the first two PCs provide a very good summary of the data. Given that the variability explained by the first PC is so high, we consider a hypothesis test to examine if a geodesic provides a good summary of the data. Kume *et al.* (2007) show that there is very strong evidence against a mean geodesic versus the alternative of a mean shape space spline, using a hypothesis test based on complex Watson distributed data. Indeed from Figure 13.3 it is clear that the mean shape-spline is a much better fit to the data than the mean geodesic. □

13.4.7   Manifold splines

Su *et al.* (2012) consider a general method for fitting smooth curves to time-indexed points p_i on Riemannian manifolds using generalizations of splines. Let M be a Riemannian manifold and $\gamma : [0, 1] \to M$ be an appropriately differentiable path on M. The goal is to find a path that minimizes the energy function:

$$E(\gamma) = \frac{\lambda_1}{2} \sum_{i=1}^{n} d^2(\gamma(t_i), p_i) + \frac{\lambda_2}{2} \int_0^1 < \frac{D^2\gamma}{dt^2}, \frac{D^2\gamma}{dt^2} > dt. \qquad (13.7)$$

The first term is referred to as the data term and the second term is referred to as the smoothing term. The asymptotic limits of the solution were investigated by Samir

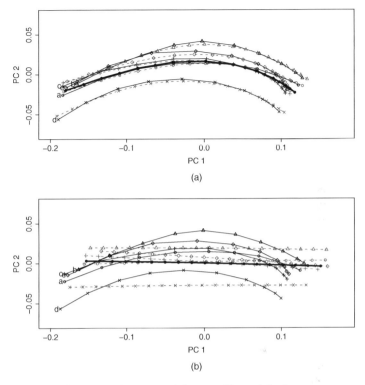

(a)

(b)

Figure 13.3   The first two PC scores of the unrolling of the human movement data paths with respect to the fitted mean path. In (a) fitted smoothing splines are shown in solid black ($\lambda = 0.00013$) with the projected data points joined by dashed lines. In (b) fitted approximate geodesics ($\lambda = 60658.8$) are shown in solid black, with the projected data points joined by dashed lines. In both plots the encircled points are knots of the mean path. Source: Kume et al. *2007. Reproduced with permission of Oxford University Press. For a colour version of this figure, see the colour plate section.*

et al. (2012). As λ_1 tends to zero, for a fixed $\lambda_2 > 0$, one obtains a geodesic curve as the optimal curve. Similarly, as λ_2 tends to zero, for a fixed $\lambda_1 > 0$, the optimal curve is analogous to a piecewise cubic polynomial that interpolates between the given points. Su *et al.* (2012) use a steepest-descent algorithm for minimizing $E(\gamma)$, where the steepest-descent direction is defined with respect to the second-order Palais metric. The algorithm is applied to several applications by Su *et al.* (2012), including planar shapes, symmetric positive definite matrices and rotation matrices. In Figure 13.4 we see the interpolated shapes from a video sequence of images of a dancer, from Su *et al.* (2012). From the initial sequence four images are taken (first row), and noise is added to the two middle figures (second row), and then a spline is fitted in a tangent space at the mean (third row), piecewise geodesics are fitted which interpolate the noisy data (fourth row), the unrolling method of Kume *et al.* (2007) (Section 13.4.6)

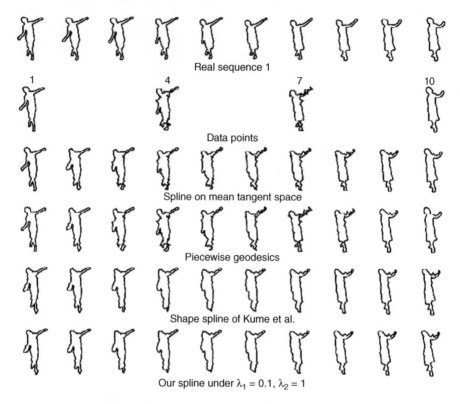

Figure 13.4   The original data, the noisy data, and interpolated and smoothed shape sequences using different techniques. Source: Su et al. 2012. Reproduced with permission of Elsevier.

is applied (fifth row) and finally the manifold spline method of Su *et al.* (2012) is applied (sixth row). Note that the last two methods are quite similar in this example.

Another promising technique is that of a principal flow (Panaretos *et al.* 2014), which is a curve on the manifold passing through the mean of the data, where a particle flowing along the principal flow attempts to move along a path of maximal variation of the data, up to smoothness constraints. The technique involves solving an ODE, and uses the Euler–Lagrange method.

13.5   Statistical shape change

Rather than describing the shape change between two individuals we may have random samples available from different populations. Comparison and explanation of population shape differences are required.

The situations under study can be categorized into two situations: independent samples; and dependent samples (matched pairs).

1. Independent samples: the objects under study are all mutually independent.

   Example: Consider the mouse vertebral data of Section 1.4.1. It is of interest to describe the differences in size-and-shape between the T2 vertebrae in the Control, Small and Large groups. All observations can be assumed to be independent here.

   Example: Bookstein and Sampson (1990) consider two independent samples of children; in one group the mothers drank excess alcohol during pregnancy; and in the other group the mothers did not. It is of interest to explore the shape differences in the faces of the children between the independent groups.

   Hypothesis tests on whether there are population shape differences in the above examples can proceed using the methods described in Chapters 9 and 10. However, we may probe further and examine, for example, whether a population shape difference is affine or has some other simple structure.

2. Dependent samples: the objects under study are related, for example the same individual before and after an operation, or describing the size-and-shape change as a particular organism grows over time.

   Example: Mardia and Walder (1994a) consider the rat calvarial growth data of Moss *et al.* (1987) in Section 1.4.11, in particular the shape difference between ages 90 and 150 days. X-ray images are taken of the head of each rat at each stage and so there is a natural pairing here, as each individual is followed through time.

Further developments that are useful for longitudinal data or the study of shape versus time include Sasaki metrics (Muralidharan and Fletcher 2012) for analysis of longitudinal data on manifolds. Sasaki metrics are defined on a cone of $M \times \mathbb{R}^+$ where M is a manifold, and so they are also appropriate for size-and-shape analysis, where M is the shape space. Also, Thompson and Rosenfeld (1994) describe deterministic and stochastic growth models for modelling shapes; Niethammer *et al.* (2011) discuss geodesic regression of image time series; and Niethammer and Vialard (2013) develop parallel transport methods for shapes.

13.5.1   Geometric components of shape change

Consider the $m = 2$ case with two figures represented in terms of the Bookstein coordinates of Equation (2.4), with $(\theta_j, \phi_j)^T$, $j = 3, ..., k$, for the first object and $(\theta_j^*, \phi_j^*)^T$, $j = 3, ..., k$, for the second object. Consider predicting the Bookstein coordinates $(\theta^*, \phi^*)^T$ at a point in the second figure given the corresponding point $(\theta, \phi)^T$

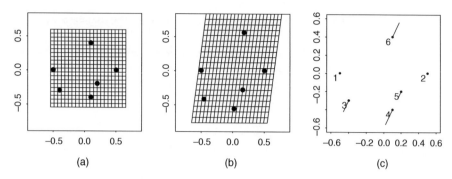

Figure 13.5   An affine deformation from (a) to (b). In (c) we see the change vectors in Bookstein coordinates, which are all parallel with length proportional to the distance from $\phi = 0$, and direction depending on the sign of the ϕ coordinate.

in the first figure. Geometric components of shape change (Bookstein and Sampson 1990) are expressions of the form:

$$\theta^* = \Phi_1(\theta, \phi), \quad \phi^* = \Phi_2(\theta, \phi) \tag{13.8}$$

subject to the constraints

$$\Phi_1(-\tfrac{1}{2}, 0) = -\tfrac{1}{2}, \ \Phi_1(\tfrac{1}{2}, 0) = \tfrac{1}{2}, \ \Phi_2(-\tfrac{1}{2}, 0) = \Phi_2(\tfrac{1}{2}, 0) = 0,$$

where $\Phi_1 : \mathbb{R}^2 \to \mathbb{R}, \Phi_2 : \mathbb{R}^2 \to \mathbb{R}$ are polynomials in two real variables. It follows that the affine transformation between two figures in Bookstein coordinates is:

$$\Phi_1(\theta, \phi) = \theta + \xi_1 \phi, \quad \Phi_2(\theta, \phi) = (1 + \xi_2)\phi, \quad (\xi_1, \xi_2 \in \mathbb{R}). \tag{13.9}$$

In Figure 13.5 we see an artificial example of a uniform transformation between two configurations in terms of Bookstein coordinates. The displacement vectors are all parallel, each vector magnitude is proportional to the vertical distance from the line $\phi = 0$ and each vector direction depends on the sign of ϕ_j. From Section 12.2.4 we see that an affine transformation always describes the shape difference for the triangle case ($k = 3$) but for $k > 3$ points the fit will in general be only approximate. Unlike the PTPS described in Section 12.3.1, this is not an interpolant in general.

Bookstein and Sampson (1990) include tests for the goodness of fit of linear and quadratic geometric components. Their motivation was a study into the differences in face shape in children whose mothers drank heavily during pregnancy and those whose mothers did not (also see the FASDs example in Section 1.4.6). Their approach was to use general multivariate normal models for the shape variables, as described in Section 9.4. If variations about the mean landmarks are small, then this approach is reasonable. The situation considered was for two independent samples. Bookstein and Sampson (1990) also consider the matched pairs situation with the rat growth data of Moss *et al.* (1987) from Section 1.4.11. The testing procedures involve the development of suitable Hotelling's T^2 tests. Since Bookstein's shape variables and Procrustes tangent coordinates are approximately linearly related for small variations (Kent 1994), it follows that the Hotelling's T^2 tests will be approximately the same as those conducted in tangent space, described in Section 9.1.

Mardia and Dryden (1989b) also considered testing for affine shape change in two independent samples using offset normal distributions with isotropic covariance structure. The fitting of second-order polynomial functions is the next extension and explicit details were given by Bookstein and Sampson (1990) for the paired samples case. For quadratic fitting we need $k > 6$ landmarks for a meaningful fit.

13.5.2   Paired shape distributions

Mardia and Walder (1994a) have investigated offset normal shape distributions for paired data (e.g. the same individual observed at two different time points). Consider two complex figures Z_1 and Z_2 which are marginally isotropic normally distributed with different means, different variances and a correlation between the two figures. Let Z_{ij} be the jth complex coordinate of the ith figure, then the model is written as:

$$\begin{pmatrix} Z_{1j} \\ Z_{2j} \end{pmatrix} \sim CN_2 \left[\begin{pmatrix} \mu_{1j} \\ \mu_{2j} \end{pmatrix}, \Omega = \begin{pmatrix} \sigma_1^2 & \rho_P \sigma_1 \sigma_2 \\ \rho_P \sigma_1 \sigma_2 & \sigma_2^2 \end{pmatrix} \right] \qquad (13.10)$$

independently for $j = 1, \ldots, k$, where ρ_P is real. Let $\tau_i^2 = 4\sigma_i^2 / \|\mu_{i2} - \mu_{i1}\|, i = 1, 2$, and let ξ be the angle between the mean baselines of the two figures, that is

$$\xi = \text{Arg}(\overline{(\mu_{22} - \mu_{21})}(\mu_{12} - \mu_{11})).$$

Let $W_i = (W_{i3}, \ldots, W_{ik})^T$ be complex Kendall coordinates for the observed figures $Z_i, i = 1, 2$, and we write $W = (W_1^T, W_2^T)^T$. Mardia and Walder (1994a) obtained the marginal shape distribution of W under this model, which has a very complicated density function. Mardia and Walder (1994b) suggest an alternative density as the rotationally symmetric bivariate complex Watson density on the pre-shape sphere, with density proportional to:

$$\exp\{\kappa_1 (z_1^* \gamma_1 \gamma_1^* z_1) + \kappa_2 (z_2^* \gamma_2 \gamma_2^* z_2)) + 2\kappa_{12} |z_1^* \gamma_1 \gamma_2^* z_2|\},$$

for pre-shapes z_1 and z_2, which in shape space is:

$$f([z_1], [z_2]) \propto \exp(\kappa_1 \cos^2 \rho_1 + \kappa_2 \cos^2 \rho_2 + \kappa_{12} \cos \rho_1 \cos \rho_2),$$

where ρ_i is the Procrustes distance from the shape $[z_i]$ to the ith mean shape $[\gamma_i]$, $i = 1, 2$. Note that κ_{12} acts as a dependence parameter between the two shapes – if $\kappa_{12} = 0$, then the shapes are independent. Prentice and Mardia (1995) have given an alternative spectral approach for paired shape data.

13.6   Robustness

Types of robustness/resistance that are of interest in landmark shape analysis include:

1. resistance to landmark outliers on specimens (i.e. some specimens have particular landmarks that are very unusually located);

2. resistance to object outliers (i.e. objects that are very different from the rest of the random sample);

3. robustness to model mis-specification.

Strictly speaking the words 'robust' and 'resistant' should be used in this manner (i.e. resistant inference can deal with outliers and robust procedures can deal with incorrect models). However, it is common in the literature to use the word 'robust' as an umbrella term to cover all cases.

Consider the situation where we have k points in m dimensions. A general regression equation for matching T to Y with additive errors is $Y = g(T) + E$, and denote $(E)_j$ as the jth row of $E, j = 1, \ldots, k$, which are the coordinates of the jth landmark, where g is a general inverse-link function.

For shape matching the model is:

$$g(T) = \beta T\Gamma + 1_k \gamma^T \qquad (13.11)$$

where $\beta > 0, \Gamma \in SO(m), \gamma \in \mathbb{R}^m$. For size-and-shape

$$g(T) = T\Gamma + 1_k \gamma^T \qquad (13.12)$$

and for affine matching

$$g(T) = TA + 1_k \gamma^T \qquad (13.13)$$

where A is a general $m \times m$ matrix. So, in the case of (13.11) estimation could be carried out by ordinary full Procrustes matching and for (13.12) by ordinary partial Procrustes matching.

Procrustes estimation involves minimizing the sum of squared norms of the errors at each point, that is using least squares:

$$s^2(E) = \|Y - g(T)\|^2 = \|E\|^2 = \sum_{j=1}^{k} \|(E)_j\|^2.$$

Several resistant methods have been proposed including the repeated median technique of Siegel and Benson (1982), also used by Rohlf and Slice (1990).

Dryden and Walker (1999) adapt the S-estimator of Rousseeuw and Yohai (1984) for shape analysis. When the response is multivariate an isotropic S-estimator could be defined as the solution to the following minimization problem: minimize $s^2\{(E)_1, \ldots, (E)_k\}$ over the unknown parameters subject to

$$\frac{1}{k} \sum_{j=1}^{k} \xi(\|(E)_j/s\|) = K. \qquad (13.14)$$

The positive continuously differentiable function $\xi(x), x \in \mathbb{R}^+$, satisfies $\xi(0) = 0$, is strictly increasing on $[0, c]$, is constant for $x > c$ for a fixed $c > 0$ and $E_\Phi(\xi) = K$, where $(E)_j/s$ is considered standard normal for calculating K.

An alternative estimator is the least median of squares (LMS) estimator (Rousseeuw 1984) with objective function

$$s^2(E) = \text{median}(\|(E)_1\|^2, \ldots, \|(E)_k\|^2).$$

The LMS residual discrepancy measure is:

$$D^2_{LMS}(Y, T) = \inf_{g \in G} \text{median} \, \|(Y - g(T))_j\|^2.$$

If we relax some of the conditions for the isotropic S-estimator to allow the choice of indicator function

$$\xi(x) = \begin{cases} 1, & \text{if } x \geq 1, \\ 0, & \text{if } x < 1, \end{cases}$$

and $K = [(k + 1)/2]/k$, then the LMS objective function leads to a solution of Equation (13.14).

The LMS procedure has a very high breakdown ϵ^* of almost 50% (the breakdown is the minimum percentage of points that can be moved arbitrarily to achieve an infinite discrepancy).

Minimization of the objective function can be difficult because the function is not smooth and there are usually local minima. An approximate procedure based on exact matching of all possible triplets of points and then choosing the triplet that minimizes the objective function leads to an approximate solution. This procedure can be speeded up and made a little more efficient by not considering very thin triplets or very small triplets, see Dryden and Walker (1999).

Many other robust regression procedures could be used for matching, such as GS-estimators, M-estimators and least absolute deviations. Verboon and Heiser (1992) use the Huber (1964) function and the bi-weight function for matching two object configurations, where reflection is also allowed. In selecting an estimator one needs to make a compromise between breakdown and efficiency, and any choice will be very much application dependent. In a simulation study Dryden and Walker (1999) found that the choice of an S-estimator with 25% breakdown leads to high efficiency when the errors are normal.

A highly resistant procedure such as LMS is very useful for identifying outliers. An approach might be to first use a resistant procedure to superimpose the objects, examine the residuals and re-investigate those landmarks with very large residuals. One possible course of action could be to ignore a suspect point and then proceed with a conventional more efficient analysis (e.g. Procrustes least squares) on the rest of the data.

Siegel and Benson (1982) considered resistant registration of objects using the technique of the repeated median. Rather than minimizing the median as in LMS the repeated median involves taking the median of the median. The algorithm for registration involves sequentially updating the scale, rotation and location parameters by univariate repeated median estimators. A major disadvantage of this technique is that it is not equivariant under affine or similarity transformations. Indeed, LMS was introduced by Rousseeuw (1984) as an alternative to the repeated median, giving equivariance as a major motivation for studying LMS.

Generalized resistant matching for random samples of objects can proceed in an analogous manner to GPA (Dryden and Walker 1998). Rohlf and Slice (1990) considered generalized matching using the repeated median technique and gave an algorithm for resistant shape matching and resistant affine matching. A practical demonstration on the shapes of mosquito wing data was compared with the usual Procrustes registration. Although the two approaches were fairly similar in that dataset, the resistant fit registration resulted in less variability at the less variable landmarks and more variability at the more variable landmarks, as expected. Dryden and Walker (1998) also examine the shape variability in a generalized matching procedure with $s(\cdot)$ given by an S-estimator and $\phi(E) = \|E\|^2$. Although plots of registered configurations look quite different in examples, the plots of the first few PCs actually look very similar when placed in the same registration.

Example 13.5   Consider matching the electrophoretic gel data of Section 1.4.14 with an affine transformation, but consider the situation where the invariant spots on gel A are located correctly but two of the invariant spots in gel B have been mislabelled. In Figure 13.6 we see the fitted points in gel A after a least squares affine transformation and after the LMS affine transformation. As expected, the least squares fit is dramatically affected by the wrongly identified points, whereas the resistant LMS fit is not affected by the two outliers.   □

13.7   Incomplete data

A practical problem that can often be encountered is that there are missing data at a subset of landmarks for some objects in a dataset. Missing data can arise for a variety of reasons, for example part of a fossil bone may be missing or an object may be occluded in an image. Distance-based methods and Euclidean Shape Tensor Analysis of Section 15.1 can deal with missing landmark data as one considers subsets of pairs or larger subsets of points, and it is not imperative that every pair/subset is

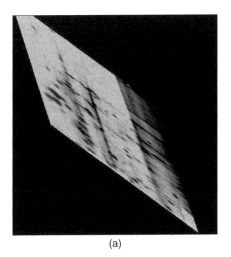 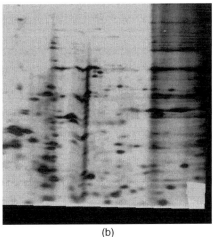

(a)   (b)

Figure 13.6   The fitted gel A registered by affine fitting using the invariant points in the gels, with two points poorly located in gel B. (a) Least squares affine transformation of gel A; (b) LMS affine transformation of gel A.

present in every object in the dataset. Obviously if a particular landmark is often missing, inferences will be less powerful concerning that landmark.

If the missing data are 'missing at random' (Rubin 1976), then one can proceed as one would do for regression analysis. We can either delete objects where landmarks are missing, delete any missing landmarks in all the objects or we can fill-in the incomplete data. The first approach is practical when just a few objects have missing landmarks. However, if a large proportion of the objects have a few (different) landmarks missing at random, then a fill-in approach may be preferable.

Procrustes analysis can be adapted using a type of EM algorithm (Bookstein and Mardia 2001). One needs to iterate between filling-in landmarks (i.e. estimating landmarks) and standard Procrustes analysis, assuming the estimated locations of the landmarks. The expectation step involves estimating the unknown landmark coordinates by taking the conditional mean of the unknown coordinates given the other landmarks in their current position through the kriging predictor, or using a thin-plate spline transformation from the mean shape for prediction. The minimization step involves the usual Procrustes registrations using the estimated missing coordinates.

Albers and Gower (2010) described a method for dealing with missing values in Procrustes analysis. Also, as described in Section 13.2, Mitchelson (2013) provided a method for handling occluded landmarks in human movement studies. Recent work dealing with missing data in studies of biological evolution of bone surfaces includes Gunz *et al.* (2009).

14

Unlabelled size-and-shape and shape analysis

The previous chapters have been mainly concerned with analysis of labelled landmarks, where there is clear, meaningful correspondence between the landmarks. Matching configurations of points where the correspondence is unknown is an important but challenging problem in many application areas, including in bioinformatics and computer vision.

In this chapter we shall consider Bayesian approaches that have been developed for matching unlabelled point sets. The matching problem, where the sets of points may be of different sizes, is relevant for the comparison of molecules and the comparison of objects from different views in computer vision. For example, if we have two protein surfaces represented by sets of amino acid locations, a question of interest is whether the two surfaces have a region of the same size-and-shape. This region may correspond to a binding site that the proteins have in common; for example they may both bind to the same protein molecule.

Some initial inferential methods to compare unlabelled shape include the MCMC simulation methodology developed by Green and Mardia (2006), Dryden *et al.* (2007) and Schmidler (2007), which themselves have connections with work stemming from Moss and Hancock (1996), Rangarajan *et al.* (1997), Chui and Rangarajan (2000, 2003) and Taylor *et al.* (2003) among others.

Consider the dataset which was described in Section 1.4.9. The active site of protein 1 contains 40 amino acids and the active site of protein 2 contains 63 amino acids. Green and Mardia (2006) developed a Bayesian model and carried out inference using an MCMC algorithm for inference about these proteins, and in particular which amino acids correspond, or match, between the two proteins.

Statistical Shape Analysis, with Applications in R, Second Edition. Ian L. Dryden and Kanti V. Mardia.
© 2016 John Wiley & Sons, Ltd. Published 2016 by John Wiley & Sons, Ltd.

14.1   The Green–Mardia model

14.1.1   Likelihood

Green and Mardia (2006) introduced a model for unlabelled size-and-shape matching which has proved successful in a number of applications. The model involves a hidden Poisson process of rate λ in a region with volume v. Two configurations X_1 and X_2 are generated as Gaussian perturbations with means given by some of the points from the hidden Poisson process, then rigid-body transformations are applied (location and rotation) to the configurations. The points in the hidden Poisson process each give rise to one of four cases: (i) points in both X_1 and X_2; (ii) a point in X_1 only; (iii) a point in X_2 only; or (iv) no points in X_1 and X_2. There is a one to one correspondence between the first type of point labels in X_1, X_2 via the hidden Poisson process. Green and Mardia (2006) then integrate out the underlying Poisson process to give a likelihood which depends on the rigid-body transformation between the configurations, a **match matrix** $\Lambda = (\lambda_{ij})$ giving the one to one correspondence between the points in X_1 and X_2, and the parameter ρ/λ, where ρ is a parameter reflecting the tendency of points to be matched.

We consider an alternative model which is very similar to that of Green and Mardia (2006), leading to a similar solution; where we assume that μ is a fixed $N \times m$ configuration and X is a $K \times m$ configuration that we apply rigid-body transformations to. In order to specify the labelling or correspondence between the points we use the match matrix Λ, which is a $K \times N$ matrix of 1s and 0s to represent a particular matching of the points in X with the points in μ. For $1 \leq j \leq N$, if $\lambda_{ij} = 1$ then the ith point of X matches to the jth point in μ. If $\sum_j \lambda_{ij} = 0$ then the ith point of X does not match to any point in μ, likewise if $\sum_i \lambda_{ij} = 0$ then the jth point of μ does not match to any point in X. If we include the constraint that each row and column can contain at most a single 1, then only one to one matches are allowed, and this is the case for Green and Mardia (2006)'s model.

The rotation matrix $\Gamma \in SO(m)$ and the translation parameter γ are also parameters in the model. The matched points in X, denoted as X^Λ, are taken as Gaussian perturbations of the matching points in μ with variance $1/\tau$, and we assume that the rows of unmatched points in X, denoted by $X^{-\Lambda}$, are distributed uniformly over a bounded region $\mathcal{A} \subset \mathbb{R}^m$ of volume $|\mathcal{A}|$, and the volume is a hyper-parameter in the model. We shall denote the points that are matched in μ as μ^Λ.

We shall concentrate on the $m = 3$ dimensional case here. Given a $K \times N$ match matrix, Λ (with p matching points), rotation matrix Γ and translation vector γ the likelihood is therefore defined as:

$$L_{GM}(X|\Lambda, \mu, \tau, \Gamma, \gamma) \propto \frac{\tau^{\frac{3p}{2}}}{|\mathcal{A}|^{K-p}} \exp\left(-\frac{\tau}{2}\text{trace}\{(\tilde{X}^\Lambda - \mu^\Lambda)^T(\tilde{X}^\Lambda - \mu^\Lambda)\}\right),$$

where $\tilde{X}^\Lambda = X^\Lambda\Gamma + 1_p\gamma^T$, $\Gamma = R_z(\theta_{12})R_y(\theta_{13})R_x(\theta_{23})$, and the rotation matrices about the x, y, z axes are:

$$R_x(\theta_{23}) = \begin{pmatrix} 1 & 0 & 0 \\ 0 & \cos\theta_{23} & \sin\theta_{23} \\ 0 & -\sin\theta_{23} & \cos\theta_{23} \end{pmatrix}, \; R_y(\theta_{13}) = \begin{pmatrix} \cos\theta_{13} & 0 & \sin\theta_{13} \\ 0 & 1 & 0 \\ -\sin\theta_{13} & 0 & \cos\theta_{13} \end{pmatrix}, \quad (14.1)$$

$$R_z(\theta_{12}) = \begin{pmatrix} \cos\theta_{12} & \sin\theta_{12} & 0 \\ -\sin\theta_{12} & \cos\theta_{12} & 0 \\ 0 & 0 & 1 \end{pmatrix}, \quad (14.2)$$

with Euler angles $\theta_{12} \in [-\pi, \pi), \theta_{13} \in [-\pi/2, \pi/2], \theta_{23} \in [-\pi, \pi)$. The uniform measure is given by:

$$\frac{1}{8\pi^2} \cos(\theta_{13}) d\theta_{12} d\theta_{13} d\theta_{23}$$

in this case (e.g. see Khatri and Mardia 1977). Further discussion of rotation matrices and a different representation were given in Section 3.2.2.

Note that Green and Mardia (2006)'s model is constructed with X_1 and X_2 as Gaussian perturbations from an underlying Poisson process. However, the likelihood is actually of the same form as the one-sided version described above (Kenobi and Dryden, 2012), where $X = X_1$ is perturbed from $\mu = X_2$, although the variance parameter is doubled and $|A| = \rho/\lambda$.

14.1.2   Prior and posterior distributions

The parameters of the model are $\tau, \Lambda, \Gamma, \gamma$, We take $\tau, \Lambda, \Gamma, \gamma$ to be mutually independent *a priori*. We use the prior distribution $\tau \sim \Gamma(\alpha_0, \beta_0)$. For the prior distribution of Λ, we assume the rows are independently distributed with the ith row having distribution

$$\pi(\lambda_{ij} = 1) = \frac{1 - \psi}{N}, \qquad 1 \le j \le N,$$

for $1 \le i \le K$ and $0 \le \psi \le 1$, where ψ is a hyperparameter to be chosen. If we insist on one to one matching then the sum of each row or column is at most 1.

We take the prior for γ as $\gamma \sim N_3(\mu_\gamma, \sigma_\gamma^2 I_3)$, and Green and Mardia (2006) use a matrix Fisher prior for Γ, which is appealing due to conjugacy. The posterior density of $(\Lambda, \tau, \Gamma, \gamma)$ conditioned on X is:

$$\pi(\tau, \Lambda, \Gamma, \gamma | X, \mu) = \frac{\pi(\tau)\pi(\Lambda)\pi(\Gamma)\pi(\gamma)L_{GM}(X|\Lambda, \tau, \mu, \Gamma, \gamma)}{\sum_\Lambda \int_0^\infty \pi(\tau)\pi(\Lambda)\pi(\Gamma)\pi(\gamma)l_{GM}(X|\Lambda, \tau, \mu, \Gamma, \gamma) d\tau}.$$

As the posterior has a complicated form, MCMC simulation is used to simulate approximately from the posterior distribution.

14.1.3 MCMC simulation

The full conditional distribution of τ is given by:

$$(\tau | X, \Lambda, \Gamma, \gamma, \mu) \sim \Gamma\left(\alpha_0 + \frac{3p}{2}, \beta_0 + \frac{\|\tilde{X}^\Lambda - \mu^\Lambda\|^2}{2}\right),$$

where p is the number of matched points in X, that is the number of non-zero rows of Λ. Hence a Gibbs update can be used for τ. The full conditional distribution of γ is given by:

$$\gamma | X, \mu, \tau, \Lambda, \Gamma \sim N\left(\frac{\mu_\gamma / \sigma_\gamma^2 + \tau \sum_{j \leq K, k \leq N, \lambda_{jk}=1}(\mu_k - x_j\Gamma)}{p\tau + 1/\sigma_\gamma^2}, \frac{1}{p\tau + 1/\sigma_\gamma^2}I_3\right), \quad (14.3)$$

and so we use a Gibbs update for γ. We update the match matrix Λ using the acceptance probability (see Section 14.2.2)

$$\alpha_\Lambda = \min\left(1, \frac{\pi(\Lambda^* | X, \mu, \tau, \Gamma, \gamma)q}{\pi(\Lambda | X, \mu, \tau, \Gamma, \gamma)q^*}\right) = \min\left(1, \frac{L(X | \Lambda^*, \mu, \tau, \Gamma, \gamma)\pi(\Lambda^*)q}{L(X | \Lambda, \mu, \tau, \Gamma, \gamma)\pi(\Lambda)q^*}\right).$$

Green and Mardia (2006) provide a Gibbs step for two of the rotation angles (due to a conjugate prior) and use a random walk Metropolis step for the other angle. Green and Mardia (2006) ensure that the matching is one to one between the points. Additional prior assumptions can be included, for example to match similar amino acids in terms of the four groups described in Section 1.4.9.

After running the MCMC algorithm for a large number of iterations the final estimate of the match matrix can be obtained using an appropriate loss function, as discussed by Green and Mardia (2006). In particular let ℓ_{01} be the loss of declaring a match when there is not one for a pair of points, and ℓ_{10} is the loss of not declaring a match when there is one. The optimal match is obtained by maximizing the sum of estimated marginal posterior probabilities minus $K = \ell_{01}/(\ell_{01} + \ell_{10})$ times the number of matched points.

The Green–Mardia model has been demonstrated to work well in a variety of situations (Mardia et al. 2007). Further examples include Green et al. (2010) for Bayesian modelling for matching and alignment of biomolecules; Ruffieux and Green (2009) alignment of multiple configurations; and additional applications are given in Mardia et al. (2011, 2013b) and Mardia (2013b) including the use of a gap prior to include positional information along the protein backbone.

Example 14.1   The active sites are given in Figure 14.1 for each of the two proteins, with 40 in 1cyd (red and yellow) and 63 active sites in 1a27 (blue and green). Following application of the Green–Mardia MCMC algorithm with 1 000 000 MCMC iterations the top 35 aligned active sites (in terms of highest probability of match) are displayed in Figure 14.1, and coloured red and blue. The non-matched amino acids

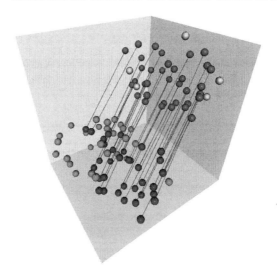

Figure 14.1   The active site locations of proteins `1cyd` *(upper point cloud) and* `1a27` *(lower point cloud) from the Protein Data Bank. The lines connect the top 35 estimated aligned active sites from Green and Mardia (2006). For a colour version of this figure, see the colour plate section.*

are yellow and green. Clearly the common binding region has a very similar size-and-shape for these 35 matches, see Green and Mardia (2006) who have 36 top matches with $K = 0.5$ whereas we chose $K = 0.75$ leading to 35 matches.   □

14.2   Procrustes model

A Procrustes size-and-shape model was developed by Dryden *et al.* (2007) and Schmidler (2007) for unlabelled size-and-shape matching. The model is similar to the Green–Mardia model, except that rotation and translation are estimated using Procrustes registration at each step of the algorithm. A MCMC algorithm is used to draw inferences about the match matrix and a concentration parameter. For the Procrustes model given a particular match matrix Λ, we use partial Procrustes registration to register the p points in X^Λ to the corresponding matched points in μ^Λ, in order to define a distance between the size-and-shapes. This aspect of the matching is present in both the Dryden *et al.* (2007) and Schmidler (2007) approaches. The Procrustes matching involves finding $\hat{\Gamma} \in SO(m)$ and $\hat{\gamma} \in \mathbb{R}^m$ such that

$$\| \mu^\Lambda - X^\Lambda\hat{\Gamma} - 1_p\hat{\gamma}^T \| = \inf_{\substack{\Gamma \in SO(m) \\ \gamma \in \mathbb{R}^p}} \| \mu^\Lambda - X^\Lambda\Gamma - 1_p\gamma^T \| = d_S(X^\Lambda, \mu^\Lambda),$$

where $d_S(X^\Lambda, \mu^\Lambda)$ is the Riemannian metric in size-and-shape space (5.5). The Procrustes estimators of rotation and translation were given in Section 7.2.3.

Let $\hat{X}^\Lambda = X^\Lambda \hat{\Gamma} + 1_p \hat{\gamma}^T$. Then $\hat{X}^\Lambda$ is the partial Procrustes fit of X^Λ onto μ^Λ. The partial Procrustes tangent coordinates of X^Λ at μ^Λ are given by the $p \times m$ matrix

$$V^\Lambda = \hat{X}^\Lambda - \mu^\Lambda = X^\Lambda \hat{\Gamma} + 1_p \hat{\gamma}^T - \mu^\Lambda$$

which is in a $pm - m(m-1)/2 - m$ dimensional linear subspace of $\mathbb{R}^{mp}$.

We denote the unmatched points in X by $X^{-\Lambda}$. We transform these points using the same transformation parameters as for X^Λ. Let $\hat{X}^{-\Lambda} = X^{-\Lambda} \hat{\Gamma} + 1_{K-p} \hat{\gamma}^T$. We consider $X^{-\Lambda}$ to lie in $\mathbb{R}^{(K-p)m}$. Given the match matrix, Λ, the size-and-shape of X^Λ lies in $S\Sigma_m^p$ and $X^{-\Lambda}$ lies in $\mathbb{R}^{(K-p)m}$.

We assume a zero mean isotropic Gaussian model for V^Λ in $q = pm - m(m-1)/2 - m$ dimensions. [There are $m(m-1)/2 + m$ linear constraints on V^Λ due to the Procrustes registration.] We assume that $X^{-\Lambda}$, the non-matching part, is uniformly distributed in a bounded region, $\mathcal{A}$, with volume $|\mathcal{A}|$ in $\mathbb{R}^m$, and this volume is a hyper-parameter in the model.

The likelihood of X given Λ and $\tau = 1/\sigma^2$, a precision parameter where σ^2 is a measure of the variability at each point, is:

$$
\begin{aligned}
L_P(X|\Lambda, \tau, \mu) &= f_{V^\Lambda}(V^\Lambda | \tau, \Lambda, \mu) f_{X^{-\Lambda}}(X^{-\Lambda} | \Lambda) \\
&= (2\pi)^{-q/2} \tau^{q/2} \exp\left(-\frac{\tau}{2} \text{trace}\{(V^\Lambda)^T V^\Lambda\}\right) \times \frac{1}{|\mathcal{A}|^{K-p}} \\
&= (2\pi)^{-q/2} \tau^{q/2} \exp\left(-\frac{\tau}{2} d_S(X^\Lambda, \mu^\Lambda)^2\right) \times \frac{1}{|\mathcal{A}|^{K-p}}.
\end{aligned}
$$

This likelihood is given by Dryden *et al.* (2007) and is essentially that of Schmidler (2007) (with $q = mp$ in the latter).

14.2.1   Prior and posterior distributions

We write $\pi(\tau)$ and $\pi(\Lambda)$ for the prior distributions of τ and Λ and assume τ and Λ are independent *a priori*.

Schmidler (2007) introduced a gap prior to include positional information along the protein backbone, so that matches are encouraged in contiguous regions along the backbone with small numbers of gaps. Mardia *et al.* (2013b) also use a gap prior in a model for unlabelled shape matching.

The posterior density of τ and Λ conditional on X is:

$$\pi(\tau, \Lambda | X, \mu) = \frac{\pi(\tau)\pi(\Lambda)L(X|\Lambda, \tau, \mu)}{\sum_\Lambda \int_0^\infty \pi(\tau)\pi(\Lambda)L(X|\Lambda, \tau, \mu)d\tau}.$$

Again to carry out inference one uses MCMC.

14.2.2   MCMC inference

The full conditional distribution of τ is available from the conjugacy of the Gamma distribution,

$$(\tau|X, \Lambda, \mu) \sim \Gamma\left(\alpha_0 + \frac{q}{2}, \beta_0 + \frac{d_S(X^\Lambda, \mu^\Lambda)^2}{2}\right),$$

so we update τ with a Gibbs step. We make updates to the match matrix using a Metropolis–Hastings step. We select a row at random and if the selected point is already matched then it becomes unmatched with probability p_{reject}, or it is matched to another point i with probability $(1 - p_{reject})/(N - 1)$. If the selected point is unmatched then it becomes matched to point i with probability $1/N$. We accept the new proposal, Λ^*, with probability

$$\alpha_\Lambda = \min\left\{1, \frac{\pi(\Lambda^*|X, \mu, \tau)q}{\pi(\Lambda|X, \mu, \tau)q^*}\right\},$$

where

$$q/q^* = \begin{cases} p_{reject}/(1/N) & \text{if making an unmatched point matched,} \\ (1/N)/p_{reject} & \text{if making a matched point unmatched,} \\ 1 & \text{if making a matched point match to a different point.} \end{cases}$$

If $p_{reject} = 1/N$ then $q/q^* = 1$, which was the value used by Dryden *et al.* (2007).

Dryden *et al.* (2007) also describe a computationally faster approximate Metropolis–Hastings update to the match matrix which does not require the use of the whole configuration in the calculation of the density. If we propose the change $(i \to l_1)$ to $(i \to l_2)$ then the alternative Hastings ratio, α_Λ^* is given by:

$$\alpha_\lambda^* = \min\{g(x_i, \mu_{l_2})q/(g(x_i, \mu_{l_1})q^*), 1\}, \tag{14.4}$$

where

$$g(x_i, \mu_j) = \begin{cases} \frac{1-\psi}{N}\left(\frac{\tau}{2\pi}\right)^{m/2} \exp\left(-\frac{\tau}{2}|x_i - \mu_j|^2\right), & \text{if } j < N + 1 \\ \psi\frac{1}{|\mathcal{A}|} & \text{if } j = N + 1. \end{cases}$$

When a new match is accepted the ordinary partial Procrustes registration is carried out on the new matching points to ensure the configuration of matching points has rotation removed. For brevity we shall refer to the size-and-shape model as the 'Procrustes model'.

Kenobi and Dryden (2012) describe various approximations to the algorithm to help prevent the MCMC routine becoming stuck at local modes, including involving some large jump proposals. Other work where large jumps have been used to help address multimodality includes that by Tjelmeland and Hegstad (2001) and Tjelmeland and Eidsvik (2004). Their scenarios are simpler where they can make use of local optimization to construct reversible jumps between modes. Kenobi and Dryden (2012) do not carry out local optimization involving gradients, but rather use the nearness of points in physical space for constructing a large jump into a different mode in this large discrete combinatorial search space of matchings. It is not clear how to make such moves reversible, and so the moves are just carried out during the initialization phase.

Kenobi and Dryden (2012) compare their implementation of the Green–Mardia model with the Procrustes model, and performance is quite similar in practice. The Green–Mardia method does appear to mix better, as the Procrustes method can become stuck at local modes. However, good matches had a higher probability of a match with the Procrustes method.

The main difference between the inference approaches is that for the Green–Mardia model one **marginalizes** over the registration parameters, that is they are integrated out, whereas in the Procrustes model one **optimizes** over the registration parameters. The two methods are often quite similar in performance. In practice it is often reasonable that a Laplace approximation holds, hence explaining the similarity of the two methods (Kenobi and Dryden 2012).

The choice as to whether to marginalize or optimize is also present in Chapter 16 in the analysis of curve shapes. The quotient space model of Section 16.4.2 involves optimization whereas the Bayesian ambient space model of Section 16.4.3 involves marginalization.

Example 14.2   Dryden *et al.* (2007) apply the Procrustes model to the steroid data of Section 1.4.4. In particular, they consider all pairwise matchings between steroids using the MCMC algorithm above, and then use the posterior modes to provide distances between each steroid molecule. Cluster analysis is then carried out, and it is demonstrated that steroids with similar binding affinities to the CBG receptor protein also have more similar size-and-shapes.   □

14.3   Related methods

Early methods for unlabelled matching include those by Moss and Hancock (1996) who use graph-based methods; Rangarajan *et al.* (1997) who developed a soft matching method (softassign) which used a very efficient numerical algorithm for optimization; Walker (1999) for matching electrophoresis gels; Chui and Rangarajan (2000, 2003); and Taylor *et al.* (2003) who developed an EM algorithm for estimating probabilities of atom matches in bioinformatics. Kent *et al.* (2010) provide an EM interpretation of the softassign algorithm of Rangarajan *et al.* (1997).

The Procrustes model is extended by Dryden *et al.* (2007) to deal with multiple observations. If there are n observations then the parameters include n matching matrices, as well as an overall unknown mean μ. The parameters are estimated again by MCMC, and the overall mean μ can be updated using an efficient Gibbs procedure. Ruffieux and Green (2009) extended the Green–Mardia model to deal with the alignment of multiple configurations. Mardia *et al.* (2011) further extended the model for estimating pharmacophores in bioinformatics, where it is of interest to find a common structure among a collection of proteins.

The above methods are all for size-and-shape comparisons, as in bioinformatics one usually wants to retain the scale. Mardia *et al.* (2013b) consider shape analysis, where isotropic scaling is also dealt with, using an adaptation of the Green–Mardia model.

An alternative but related method for unlabelled matching was provided by Czogiel *et al.* (2011) where the molecules were regarded as continuous fields of some quantity (e.g. partial charge or atomic radius), and observations from the field are available at a finite set of atoms. A Hilbert space representation is used to calculate a distance between the fields, and again a Bayesian model is developed for inference. The advantage with Czogiel *et al.* (2011)'s procedure is that there is no need to estimate the combinatorially expensive matching matrix. The methodology was applied to the steroids data of Section 1.4.4 and the procedure provided good clustering of the molecules in terms of size-and-shape being associated with binding affinity to the CBG receptor protein. The MCMC methodology has also been extended to outline matching with occlusions with applications in computer vision (Cao *et al.* 2011). Yu *et al.* (2014) also consider hierarchical Bayesian modelling for multigroup shape analysis where correspondences need to be inferred, through a joint distribution of shape variables and object boundary data. Posterior inference is via MCMC simulation.

An alternative method for matching unlabelled objects is to use Minimum Description Length (MDL), which is a method for choosing between models using the shortest description of a training dataset. There are strong connections to model choice methods, including the Bayesian Information Criterion (BIC) (Schwarz 1978). Davies *et al.* (2002) introduce an MDL approach to statistical shape modelling, and many applications including to image and surface matching are given by Davies *et al.* (2008b). Generalized matching of a group of surfaces is discussed by Davies *et al.* (2008a) and Twining *et al.* (2011).

Many of the applications of unlabelled shape analysis have been in bioinformatics, which motivated the methodology of Green and Mardia (2006); Dryden *et al.* (2007) and Schmidler (2007). Some other examples where shapes of protein structures are compared can be found in Liu *et al.* (2011) and Rodriguez and Schmidler (2014), who use a gap prior. Schmidler *et al.* (2002) provided an early method of Bayesian protein structure prediction; Schmidler *et al.* (2007) considered models of helices in bioinformatics; and Panaretos and Konis (2011) develop sparse approximations of protein structure from noisy random projections from single molecule experiments. For further relevant work, a collection of papers on the topic of Bayesian structural bioinformatics is found in Hamelryck *et al.* (2012).

14.4   Unlabelled points

In the preceding sections it makes sense to estimate a correspondence between points, and so we assume that there is an underlying unknown labelling. However, in other applications we want to take into account all possible labellings – there is no sensible correspondence that can be estimated. One particular area of development has been that of unlabelled triangles, which work with a subset of the shape space. For example, for the triangle case where reflections and labelling are not important we work with a half-lune which is one-twelfth of the shape space (see Section 2.6.2). Procrustes methods can be adapted to the unlabelled case, and the minimizations are carried out over the similarity transformations **and** permutations of the landmark labels.

14.4.1   Flat triangles and alignments

Kendall and Kendall (1980), Kendall (1984), Small (1988, 1996, p. 158) and Stoyan *et al.* (1995, Chapter 8) describe the analysis of unlabelled shapes of triplets formed from the megalithic standing stones of Cornwall (see Section 1.4.18). The expected number and variance of nearly collinear triangles are obtained and used to test whether the stones could have been deliberately placed in straight lines or whether the stones were placed at random. The set of collinear shapes is denoted by *Coll* (Kendall 1984). Le (1991a, 1992) obtains Procrustes distances of configurations to the collinearity set *Coll*, and further applications to testing for collinearities are given by Kendall (1991a) and Le (1992). With the motivation of understanding human vision, Bhavnagri (1995a) also considers distances to *Coll* and to the set of regular polygons *Sph*, which was also defined by Kendall (1984). Le and Bhavnagri (1997) also derive detailed results based on distances to *Coll*.

14.4.2   Unlabelled shape densities

We can also consider shape distributions for unlabelled i.i.d. points in a region. Kendall (1985), Kendall and Le (1987) and Le (1987) consider shape distributions for triangles generated from i.i.d. triplets in a convex polygon K. A particularly simple result is that if three points are randomly placed inside a square, then the resulting shape density of Kendall's shape variables (u^K, v^K), with respect to the uniform shape measure $d\gamma$, is given by (Kendall and Le 1987):

$$f_1(u^K, v^K) = \frac{1}{4} - \frac{1}{40}(v^K\sqrt{3})^2 \left[\frac{1}{1 - u^K\sqrt{3}} + \frac{1}{1 + u^K\sqrt{3}} \right]$$

$$+ \frac{1}{240}(v^K\sqrt{3})^4 \left[\frac{1}{(1 - u^K\sqrt{3})^3} + \frac{1}{(1 + u^K\sqrt{3})^3} \right].$$

If the convex polygon is a rectangle of sides $s \times 1$, then the density is:

$$f_2(u^K, v^K) = \frac{s^3 + s^{-3}}{2}(f_1(u^K, v^K) - \frac{1}{4}) + \frac{s + s^{-1}}{8}.$$

Another important result for three i.i.d. points in a general convex region is that the shape density is constant for values of $v^K = 0$ (Small 1982), as is simply seen in the rectangle case where $f_1(u^K, 0) = \frac{1}{4}$ and $f_2(u^K, 0) = (s + s^{-1})/8$.

14.4.3 Further probabilistic issues

Further work has been carried out on probabilistic issues for unlabelled shapes. For example, Watson (1986) and Mannion (1988, 1990a,b) have considered Markovian sequences of shapes of triangles and Gates (1994) has considered the shapes of triangles formed by random lines in a plane. D.G. Kendall (1977), W.S. Kendall (1990a, 1998) and Le (1991b, 1994) have considered the diffusion of shape, and relationships to the offset normal shape distributions of Section 11.1 have been made. In particular, consider $k \geq 3$ points $(X_j, Y_j)^T, j = 1, \ldots, k$, that are in Brownian motion having started at time $t = 0$ at $(\mu_j, \nu_j)^T, j = 1, \ldots, k$. The distribution of the points at time $t = \sigma^2$ is the isotropic model of Equation (11.10), and so the diffused shape at this time has the offset normal shape distribution of Section 11.1.2.

14.4.4 Delaunay triangles

Another application involves the shapes of Delaunay triangles in Dirichlet tessellations. Miles (1970) gave the basic result that the shapes of typical Delaunay triangles from a Poisson process are independent of the circumradius, and the shape distribution is particularly simple. From Kendall (1983), when shape is measured in terms of two of the internal angles of the triangle α and β, the density with respect to the uniform shape measure $d\gamma$ is:

$$f(\alpha, \beta) = \frac{4}{9} \left[\sin^2 \alpha + \sin^2 \beta + \sin^2(\alpha + \beta) \right]^2.$$

This distribution is known as the Miles distribution and the mode is the equilateral shape. See Miles (1970) and Kendall (1983) for further discussion.

Example 14.3   Mardia *et al.* (1977) and Mardia (1989b) have examined whether Delaunay triangles from the Iowa data of Section 1.4.17 could be more equilateral than expected by chance. In Figure 14.2 we see a plot of a spherical blackboard corresponding to $AB \geq AC \geq BC$. Points have been plotted on the bell according to the shapes of triangles from the Delaunay triangles of the Iowa data in Section 1.4.17. There is a tendency for shapes to concentrate towards the top of the bell (i.e. to be more equilateral), although there was no evidence for central place theory in the test of Mardia (1989b).   □

Dryden *et al.* (1997) also consider a statistic based on average squared partial Procrustes distance to the equilateral triangle, where the fact that neighbouring triangles are correlated has been taken into account. The application that Dryden *et al.* (1995, 1997), Taylor *et al.* (1995) and Faghihi (1996) consider is the examination of regularity ('equilateralness') in an image of the cross-section of muscle fibres. Also, Dryden

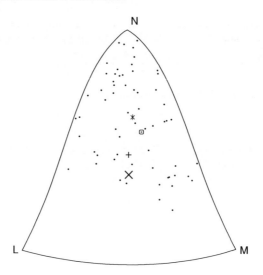

Figure 14.2   The Iowa central place data on Kendall's spherical blackboard. There are some special points marked by symbols representing a full Procrustes mean shape (*), the mean of the uniform distribution (X), the centre of the bell (+), and the circle is the mean of 63 simulated values from the Miles distribution. Source: Mardia 1989b. Reproduced with permission from John Wiley & Sons.

and Zempléni (2006) investigate extreme unlabelled triangle shapes in muscles, using the generalized extreme value and generalized Pareto distributions. Here 'extreme' relates to being a long way in shape distance from the equilateral shape. Differences between diseased and healthy muscles in triangle shapes were observed with these methods.

15

Euclidean methods

15.1 Distance-based methods

An alternative to working with geometrical configurations directly is to work with inter-landmark distances. Consider the **squared Euclidean distance matrix** D from the configuration X ($k \times m$ matrix) given by:

$$(D)_{rs} = \|(X)_r - (X)_s\|^2 \ , \ r, s = 1, \dots, k, \tag{15.1}$$

where $(X)_r$ are the coordinates of the rth point ($r = 1, \dots, k$). We consider methods for shape and size-and-shape analysis that involve working with the full collection of such distance matrices, and in some cases the estimates can be similar to Procrustes techniques. Traditional morphometrics studying lengths, ratios of lengths or angles usually considers just a subset of the inter-landmark distances, and was summarized in Section 2.3.

15.2 Multidimensional scaling

15.2.1 Classical MDS

Multidimensional scaling (MDS) is concerned with constructing a configuration of k points in Euclidean space from information about the distances between the k points (see Mardia *et al.* 1979, pp. 394–398). Consider X to be a $k \times m$ configuration with $k \times k$ squared Euclidean distance matrix D, as in Equation (15.1). It can be shown that D is a squared Euclidean distance matrix if and only if

$$B = -\tfrac{1}{2} CDC$$

Statistical Shape Analysis, with Applications in R, Second Edition. Ian L. Dryden and Kanti V. Mardia.
© 2016 John Wiley & Sons, Ltd. Published 2016 by John Wiley & Sons, Ltd.

is positive semi-definite, where C is the $k \times k$ centring matrix of Equation (2.3). We can interpret B as the centred inner product matrix of the $k \times m$ configuration X.

If B is positive semi-definite with rank $p \leq k - 1$, then a configuration can be constructed which has the same squared Euclidean distance matrix as D, for example

$$Y = [f_1, f_2, \ldots, f_p],$$

where f_j are the eigenvectors of B scaled so that $f_j^T f_j = a_j$, where the a_j are the eigenvalues of B in descending order. Note that any translated, rotated and reflected version of Y will also have the same squared Euclidean distance matrix.

In practice it is often useful to find a configuration Y^* in $m < p$ dimensions with squared Euclidean distance matrix approximately equal to D. A sensible approximation is to take

$$MDS_m(D) = [f_1, f_2, \ldots, f_m].$$

This configuration is called the classical solution to the MDS problem. It will be reasonable provided the first m eigenvalues of B are large compared with the rest. Any rotation and reflection of $MDS_m(D)$ will also be an equivalent solution.

15.2.2   MDS for size-and-shape

Consider a sample of n configurations of k landmarks. Let $(W)_{h_1, h_2} = \frac{1}{n} \sum_{i=1}^n d_i^2(h_1, h_2)$ be the average squared Euclidean distance matrix over the n configurations, where $d_i^2(h_1, h_2)$ is the squared Euclidean distance between the landmarks $h_1, h_2 \in \{1, \ldots, k\}$ for the ith observation, $i = 1, \ldots, n$. Then following the classical MDS method an estimate of the mean reflection size-and-shape is:

$$\hat{\mu}_{MDS} = MDS_m(W). \tag{15.2}$$

In such a distance-based method we cannot distinguish between an estimate and its reflection, but for concentrated data, the appropriate choice is not difficult. This method is described in Kent (1994) which applies for both size-and-shape and shape data. For size-and-shape data no pre-scaling is necessary. For shape data the configurations are pre-scaled to unit size. In the planar case ($m = 2$) we can plot a 2D solution by using $f_1 + if_2$ or $f_1 - if_2$.

15.3   Multidimensional scaling shape means

Several types of MDS means have been suggested in the literature and the following summary is taken from Dryden et al. (2014). The reflection shape of an object may be considered to be the geometric information remaining when all location, scale, rotation and reflection information has been removed. There are various ways of defining the mean reflection shape of an object, including the approach proposed independently by Bandulasiri and Patrangenaru (2005) and Dryden *et al.* (2008a), and the alternative definition proposed by Bhattacharya (2008). There is in fact a family of related definitions of mean reflection shape and details were given by Dryden *et al.* (2014) and in an unpublished ISI 2009 conference paper by Preston and Wood (2009).

Using the notation in Dryden *et al.* (2008a), the MDS mean reflection shape of an object in m dimensions described by k landmarks is defined as follows. Let X denote an $m \times (k - 1)$ Helmertized transposed configuration matrix, scaled so that the trace $\text{trace}(X^\top X) = 1$. We assume throughout that $1 \leq m < k - 1$. Then $X^\top X$ lies in the space $P_m(k - 1)$, where we define $P_m(k - 1)$ to be the space of $(k - 1) \times (k - 1)$ symmetric matrices Y with $1 \leq \text{rank}(Y) \leq m$ and $\text{trace}(Y) = 1$.

Suppose that the population mean $\Xi = E(X^\top X)$ has spectral decomposition $\sum_{i=1}^{k-1} \lambda_i u_i u_i^\top$, where $\lambda_1 \geq \cdots \geq \lambda_{k-1} \geq 0$ are the eigenvalues of Ξ, with corresponding unit eigenvectors $u_1, \ldots, u_{k-1}$. The mean reflection shape, as defined by Bandulasiri and Patrangenaru (2005) and Dryden *et al.* (2008a) is given by:

$$\phi(\Xi) = \frac{1}{\lambda_1 + \cdots + \lambda_m} \sum_{i=1}^{m} \lambda_i u_i u_i^\top. \tag{15.3}$$

Note that, when the distribution of $X^\top X$ is non-degenerate, Ξ will not lie in $P_m(k - 1)$ but, by construction, $\phi(\Xi)$ always does.

An alternative definition of the mean reflection shape, proposed by Bhattacharya (2008, Section 6), is given by:

$$A(\Xi) = \sum_{i=1}^{m} (\lambda_i - \bar{\lambda} + m^{-1}) u_i u_i^\top, \tag{15.4}$$

where the λ_i and u_i are as before and $\bar{\lambda} = m^{-1} \sum_{i=1}^{m} \lambda_i$. Note that $A(\Xi)$ also lies in $P_m(k - 1)$.

Observe that the adjustments for the MDS estimators to lie in $P_m(k - 1)$ are different: for Equation (15.3) there is a multiplicative adjustment and for Equation (15.4) an additive adjustment. In fact, Dryden *et al.* (2014) showed that $\phi(.)$ and $A(.)$ are both members of a one-parameter family of projections indexed by $\alpha \geq 1/2$ of the symmetric non-negative definite matrices onto $P_m(k - 1)$, using distance

$$d_\alpha(A, B) = \|A^\alpha - B^\alpha\|,$$

where A and B are non-negative definite square matrices. The projections used in Equation (15.3) and Equation (15.4) are for $\alpha = 1/2$ and $\alpha = 1$, respectively, although in both cases the averaging is done using $\alpha = 1$. We use the abbreviation MDS for the mean in Equation (15.3) and MDS($\alpha = 1$) for the estimate in Equation (15.4).

Example 15.1   An example comparing MDS and MDS($\alpha = 1$) means to a family of extrinsic means and the intrinsic mean was given in Example 6.2 applied to the male gorilla data of Section 1.4.8. To obtain the MDS means in R we can use the following code:

```
data(gorm.dat)
> A<-MDSshape(gorm.dat,alpha=1,projalpha=1/2)
> A
        [,1] [,2]
[1,]  0.5002587  0.0255867592
[2,] -0.4369590 -0.0676547676
[3,] -0.3239341  0.1445435828
[4,] -0.1997595  0.1625496138
[5,]  0.1127959  0.1400977827
[6,]  0.4304058 -0.0005215325
[7,]  0.1291547 -0.1888126195
[8,] -0.2119625 -0.2157888189
> B<-MDSshape(gorm.dat,alpha=1,projalpha=1)
> B
        [,1] [,2]
[1,]  0.5000616  0.0256419012
[2,] -0.4367869 -0.0678005703
[3,] -0.3238065  0.1448550885
[4,] -0.1996808  0.1628999242
[5,]  0.1127515  0.1403997073
[6,]  0.4302363 -0.0005226565
[7,]  0.1291038 -0.1892195293
[8,] -0.2118790 -0.2162538651
> riemdist(A,B)
[1] 0.0009210365
```

Here we compute the MDS mean shape followed by the MDS($\alpha = 1$) mean shape, for the male gorilla data of Section 1.4.8. Here the two mean shapes are very similar, with Riemannian shape distance just 0.0009 apart. □

It is not clear which definition of mean reflection shape is to be preferred. Bhattacharya (2008) points out that (15.4) is a Fréchet mean with respect to a particular metric d_1 and indicates a preferance for that reason, but it is not clear why a projection involving an additive adjustment should be preferred to projection involving a multiplicative adjustment or something else.

Some inferential procedures for MDS shape were given by Dryden et al. (2008a), including central limit theorems, and Preston and Wood (2010, 2011) employed the MDS definition of the mean (15.3) to develop bootstrap approaches for one- and two-sample problems. From numerical investigations inference based on the null asymptotic distributions is rarely likely to be reliable; but using the bootstrap typically offers a substantial improvement in performance. For settings in which n is only moderately large, there is often a need to regularize the test statistic used within the bootstrap. Preston and Wood (2010) considered three different adjustments: (i) leaving the first p eigenvalues unchanged, $\tilde{\lambda}_i = \lambda_i$ for $i = 1, \ldots, p$, and replacing the rest with their mean, that is, $\tilde{\lambda}_i = (d - p)^{-1} \sum_{j=p+1}^{d} \lambda_j$ for $i = p + 1, \ldots, d$; (ii) leaving the first p eigenvalues unchanged and replacing the rest with the $(p + 1)$th eigenvalue, that is $\tilde{\lambda}_i = \lambda_{p+1}$ for $i = p + 1, \ldots, d$; and (iii) adding the $(p + 1)$th eigenvalue to each eigenvalue, that is $\tilde{\lambda}_i = \lambda_i + \lambda_{p+1}$ for $i = 1, \ldots, d$. Preston and Wood (2010) carried out simulation studies which demonstrated that all three regularizations worked well, but (i) was the best in their examples.

If variations are small, then the MDS mean reflection shape will be very similar to that of the full Procrustes mean shape or the Bookstein mean shape. The reason for the similarity of the approaches is that under a perturbation model pre-scaled distances are approximately linear transformations of errors at landmarks. From Example 6.2 we see that the MDS means are less resistent to outliers than the Procrustes and intrinsic mean.

15.4 Euclidean distance matrix analysis for size-and-shape analysis

15.4.1 Mean shape

The classical MDS approach leads to a biased estimate of mean shape or size-and-shape under normal errors. In order to correct for this bias the method of Euclidean distance matrix analysis has been proposed for size-and-shape analysis. An overall summary of this area of work was given by Lele and Richtsmeier (2001).

The method of Euclidean distance matrix analysis (EDMA) (Lele 1991, 1993) also involves the classical MDS solution from an estimated $k \times k$ matrix of all inter-landmark distances, but a correction is made for bias under normal models. Modelling assumptions are made in EDMA whereas there were no modelling assumptions in MDS. Euclidean distance matrix analysis is intended for size-and-shape analysis, whereas MDS is appropriate for size-and-shape or shape analysis.

Let $F(X)$ be the **form distance matrix**, which is the $k \times k$ matrix of all pairs of inter-landmark distances in the configuration X. An estimate of the population form distance matrix $F(\mu)$ can be obtained using normal models. Stoyan (1990) and Lele (1991, 1993) use inter-landmark distances and then use MDS to estimate reflection size-and-shape. Their method involves estimating population distances using a method of moments under a normality assumption, as in the following.

We shall concentrate on the $m = 2$ dimensional case here, although calculations for $m = 3$ dimensions can also be derived (Lele and Richtsmeier 2001). Let (x_j, y_j) be the coordinates of the jth landmark in two dimensions, distributed independently as:

$$(x_j, y_j)^T \sim N_2 \left((\mu_j, \nu_j)^T, \frac{\sigma^2}{2} I_2 \right), j = 1, \ldots, k. \tag{15.5}$$

Then

$$(x_r - x_s)^2 + (y_r - y_s)^2 = D_{rs}^2 \sim \sigma^2 \chi_2^2 (\delta_{rs}^2 / \sigma^2), \tag{15.6}$$

where $r = 1, \ldots, k$, $s = 1, \ldots, k$, $r \neq s$, χ_2^2 denotes a non-central chi-squared distribution with two degrees of freedom, and $\delta_{rs}^2 = (\mu_r - \mu_s)^2 + (\nu_r - \nu_s)^2$. We have

$$E(D_{rs}^2) = 2\sigma^2 + \delta_{rs}^2, \quad \text{var}(D_{rs}^2) = 4\sigma^4 + 4\delta_{rs}^2\sigma^2, \tag{15.7}$$

so that

$$\delta_{rs}^4 = \{E(D_{rs}^2)\}^2 - \text{var}(D_{rs}^2). \tag{15.8}$$

Important point: Consider a random sample of squared distances between the two landmarks labelled r and s, for n objects, given by $d_1^2, \ldots, d_n^2$. From Equation (15.7) we see that the sample average of the squared distances between landmarks r and s is **biased**, and the bias is $2\sigma^2$.

We can remove the bias by substituting the sample moments into Equation (15.8) to obtain a moment estimate of the population squared Euclidean distance δ_{rs}^2,

$$\hat{\delta}_{rs}^2 = \left\{ (\bar{d}_{sq})^2 - \frac{1}{n}\left[\sum_{i=1}^n (d_i^2 - \bar{d}_{sq})^2 \right] \right\}^{1/2},$$

where $\bar{d}_{sq} = n^{-1}\sum_{i=1}^n d_i^2$. Anisotropy can also be accommodated (Lele 1993). The estimate of mean reflection size-and-shape

$$MDS_2(\Delta), \quad (\Delta)_{rs} = \hat{\delta}_{rs},$$

is consistent for mean reflection size-and-shape μ under model (15.5) with isotropic multivariate normal errors (Lele, 1993), but is not robust since $\text{var}(d^2)$ involves fourth moments. Thus, EDMA is closely related to MDS, but the additional modelling assumptions allow any bias to be removed. As with any size-and-shape study, the objects have to be recorded to the same scale. Any rescaling will affect the distributional assumptions and the bias corrections will no longer be appropriate. Stoyan and Stoyan (1994) also discuss similar distance-based methods in some detail, concentrating on the triangle case.

If variations are small, then the EDMA mean reflection shape will be very similar to that of MDS, which in turn is very similar to the full Procrustes mean shape or the Bookstein mean shape. For further details of EDMA methods see Lele and Richtsmeier (2001); Heo and Small (2006) also provide a detailed review.

Recent work on estimating Euclidean distance matrices with applications to protein size-and-shape estimation includes that by Zhang *et al.* (2016) who use a regularized kernel-based method for consistent estimation.

15.4.2   Tests for shape difference

We consider two tests for shape difference between mean shape or size-and-shape in two independent groups, using distance-based methods.

15.4.2.1   EDMA-I

Lele and Richtsmeier (1991) consider the EDMA-I test for difference in reflection size-and-shape between means in two groups. In order to examine differences in reflection size-and-shape consider the form ratio distance matrix

$$D_{ij}(X, Y) = F_{ij}(X)/F_{ij}(Y). \tag{15.9}$$

In order to test for mean reflection size-and-shape differences Lele and Richtsmeier (1991) advise the use of the test statistic

$$T = \max_{i,j} D_{ij}(\hat{\mu}, \hat{v})/\min_{i,j} D_{ij}(\hat{\mu}, \hat{v}), \tag{15.10}$$

where $\hat{\mu}$ and $\hat{v}$ are suitable estimators of mean size-and-shape (such as the EDMA estimator). Bootstrap procedures are used to estimate the null distribution of the test statistic.

Example 15.2   Taylor (1996) used EDMA to investigate the size-and-shape of oxygen masks produced for the US Air Force. A set of 3D coordinates between landmarks were taken on 30 male and 30 female subjects – 16 anatomical landmarks on the face and 14 landmarks on the mask, using a laser scanner. An assessment was made as to how the mask fitted, and the fit was rated as a Pass or Fail. Tests using EDMA were carried out to assess whether or not there were reflection size-and-shape differences between the Passes and Fails. There were no significant differences in facial reflection size-and-shape, but there were significant differences in how the mask was placed on the face in the Passes and Fails. Thus, the use of EDMA provided mask placement specifications for a good fit.   □

15.4.2.2   EDMA-II

Lele and Cole (1995) propose an alternative, more powerful test, EDMA-II, which is appropriate for testing for different mean reflection shapes or reflection size-and-shapes in two groups. First of all estimates of the average form distance matrices are obtained for each group $\hat{F}_\mu$ and $\hat{F}_v$. Each entry of an average form distance matrix is then scaled by an overall size measure for the group. The 'shape difference matrix' is then defined by the arithmetic difference of the two scaled average form distance matrices. The test statistic is given by the value of that entry in the shape difference matrix which is farthest from zero. Using the normal model and estimated mean size-and-shape and estimated covariance matrices, a parametric Monte Carlo confidence interval for the test statistic is obtained. The null hypothesis would be rejected at the appropriate level if the interval does not contain zero.

Lele and Cole (1995) consider a power study comparing EDMA-II and Hotelling's T^2 test using Bookstein coordinates. The relative power of the two tests depends on the particular mean shape difference in the alternative hypothesis.

EDMA-II can be more powerful than Hotelling's T^2 for some alternatives, whereas Hotelling's T^2 can be more powerful for other alternatives. Kent (1995, MORPHMET electronic discussion list) gives insight into this phenomenon. Hotelling's T^2 test can be expressed as a union-intersection test where the test statistic is:

$$\sup_{a:\|a\|=1} \frac{n_1[a^{\mathrm{T}}(\bar{u}_1 - \bar{u})]^2 + n_2[a^{\mathrm{T}}(\bar{u}_2 - \bar{u})]^2}{n_1 a^{\mathrm{T}} S_1 a + n_2 a^{\mathrm{T}} S_2 a}, \tag{15.11}$$

where $\bar{u}_i$ and S_i are the sample means and covariance matrices for the ith group, and $\bar{u}$ is the pooled mean. For small variability a particular difference of scaled Euclidean distances is approximately a linear function of Bookstein coordinates. Hence, EDMA-II involves examining a similar statistic to that of Equation (15.11) except that the maximum is carried out over $k(k-1)/2$ fixed directions. If the mean difference in the alternative lies along one of the fixed directions, then EDMA-II will have slightly higher power than Hotelling's T^2. However, if the alternative difference lies away from the fixed directions, then Hotelling's T^2 will have higher power. Other power studies are given by Rohlf (2000).

15.5   Log-distances and multivariate analysis

An alternative approach to shape analysis based on inter-landmark distances has been proposed by Mardia *et al.* (1996b) and Rao and Suryawanshi (1996), using logarithms of distances. Estimates of mean reflection size-and-shape and reflection shape are obtained using average log-distances and MDS.

Let $G(X)$ be the **form log-distance matrix**, which is the $k \times k$ matrix of all pairs of inter-landmark log-distances in the configuration X. The **shape log-distance matrix** is:

$$G^*(X) = G(X) - \bar{G}1_k 1_k^{\mathrm{T}}, \quad \bar{G} = \frac{2}{k(k-1)} \sum_{i=1}^{k-1} \sum_{j=i+1}^{k} [G(X)]_{ij}.$$

If $d_i(h_1, h_2)$ is the distance between landmarks h_1 and h_2 for the ith object X_i, $i = 1, \ldots, n$, then the average form log-distance matrix is:

$$G(\hat{\mu}) = \frac{1}{n} \sum_{i=1}^{n} \log d_i(h_1, h_2).$$

An average form matrix can be obtained by exponentiating, and then MDS can be used to obtain an estimate of the mean reflection size-and-shape

$$[MDS_m(\exp(G(\hat{\mu})))]_{RS}.$$

An average shape log-distance matrix is given by $G^*(\hat{\mu})$ and an average shape can be obtained by exponentiating, and then using MDS, that is the estimate of the mean reflection shape is:

$$[MDS_m(\exp(G^*(\hat{\mu})))]_R.$$

Any arbitrary scalings of individual objects will appear as an overall constant added to each inter-landmark log-distance, and so the same estimate of reflection mean shape will be obtained regardless of the arbitrary scaling of objects.

If variations are small, then the estimate of mean reflection shape using the logarithm of distances will be very similar to using MDS directly without taking logs, which in turn will be very similar to the Procrustes and Bookstein means.

Rao and Suryawanshi (1996) suggested the adaptation of multivariate analysis techniques for inference. The analysis of size and shape through functions of distances is closely related to the traditional multivariate morphometric approach, briefly described in Section 2.1. Consider multivariate procedures on the $q = k(k-1)/2$-vector of inter-landmark log-distances

$$d = [\log d(1,2), \log d(1,3), \dots, \log d(k-1,k)]^T.$$

In particular, one example of Rao and Suryawanshi (1996)'s procedures is a two-sample test for mean size and shape difference. If $\bar{d}_1$ and $\bar{d}_2$ are the sample means of this inter-landmark log-distance vector in two independent samples, with pooled sample covariance matrix S, then the Mahalanobis distance can be split into two parts:

$$
\begin{aligned}
D_0^2 &= (\bar{d}_1 - \bar{d}_2)^T S^{-1}(\bar{d}_1 - \bar{d}_2) \\
&= (\bar{d}_1 - \bar{d}_2)^T H^T (HSH^T)^{-1} H(\bar{d}_1 - \bar{d}_2) + \frac{[1^T S^{-1}(\bar{d}_1 - \bar{d}_2)]^2}{1^T S^{-1} 1} \\
&= D_{sh}^2 + D_{si}^2,
\end{aligned}
$$

where H is a $(q-1) \times q$ matrix of rank $q-1$ such that $H1_q = 0$ (e.g. the Helmert submatrix). The expression D_{sh}^2 reflects the difference in reflection shape and D_{si}^2 reflects the difference in size. The objects will need to be recorded to the same scale for D_{si}^2 to be meaningful.

15.6   Euclidean shape tensor analysis

Cooper et al. (1995) and Goodall (1995) have introduced Euclidean shape tensor analysis (ESTA) which is a generalization of distance-based methods. One works with the size-and-shape of subsets of points (say 3 or 4 points) rather than just lengths between pairs of points. An average size-and-shape for each subset is obtained by performing Procrustes analysis, and then an overall average size-and-shape for the k point configuration is obtained by a metrical scaling method. Some related distributional results are also given by Dryden *et al.* (1997), where statistics based on subsets of triangles in nearly regular spatial patterns have been investigated. Dryden *et al.* (1997, 1999) and Faghihi *et al.* (1999) developed procedures for detecting abnormalities in muscle fibres using the shapes and sizes of subsets of Delaunay triangles.

15.7   Distance methods versus geometrical methods

There has been a great deal of discussion about the advantages and disadvantages of distance-based methods over geometrical shape methods, such as Procrustes analysis and edge registration. If one is interested in a few specific length measurements on an organism, then it would seem most suitable to use distance-based methods in such applications. However, if one is interested in the complete geometrical configuration, then one must consider the merits of the various methods, always bearing in mind that for small variations distance-based methods and geometrical shape methods give similar results. Distance, area, volume, weight and energy are examples of extensive measurements which are dependent on scale. These are in contrast to intensive measurements, such as angles and proportions. Bookstein (2015a)'s study of self-similarity is appropriate for intensive rather than extensive measurements, as one is interested in phenomena at every scale.

An advantage of distance-based methods is that they can be applied to distances that do not require the location of landmarks, for example the maximum and minimum widths of an organism. Also dealing with missing landmarks is straightforward. Finally the mean size-and-shape can be estimated consistently under normal models, with appropriate adjustments.

A disadvantage with distance-based methods is that they are invariant under reflections (which may not be desirable for certain applications). Also a form difference matrix is difficult to interpret, and visualization of shape variability and shape differences is difficult.

Important point: However, we would like to emphasize that if variations are small, then registration methods and distance-based methods can give similar conclusions about shape (Kent 1994) when the coordinates are approximately linearly related, although there may be practical differences in efficiency.

An important question is: when are variations small? The answer will depend on the curvature of the space, and also on the number of landmarks k and number of dimensions m. If all the data are away from the singularities of the shape space and if all the data lie within $d_F = 0.2$ of an average shape, then the tangent space approximations have seemed fine in many examples and so this provides a useful rule-of-thumb.

16

Curves, surfaces and volumes

16.1   Shape factors and random sets

Shape analysis is a much wider topic than just landmark analysis. We have concentrated most of our work on point set configurations, but the extension to more general objects is widespread.

Some simple measures for describing the shapes of sets include shape factors and other moment-based measures. These shape factors are dimensionless and are often used in microscopy (Exner and Hougardy 1988) and image analysis (Glasbey and Horgan 1995). Commonly used measures include the area–perimeter ratio

$$f_{AP} = \frac{4\pi A}{P^2}, \tag{16.1}$$

which has maximum value 1 for a circle and minimum value 0, with A the area and P the perimeter. The elongation shape factor is:

$$f_E = \frac{B}{L},$$

where B is the breadth of the object measured orthogonally to the maximum length L between the two most extreme points. See Stoyan and Stoyan (1994, Chapter 8) for further examples of simple shape measures.

Stoyan and Molchanov (1995) have developed theory for the shapes of set-valued data, and include methodology for the calculation of set-valued means. Distance measures between sets, such as the Hausdorff distance, can be used to develop estimated

---

Statistical Shape Analysis, with Applications in R, Second Edition. Ian L. Dryden and Kanti V. Mardia.
© 2016 John Wiley & Sons, Ltd. Published 2016 by John Wiley & Sons, Ltd.

mean sets in a manner analogous to Fréchet means for shapes discussed in Chapter 6. The distance between a point x and a set C is:

$$d(x, C) = \inf_{y \in C} d(x, y),$$

and the Hausdorff distance between two sets A and B is:

$$d_H(A, B) = \max(\sup_{x \in A} d(x, B), \sup_{x' \in B} d(A, x')). \tag{16.2}$$

Stoyan and Molchanov (1995) develop algorithms for estimating mean sets (Aumann and Vorob'ev means) from a random sample of random sets, where invariances such as rotation and translation are taken into account. They provide an algorithm in the spirit of GPA. Applications to random sets of particles are considered in Stoyan and Molchanov (1995), and a further application of interest is that of describing the shape variability of sand particles in Stoyan and Stoyan (1994) and Stoyan (1997), as described in Section 1.4.10. Charpiat *et al.* (2005) use the Hausdorff metric for averaging set objects in computer vision. Also motivated by image analysis applications, Dubuisson and Jain (1994) develop various modifications of Hausdorff distances. Note only the Hausdorff distance in the variants considered by Dubuisson and Jain (1994) is actually a metric.

16.2   Outline data

16.2.1   Fourier series

Several approaches have been suggested for use on objects that do not have landmarks. The shape of outlines in a plane can be considered using Fourier series representations (Zahn and Roskies 1972). A Fourier series decomposition of a polar representation or a tangent angle representation of the x–y coordinates around the outline leads to a new set of variables (the Fourier coefficients). In particular, for a 2D outline a polar representation of the radius $r(\theta)$ as a function of angle θ with the horizontal axis is given by:

$$r(\theta) = \sum_{j=0}^{p} a_j \cos(j\theta) + b_j \sin(j\theta),$$

where a_j, b_j are the Fourier coefficients. If the object has been translated and rotated into a standard reference frame, and normalized to unit size in some way, then the Fourier coefficients represent the shape of the object. The Fourier coefficients can be estimated for each outline using least squares techniques, and an important consideration is to choose p. Standard multivariate analysis on the Fourier coefficients provides a suitable way of analysing such data. There are numerous applications of this technique in the literature including in engineering and biology – for some examples see Zahn and Roskies (1972); Rohlf and Archie (1984); Johnson *et al.* (1985); Rohlf (1990); Stoyan and Stoyan (1994); Renaud (1995); and Lestrel (1997). This Fourier approach is most often used when there is some approximate underlying correspondence between parts of the outlines. The technique can be extended to three dimensions using spherical harmonics (Chung *et al.* 2007; Dette and Wiens 2009). If the outlines have discontinuities or sharp edges then a wavelet representation is more appropriate. Other techniques include that of Parui and Majumder (1983) who consider shape similarity measures for open curves. Note that open curves (with distinct start and end points) are a little simpler to describe than closed curves (where the start point is an unknown parameter that needs to be factored out).

16.2.2 Deformable template outlines

In modelling the outline of an object, Grenander and co-workers (Grenander 1994; Grenander and Miller 2007) specify a series of points around the outline connected by straight line segments. In their method, variability in the template is introduced by pre-multiplying the line segment by random scale-rotation matrices. Let us assume that the outline is described by the parameters $\theta^T = \{\theta_1^T, \theta_2^T\}$, where θ_1 denotes a vector of similarity parameters (i.e. location, size and rotation) whereas θ_2 denotes say k landmarks of the outlines. For θ_1, we can construct a prior with centre of gravity parameters, a scale parameter and a rotation by an angle α. Conditional on θ_1, we can construct a prior distribution of θ_2 as follows. Let η_j and e_j be the template edge and object edge, respectively, in complex coordinates, $j = 1, \ldots, k$, where the edges are formed from the vertices (landmark points) $v_j \in \mathbb{C}$ by taking differences $e_j = v_j - v_{j-1}$, with $v_0 = v_k$ and the edges satisfy the constraint $\sum_{j=1}^{k} e_j = 0$. A simple model is the Gaussian Markov Random Field (MRF) of order 1

$$e_j = (1 + t_j)\eta_j,$$

where $\{t_j\}$ are random Gaussian rotations and

$$E(t_j | \text{rest}) = \lambda(t_{j-1} + t_{j+1}), \quad 0 < \lambda < \tfrac{1}{2}, \quad \text{var}(t_j | \text{rest}) = \sigma^2.$$

The process $\{\theta_2\}$ is an improper Gaussian MRF of second order. The prior on $\{\theta_2\}$ is invariant under location but not scale or rotation. Kent *et al.* (1996b) and Mardia (1996a) give full details and also provide a link with the snake (see Section 17.3.5).

Grenander and Manbeck (1993) use edge models in an application involving automatic testing of defective potatoes. Using real notation consider a template with edges $g_1^0, \ldots, g_k^0$ with $\sum_{j=1}^{k} g_j^0 = 0$. Define the random edges g_j by

$$g_j = r \begin{pmatrix} \cos\psi & \sin\psi \\ -\sin\psi & \cos\psi \end{pmatrix} \begin{pmatrix} \cos\psi_j & \sin\psi_j \\ -\sin\psi_j & \cos\psi_j \end{pmatrix} g_j^0,$$

where r is global size, ψ is global rotation and ψ_j are individual rotations. We can take the distribution of r to be log normal, ψ to be uniform on $(0, 2\pi)$, the joint distribution of $\psi_1, \ldots, \psi_k$, to be multivariate von Mises such as (Mardia 1975):

$$c \exp \left\{ \beta_1 \sum_{i=1}^{k} \cos(\psi_{i+1} - \psi_i) + \beta_2 \sum_{j=1}^{k} \cos \psi_j \right\},$$

where $0 < \psi_j < 2\pi, j = 1, \ldots, k; \psi_{k+1} = \psi_1$, and $\beta_1, \beta_2 > 0$. Thus β_1 and β_2 are deformation parameters, β_1 allows similar neighbouring rotations whereas β_2 maintains fidelity to the template. Since $\sum g_j = 0$, there are constraints on $\{\psi_j\}$. Let us take $r = 1$ and $\psi = 0$. The constraints $\sum g_j = 0$ and $\sum g_j^0 = 0$ imply the following constraints on $\psi_1, \ldots, \psi_k$:

$$\sum g_{ix}^0 \cos \psi_i + \sum g_{iy}^0 \sin \psi_i = 0, \qquad -\sum g_{ix}^0 \sin \psi_i + \sum g_{iy}^0 \cos \psi_i = 0,$$

where $\sum g_{jx}^0 = \sum g_{jy}^0 = 0$. These can be simplified further for small values of $\{\psi_j\}$ leading to linear constraints on ψ_j. The field of view encloses a single position and therefore location parameters do not matter. Grenander and Manbeck (1993) also use information from a colour model in the defective potato testing application.

16.2.3   Star-shaped objects

Consider star-shaped objects in $\mathbb{R}^d$ which can be represented by a radial function $R(u)$, where $R(u) > 0$ and $u \in S^{d-1}$ are suitable spherical coordinates. An example of a star-shaped object suitable for such analysis is the leaf outline of Figure 16.1, which was modelled by Mardia and Qian (1995).

In general the radial function $R(u)$ contains size, shape, location and rotation information of the surface. However, location and rotation information is not present if the

Figure 16.1   A leaf outline which is a star-shaped object. Various radii are drawn from the centre to the outline.

object can be regarded as being in a fixed location and orientation, for example from preliminary registration. In some applications, it is of interest to restrict u to part of the sphere $\mathcal{D} \subseteq S^{d-1}$. The size of object the could be taken as the geometric mean

$$S = \exp\{\frac{1}{|\mathcal{D}|} \int_{\mathcal{D}} \log R(u) du\},$$

and, assuming the registration is fixed, the shape of surface is taken as $X(u) = R(u)/S$. A possible model for $\log R(u)$ is a Gaussian process with mean $\log \mu(u)$ and covariance function $\sigma(u_i, u_j), u_i, u_j \in \mathcal{D}$.

An alternative procedure is to take the size as $A = \frac{1}{|\mathcal{D}|} \int_{\mathcal{D}} R(u) du$ with shape $X_A(u) = R(u)/A$. A suitable model here could be a Gaussian process for $R(u)$ (Mancham and Molchanov 1996). The two procedures will be similar for small variations about the mean $\mu(u)$.

In practice a discrete set of points will be available, and we assume that the points are identified on the surface for k values of $u \in \mathcal{D}$ (e.g. on regularly spaced rays from the origin). Assuming an initial registration has been determined, the collection of radii $R = (R_1, \ldots, R_k)^T$ measures the size and shape of the surface. Let $X^* = (\log R_1, \ldots, \log R_k)^T$ be the vector of log-radii. The size of the surface is measured by:

$$\exp\{\bar{X}\} = (\prod R_i)^{1/k},$$

and the shape of the surface is measured by the vector

$$\exp(X) = \exp(X^* - \bar{X} 1_k). \tag{16.3}$$

A possible model for log-shape is a (singular) multivariate normal distribution, and PCA can be used to reduce the dimensionality for inference.

Example 16.1   Dryden (2005b) and Brignell *et al.* (2010) consider the brain surface data of Section 1.4.15. Estimates of the mean surface shape were obtained using (16.3), and shape variability was investigated using PCA. In addition surface symmetry measures were calculated, and the schizophrenia patients were found to be more symmetrical than the controls – an effect known as brain torque.   □

16.2.4   Featureless outlines

In some situations there is no sensible correspondence between outlines, for example in the sand particles of Section 1.4.10. Miller *et al.* (1994) develop an approach where the complex edge transformations $\{t_i\}$ are assumed to have a circulant Toeplitz covariance matrix in constructing a complex multivariate normal model. The model has been applied successfully to images of mitochondria (Grenander and Miller 1994). This model with circulant symmetry has also been studied by Kent *et al.* (2000), who consider additional constraints to make $\{\theta_2\}$ invariant under the similarity transformations. Kent *et al.* (2000) also explore the eigenstructure of the circulant covariance matrix. Kent *et al.* (2000) and Hobolth *et al.* (2002) apply their circulant models to the sand particles of Section 1.4.10, where the mean is a circle and the variability about the mean is of particular interest. In this application the shape variability of the river particles is significantly higher than the sea particles. The circulant models build on those developed by Grenander and Miller (1994), and Hobolth *et al.* (2002) make an explicit connection between edge-based and vertex-based models, with a focus on Markov random field models around the outline. Studies where the objects under study are featureless (i.e. no obvious correspondence in parts of the outline) are common, for example Grenander and Manbeck (1993); Grenander and Miller (1994); Mardia *et al.* (1996f); Stoyan (1997); Rue and Syversveen (1998); Hobolth and Vedel Jensen (2000); Kent *et al.* (2000); Hobolth *et al.* (2003) and Gardner *et al.* (2005) study potatoes, mitochondria, sand grains and ceramic particles, mushrooms, cells, cell nuclei, and tumour classification. Several of the models have continuous representations using stochastic processes, and these models are not dependent on the particular discretization of the curves. Parameter estimation is often carried out using likelihood-based methods (e.g., Kent *et al.* 2000; Hurn *et al.* 2001; Hobolth *et al.* 2002).

16.3   Semi-landmarks

Bookstein (1996a,c) has introduced the shape analysis of outlines based on semi-landmarks, which are allowed to slip around an outline. This approach is appropriate when there is some approximate notion of correspondence, which can be indicated by pseudo-landmarks that are allowed to slide along the outline.

First of all some landmarks are chosen and these are free to slide along the tangent direction of a curve. Let $X_i = (X_{ix}, X_{iy})^T$ be the 'source' or old landmarks and let $Y_i = (Y_{ix}, Y_{iy})^T$ be the 'target' or new landmarks ($i = 1, \dots, k$). Let us write $Y = (Y_x, Y_y)^T$ where $Y_x = (Y_{1x}, \dots, Y_{kx})^T$ and $Y_y = (Y_{1y}, \dots, Y_{ky})^T$. Now suppose that $Y_j, j = 1, \dots, q < k$, are free to slide away from their old position $X_j, j = 1, \dots, q$, along the directions $u_j = (u_{jx}, u_{jy})^T$ with $\|u_j\| = 1$. Thus the new position of Y_j is:

$$Y_j = X_j + t_j u_j, \quad j = 1, \dots, q.$$

Note that $Y_{q+1}, \dots, Y_k$ are fixed in advance. We know that the bending energy is given by:

$$Y^T \mathrm{diag}(B_e, B_e) Y = Y^T B_2 Y$$

where B_e depends only on the old configuration $X_i, i = 1, \dots, k$. Now we can write:

$$Y = Y^0 + U t^T,$$

where

$$Y^0 = \begin{pmatrix} X_{1x}, \ldots, X_{qx}, Y_{q+1,x}, \ldots, Y_{kx} \\ X_{1y}, \ldots, X_{qy}, Y_{q+1,y}, \ldots, Y_{ky} \end{pmatrix}^{\mathrm{T}}$$

and

$$U = \begin{pmatrix} \mathrm{diag}(u_{1x}, \ldots, u_{qx}) \\ \mathrm{diag}(u_{1y}, \ldots, u_{qy}) \\ 0_{2k-2q,q} \end{pmatrix}$$

is a $2k \times q$ matrix and $t = (t_1, \ldots, t_q)^{\mathrm{T}}$. We need to minimize the bending energy

$$(Y^0 + Ut^{\mathrm{T}})^{\mathrm{T}} B_2 (Y^0 + Ut^{\mathrm{T}})$$

with respect to t. The problem is similar to generalized least squares linear regression, with solution

$$t = -(U^{\mathrm{T}} B_2 U)^{-1} U^{\mathrm{T}} B_2 Y^0.$$

Bookstein (1996a,c) advocates an iterative procedure where the initial value of $u_j = (X_{j+1} - X_{j-1})/\|X_{j+1} - X_{j-1}\|$ (or an alternative procedure which iterates through $u_j = (u_{j+1} - u_{j-1})/\|u_{j+1} - u_{j-1}\|$). So, one obtains U then calculates t, then iterates. Once the semi-landmarks have been obtained we can consider Procrustes and other shape analysis techniques, to obtain a suitable shape prior distribution for object recognition. Alternatively, one may be interested in shape analysis of the outlines, and for example differences in average shape between groups. Bookstein (1996c) has investigated the shape of the corpus callosum in the midsagittal sections of MR scans of the brain in control subjects and schizophrenia patients, as in Figure 1.14. An important practical aspect of the work is how to sample the points: one needs to achieve a balance between coarse sampling of landmarks where information might be lost and over-fine sampling of landmarks where power can be lost. For an alternative but related approach, see Green (1996).

16.4   Square root velocity function

16.4.1   SRVF and quotient space for size-and-shape

Srivastava *et al.* (2011a) have introduced a very promising method for the size-and-shape analysis of elastic curves, building on earlier work that started with Klassen *et al.* (2003). The method involves the analysis of curve outlines or functions and is invariant under re-parameterization of the curve.

Let f be a real valued differentiable curve function in the original space, $f(t)$: $[0, 1] \to \mathbb{R}^m$. From Srivastava *et al.* (2011a) the square root velocity function (SRVF) or normalized tangent vector of f is defined as $q : [0, 1] \to \mathbb{R}^m$, where

$$q(t) = \frac{\dot{f}(t)}{\sqrt{\|\dot{f}(t)\|}},$$

and $\|f(t)\|$ denotes the standard Euclidean norm. After taking the derivative, the q function is now **invariant under translation** of the original function. In the 1D functional case the domain $t \in [0, 1]$ often represents 'time' rescaled to unit length, whereas in two and higher dimensional cases t represents the proportion of arc length along the curve.

Let f be warped by a re-parameterization $\gamma \in G$, that is $f \circ \gamma$, where $\gamma \in G$: $[0, 1] \to [0, 1]$ is a strictly increasing differentiable warping function. Note that γ is a diffeomorphism. The SRVF of $f \circ \gamma$ is then given as:

$$q^*(t) = \sqrt{\dot{\gamma}(t)} q(\gamma(t)),$$

using the chain rule. We would like to consider a metric that is invariant under re-parameterization transformation G. If we define the group G to be domain re-parameterization and we consider an equivalence class for q functions under G, which is denoted as $[q]$, then we have the equivalence class $[q] \in Q$, where Q is a **quotient space** after removing arbitrary domain warping. Inference can be carried out directly in the quotient space Q.

An elastic distance (Srivastava *et al.* 2011a) defined in Q is given as:

$$d(q_1, q_2) = d([q_1], [q_2]) = \inf_{\gamma \in G} \|q_1 - \sqrt{\dot{\gamma}} q_2(\gamma)\|_2 = d_{\text{Elastic}}(f_1, f_2),$$

where $\|q\|_2 = \{\int_0^1 q(t)^2 dt\}^{1/2}$ denotes the L^2 norm of q.

There are several reasons for using the q representation instead of directly working with the original curve function f. One of the key reasons is that we would like to consider a metric that is invariant under re-parameterization transformation G. The elastic metric of Srivastava *et al.* (2011a) satisfies this desired property,

$$d_{\text{Elastic}}(f_1 \circ \gamma, f_2 \circ \gamma) = d_{\text{Elastic}}(f_1, f_2),$$

although it is quite complicated to work with directly on the functions f_1 and f_2. However, the use of the SRVF representation simplifies the calculation of the elastic metric to an easy-to-use L^2 metric between the SRVFs, which is attractive both theoretically and computationally. In practice q is discretized using p points.

If q_1 can be expressed as some warped version of q_2, that is they are in the same equivalence class, then $d([q_1], [q_2]) = 0$ in quotient space. This elastic distance is a proper distance satisfying symmetry, non-negativity and the triangle inequality. Note that we sometimes wish to **remove scale** from the function or curve, and hence we can standardize so that

$$\left\{ \int_0^1 q(t)^2 dt \right\}^{1/2} = 1, \tag{16.4}$$

and then the SRVF measures shape difference.

In the $m \geq 2$ dimensional case it is common to also require **invariance under rotation** of the original curve. Hence we may also wish to consider an elastic distance (Joshi *et al.* 2007; Srivastava *et al.* 2011a) defined in Q given as:

$$d([q_1],[q_2]) = \inf_{\gamma \in G, \Gamma \in SO(m)} \| q_1 - \sqrt{\dot{\gamma}} q_2(\gamma) \Gamma \|_2.$$

The $m = 2$ dimensional elastic metric for curves was first given by Younes (1998).

16.4.2 Quotient space inference

Inference can be carried out directly in the quotient space Q, and in this case the population mean is most naturally the Fréchet/Karcher mean μ_Q. Given a random sample $[q_1], \ldots, [q_n]$ we obtain the sample Fréchet mean by optimizing over the warps for the 1D function case (Srivastava *et al.* 2011b):

$$\hat{\mu}_Q = \arg \inf_{\mu \in Q} \sum_{i=1}^n \inf_{\gamma_i \in G} \| \mu - \sqrt{\dot{\gamma}_i}(q_i \circ \gamma_i) \|_2^2.$$

In addition for the $m \geq 2$ dimensional case (Srivastava *et al.* 2011a) we also need to optimize over the rotation matrices Γ_i where

$$\hat{\mu}_Q = \arg \inf_{\mu \in Q} \sum_{i=1}^n \inf_{\gamma_i \in G, \Gamma_i \in SO(m)} \| \mu - \sqrt{\dot{\gamma}_i}(q_i \circ \gamma_i) \Gamma_i \|_2^2.$$

This approach can be carried out using dynamic programming for pairwise matching, then ordinary Procrustes matching for the rotation, and the sample mean is given by:

$$\hat{\mu}_Q = \frac{1}{n} \sum_{i=1}^n \sqrt{\dot{\hat{\gamma}}_i}(q_i \circ \hat{\gamma}_i) \hat{\Gamma}_i.$$

Each of the parameters is then updated in an iterative algorithm until convergence.

16.4.3 Ambient space inference

Cheng *et al.* (2016) introduce a Bayesian approach for analysing SRVFs, where the prior re-parameterization for γ is a Dirichlet distribution. This prior distribution is uniform when the parameters in the Dirichlet are all $a = 1$, and a larger value of a leads to a transformation more concentrated on $\dot{\gamma} = 1$ (i.e. translations). In the $m \geq 2$ dimensional case, we consider a Gaussian process for the difference of two vectorized q functions in a relative orientation Γ, that is $\{\text{vec}(q_1 - q_2^*)|\gamma, \Gamma\} \sim GP$, where $q_2^* = \sqrt{\dot{\gamma}} q_2(\gamma)\Gamma$. The matrix $\Gamma \in SO(m)$ is a rotation matrix with parameter vector θ. If we assign a prior for rotation parameters (Eulerian angles) θ corresponding to rotation matrix Γ, then the joint posterior distribution of $(\gamma([t]), \theta)$, given $(q_1([t]), q_2([t]))$ is:

$$\pi(\gamma, \theta | q_1, q_2) \propto \kappa^{p/2} e^{-\kappa \|q_1([t]) - q_2^*([t])\|^2} \pi(\gamma)\pi(\theta)\pi(\kappa),$$

where γ, θ, κ are independent *a priori*. Cheng *et al.* (2016) carried out inference by simulating from the posterior using MCMC methods. For large samples estimation is consistent, which can be shown using the Bernstein–von Mises theorem (van der Vaart 1998, p. 141). Cheng *et al.* (2016) provide a joint model for the SVRF and the warping function, and hence it is in the **ambient space**.

Example 16.2   Consider a pairwise comparison of two mouse vertebrae from the dataset in Section 1.4.1, where there are 60 landmarks/pseudo-landmarks on each outline. We use Cheng *et al.* (2016)'s MCMC algorithm for pairwise matching with 50 000 iterations. The q-functions are obtained by initial smoothing, and then normalized so that $\|q\|_2 = 1$. The registration is carried out using rotation through an angle θ about the origin, and a warping function γ. The original and registered pair (using a posterior mean) are shown in Figure 16.2 and the point-wise correspondence between the curves and a point-wise 95% credibility interval for $\gamma(t)$ are shown in Figure 16.3. The start point of the curve is fixed and is given by the left-most point on the curve in Figure 16.3 that has lines connecting the two bones. The narrower regions in the

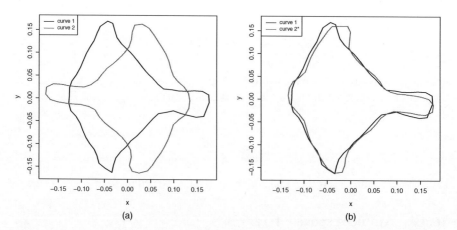

Figure 16.2   (a) Unregistered curves and (b) registration through $\hat{\gamma}(t)_A$. Source: Cheng et al. 2016. For a colour version of this figure, see the colour plate section.

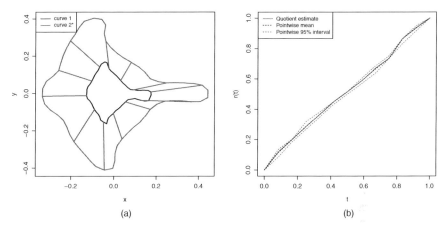

(a)   (b)

Figure 16.3   (a) Correspondence based on $\hat{\gamma}(t)_A$ and (b) 95% credibility interval for $\gamma(t)$. In (a) one of the bones is drawn artificially smaller in order to better illustrate the correspondence. Source: Cheng et al. 2016. For a colour version of this figure, see the colour plate section.

credibility interval correspond well with high curvature regions in the shapes. We also applied the multiple curve registration, as shown in Figure 16.4.   □

There are many applications where the SRVF methods can be used, for example Srivastava *et al.* (2012) develop techniques for 2D and 3D shape analyses in gait-based biometrics, action recognition, and video summarization and indexing. The

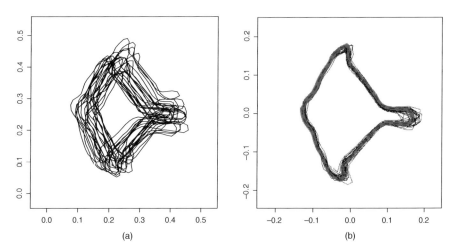

(a)   (b)

Figure 16.4   The original curves from the Small group, (a) without and (b) with registration. The dashed curve in (b) is the estimated μ_A and grey colour shows the credible region given by 10 000 samples of mean. Source: Cheng et al. 2016. For a colour version of this figure, see the colour plate section.

shape of curves that are invariant to re-parameterization has been the subject of deep mathematical work on different types of quotient spaces, with connections to elastic metrics (Michor and Mumford 2006, 2007; Younes *et al.* 2008a).

The choice as to whether to use quotient space inference or ambient space inference was discussed by Cheng *et al.* (2016), and in practice the two methods are often similar due to a Laplace approximation to the posterior density. This situation is similar to that of Chapter 14 in the analysis of unlabelled shape, where the choice between optimization (quotient space model) or marginalization (ambient space model) needs to be considered (see Section 14.2.2).

16.5   Curvature and torsion

Mardia *et al.* (1999) have approached the problem of analysing curves in three dimensions through the Frenet–Serret framework, in particular to estimate curvature and torsion. Recall from differential geometry that curvature measures the rate of change of the angle of tangents and it is zero for straight lines. Torsion measures the twisting of a curve, and it is zero for the curves lying in two dimensions. For a helix, curvature and torsion are constant. Curvature and torsion are invariant under a rigid transformation and thus useful for studying the size-and-shape of curves. Given a curve $\{x(t), y(t), z(t)\}$ the torsion function is given by $\tau(t) = \tau_1(t)/\tau_2(t)$, where

$$\tau_1 = x'''(y'z'' - y''z') + y'''(x''z' - x'z'') + z'''(x'y'' - x''y')$$
$$\tau_2 = (y'z'' - y''z')^2 + (x''z' - x'z'')^2 + (x'y'' - x''y')^2$$

and the curvature function is $\kappa = \sqrt{\tau_2}/\|(x', y', z')\|^{3/2}$. Note that the amount of smoothing required to estimate torsion can be large given that third derivatives need to be estimated, which in turn may bias the estimation. Recently, by using splines, Kim *et al.* (2013) have provided consistent estimators of curves in three dimensions which in particular lead to consistent estimators of curvature and torsion. One of the applications is to infer if torsion is zero or not, for example, in spinal deformity (Mardia *et al.* 1999). Cheng *et al.* (2014) use curvature and torsion to study the size-and-shapes of carotid arteries. Another application related to change in curvature, for example, is finding a kink in helices related to drug discovery (Deane *et al.* 2013).

16.6   Surfaces

Comparing surfaces is more complicated than comparing curves, as the natural ordering around an outline is not present for a surface. Extensions of the SRVF method to surfaces have been presented by Kurtek *et al.* (2010) and Jermyn *et al.* (2012). In particular, the SRVF is replaced by a **square root normal field** by Jermyn *et al.* (2012) and then distances between the shapes of surfaces need to be invariant to parameterization (common diffeomorphic transformations) of the surfaces. The methodology is applied to a wide variety of examples, including the comparisons of organs from medical images and 3D surface images from computer vision.

The 'landmark-free' matching procedure called the iterative closest point (ICP) algorithm uses a procedure where a set of points is located on one object and the corresponding closest points on the second object are used to update the registration (Besl and McKay 1992). A new set of closest corresponding points is then chosen and the registration is updated again, and so on.

In image analysis a method of representing shapes is via distance functions between points on the surface, which are isometrically embedded into Euclidean space. Biasotti *et al.* (2014) provide a review of 3D shape similarity using maps; Rosman *et al.* (2010) consider nonlinear dimension reduction by using a topologically constrained isometric embedding; Bronstein *et al.* (2010) consider metrics within a shape object for shape matching, and the methods are invariant under inelastic matching. Elad and Kimmel (2003) also use Euclidean embedding, and shape is regarded as a metric space. Further matching procedures have been proposed by minimizing pixel-based measures (e.g. Glasbey and Mardia 2001).

Other methods for surface shape comparison include using conformal Wasserstein distances (Lipman and Daubechies 2010; Lipman *et al.* 2013) with an emphasis on practical computation. Finally, in order to deal with continuous (infinite dimensional) surface shapes it can be useful to employ kernel methods in Hilbert spaces (Czogiel *et al.* 2011; Jayasumana *et al.* 2013b; 2015, Awate *et al.* 2014).

16.7   Curvature, ridges and solid shape

It is often of great interest to study curvature and boundary information, especially when landmarks are not available, but also in addition to landmarks. Curvature-based methods can be used for describing shape or for comparing object shapes for two dimensions or three dimensions. Other related topics of interest include the use of ridge curves (Kent *et al.* 1996a) and crest lines (Subsol *et al.* 1996) which could be used for object matching. Methods for describing solid shape, that is shapes of whole object volumes, have been summarized by Koenderink (1990). Also, Koenderink and van Doorn (1992) consider surface shape and curvature at different scales. A typical 3D application is the following from a study of human faces before and after operations.

Example 16.3   Range images are obtained of human faces before and after plastic surgery and the data are acquired using a laser scanner and charge-coupled device (CCD) camera (Coombes *et al.* 1991). A laser beam is fanned into a line and directed at a person sitting in a chair. Two mirrors reflect the image of the resulting profile into a CCD camera. The scan lines are recorded for the profiles and the chair is rotated under computer control. The result is that approximately 24 000 (3D) coordinates are obtained on the surface of the face, see Figure 16.5.

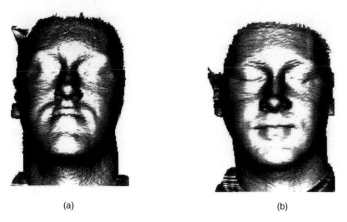

(a)   (b)

Figure 16.5   Range data surface of a human face: (a) before surgery; and (b) after surgery to correct a cleft palate.

Consider a patient who has a cleft palate and therefore requires surgery to improve comfort and appearance. It is of interest to describe the differences in the shape of the face before and after surgery, for the benefit of the patient and surgical planning.

Mardia *et al.* (1995) provide maps on a patient scan, coloured according to different surface types, which are evaluated by curvature (e.g. saddle point, pit, valley, etc.). These maps are helpful for describing surface shape and can highlight abnormalities or differences in shape.   ☐

Note that in all examples of this chapter there is a large step from the object in the real world and the data that are obtained for analysis. How the data are acquired will feed in strongly to the type of statistical analysis carried out. Koenderink (1990) discusses apertures for instruments that lie between a natural phenomenon and the measured data. It is important that apertures, e.g. image pixels or sampling points, should not become a feature of the analysis. This is highly relevant for the study of discrete differential properties of curves, surfaces and volumes.

17

Shape in images

17.1 Introduction

There has been an explosion in the collection of image data in the last decade, especially with the availability of cheap high-quality cameras and smart-phones. An important area of study is the interpretation of images, and shape is a very important component. A wide ranging summary of measuring shape in images was given by Neale and Russ (2012).

A **digital image** is an $r \times c$ grid of pixels (picture elements) each of which is assigned an integer representing the brightness (or intensity) at that pixel. The pixel is coloured a particular shade of grey depending on the brightness at that position, and hence the integer representing brightness is called the grey level.

A common scale used is for the grey levels to range from 0 (black) through to 255 (white). So, for example, a grey level of 50 would be dark grey, 128 mid grey and 200 light grey. Such images are called grey-level or grey-scale images. The range of grey levels is usually on the scale 0 to $(2^g - 1)$, where $g = 1$ for binary images (black/white), $g = 8$ for 256 grey-scale (8 bit) images and $g = 12$ for 4096 grey-scale (12 bit) images. Colour images can be represented as three grey-level images – each image for the red, green and blue colour bands. For a 256×256 image a scene is represented by $2^{16} = 65\,536$ integers, and hence the data are very high dimensional. Image analysis is involved with all aspects of analysing such image data. For an introduction to statistical image analysis see, for example, Ripley (1988) and Winkler (1995). Other reviews include Mardia and Kanji (1993); Mardia (1994); Grenander and Miller (2007); Davies *et al.* (2008b) and Sotiras *et al.* (2013).

Statistical Shape Analysis, with Applications in R, Second Edition. Ian L. Dryden and Kanti V. Mardia.
© 2016 John Wiley & Sons, Ltd. Published 2016 by John Wiley & Sons, Ltd.

Figure 17.1   A simple digital image of the letter H.

Example 17.1   In Figure 17.1 we see a simple $(r = 7) \times (c = 6)$ image of the letter H with grey levels in the matrix

$$
\begin{bmatrix}
0 & 0 & 0 & 0 & 0 & 0 \\
0 & 211 & 2 & 1 & 243 & 0 \\
0 & 200 & 5 & 8 & 251 & 0 \\
0 & 210 & 241 & 251 & 254 & 0 \\
0 & 236 & 2 & 12 & 204 & 0 \\
0 & 251 & 7 & 12 & 218 & 0 \\
0 & 0 & 0 & 0 & 0 & 0
\end{bmatrix}
\qquad \square
$$

Image analysis tasks can be broadly divided into two distinct areas – low and high level image analysis. Low level image analysis involves techniques at a pixel by pixel level, such as cleaning up noisy images, deblurring images, classifying each pixel into a few classes (segmentation) and modelling of textures. High level image analysis involves direct modelling of the objects in the images and tasks such as object recognition and location of objects in an image. It is particularly in the area of high level image analysis that shape analysis has a prominent rôle to play.

In image analysis the registration parameters (location, rotation and scale) will usually need to be modelled, although a partition of variables into shape and registration parameters is often helpful.

17.2   High-level Bayesian image analysis

Since the early 1980s statistical approaches to image analysis using the Bayesian paradigm have proved to be very successful. Initially, the methodology was primarily developed for low level image analysis but then widely used for high level tasks.

To use the Bayesian framework one requires a prior model which represents our initial knowledge about the objects in a particular scene, and a likelihood or noise model which is the joint probability distribution of the grey levels in the image, dependent on the objects in the scene. By using Bayes' theorem one derives the posterior distribution of the objects in the scene, which can be used for inference. Examples of the tasks of interest include segmentation of the scene into 'object' and 'background', and object recognition. The computational work involved is generally intense because of the large amount of data in each image.

An appropriate method for high-level Bayesian image analysis is the use of deformable templates, pioneered by Grenander and colleagues (Grenander and Keenan 1993; Grenander 1994; Grenander and Miller 2007). Our description follows the common theme of Mardia *et al.* (1991, 1995). We assume that we are dealing with applications where we have prior knowledge on the composition of the scene and we can formulate parsimonious geometric descriptions for objects in the images. For example, in medical imaging we can expect to know a priori the main subject of the image, for example a heart or a brain. Consider our prior knowledge about the objects under study to be represented by a parameterized ideal prototype or template S_0. Note that S_0 could be a template of a single object or many objects in a scene. A probability distribution is assigned to the parameters with density (or probability function) $\pi(S)$, which models the allowed variations S of S_0. Hence, S is a random vector representing all possible templates with associated density $\pi(S)$. Here S is a function of a finite number of parameters, say $\theta_1, ..., \theta_p$.

In addition to the prior model we require an image model. Let the observed image I be the matrix of grey levels x_i, where $i = (i_1, i_2) \in \{1, ..., r\} \times \{1, ..., c\}$ are the $r \times c$ pixel locations. The image model or likelihood is the joint p.d.f. (or probability function for discrete data) of the grey levels given the parameterized objects S, written as $L(I|S)$. The likelihood expresses the dependence of the observed image on the deformed template. It is often convenient to generate an intermediate synthetic image but we will not need it here (Mardia *et al.* 1995).

By Bayes' theorem, the posterior density $\pi(S|I)$ of the deformed template S given the observed image I is:

$$\pi(S|I) \propto L(I|S)\pi(S). \tag{17.1}$$

An estimate of the true scene can be obtained from the posterior mode (the MAP estimate) or the posterior mean. The posterior mode is found either by a global search, gradient descent (which is often impracticable due to the large number of parameters) or by techniques such as simulated annealing (Geman and Geman 1984), iterative conditional modes (ICM) (Besag 1986) or ICM for image sequences (Hainsworth and Mardia 1993). Alternatively, MCMC algorithms (e.g. see Besag *et al.* 1995; Gilks *et al.* 1996; Gelman *et al.* 2014; Green *et al.* 2015) provide techniques for simulating from a density. MCMC has the advantage that it allows a study of the whole posterior density itself, and so credibility or confidence regions can be easily obtained.

17.3   Prior models for objects

The key to the successful inclusion of prior knowledge in high level Bayesian image analysis is through specification of the prior distribution. Many approaches have been proposed, including methods based on outlines, landmarks and geometric parameters. The prior can be specified either through a model with known parameters or with parameters estimated from training data.

17.3.1   Geometric parameter approach

One approach is to consider a geometric template for S consisting of parametric components, for example line segments, circles, ellipses and arcs. Examples include a circle of random radius to represent the central disc of galaxies (Ripley and Sutherland 1990); simple geometric shapes for facial features such as eyes and mouths (Phillips and Smith 1993, 1994); circular templates to represent pellets in an image, where the number of pellets is unknown (Baddeley and van Lieshout 1993); and ellipses for leaf cell shape (Hurn 1998).

In these models, distributions are specified for the geometrical parameters, and the hierarchical approach of Phillips and Smith (1993, 1994) provides a simple method. Often templates are defined by both global and local parameters. The global parameters represent the object on a coarse scale and the local parameters give a more detailed description on a fine scale. The idea of a hierarchical model for templates is to specify the marginal distribution of the global parameters and a conditional distribution for the local parameters given the global values. This hierarchical division of parameters can be extended to give a complete description of the local dependence structure between variables. Hence, conditionally, each parameter depends only on variables in the levels immediately above and below it.

In general, we assume that templates can be summarized by a small number of parameters $\theta = (\theta_1, ..., \theta_p)$ say, where variation in θ will produce a wide range of deformations of the template. By explicitly assigning a prior distribution to θ, we can quantify the relative probability of different deformations. The prior can be based on training data which need not be a large dataset. By simulation, we can check the possible shapes that could arise.

17.3.2   Active shape models and active appearance models

Consider a single object in $\mathbb{R}^m$ with k landmarks. One approach to specifying the template is to work with landmarks on a template directly. Denote the coordinates of the landmarks in $\mathbb{R}^m$ as X, a $mk \times 1$ vector. The configuration can be parameterized as an overall location (e.g. centroid), a scale β, a rotation Γ and some suitable shape coordinates, such as shape PC scores. A shape distribution from Chapters 10 or 11 could be chosen as a prior model together with more vague priors for location, scale and rotation.

A point distribution model (PDM) is a PC model prior model for object recognition and was suggested by Cootes et al. (1992) (see Section 7.7.4). In Figure 7.17

we saw a model for hand shapes which uses three PC scores. The PDM is a template model which is estimated from a training dataset of images, where landmarks are usually located on each training image by an observer. The PDM forms a shape prior distribution in the set of models known as active shape models (ASMs) (Cootes *et al.* 1994; Cootes and Taylor 1995; Davies *et al.* 2008b). The ASM combines a prior shape model with a likelihood type of model for the pixels based on local grey level intensities in the image at normal directions to the template. The ASM can be used to for detecting objects in a new image, making use of the prior information in a training set.

An extension to the ASM is the active appearance model (AAM) which also includes full information of the grey levels of the training images (rather than in a few local regions as in ASMs). The AAMs also use PCA for reducing the dimension of the variability of the image grey levels, and it is important to register the grey level images in the training set in order to estimate the image PCs in the AAM, see Cootes *et al.* (2001). The AAM can then be used for detecting objects in a new image, making use of the prior information in a training set of images. Both AAMs and ASMs have been extremely successful in a wide variety of medical and engineering image applications.

17.3.3 Graphical templates

Amit and Kong (1996) and Amit (1997) use a graphical template method which is based on comparisons between triangles in an image and a template. Cost functions are given for matching triangles in the deformed template to triangles in the observed template and the total cost gives a measure of discrepancy. The cost functions involve hard constraints limiting the range in which the observed angles can deviate from the template angles and soft constraints penalizing the deviations from template angles.

The procedure had been implemented into a fast algorithm which finds an optimal match in an image in polynomial time, provided the template graph is decomposable (Amit 1997; Amit and Geman 1997).

17.3.4 Thin-plate splines

The thin-plate spline introduced in Section 12.3 can be used in a prior model. The PTPS transformation from the ideal template S_0 to a deformed version S has an energy function associated with it. The total minimum bending energy $J(\Phi)$ from a PTPS is given in Equation (12.16). Thus if $E_{int}(S, S_0) = J(\Phi)$, then a prior distribution of the deformation could be obtained using the Gibbs distribution with density proportional to:

$$\exp\{-\kappa J(\Phi)\}.$$

This would inhibit landmark changes that require a large bending energy. If S is an affine transformation of S_0, then the total bending energy is zero. This prior was suggested by Mardia *et al.* (1991) and further applications were given by Mardia and Hainsworth (1993).

17.3.5   Snake

The snake (Kass *et al.* 1988) in its simplest form is a spline with a penalty on the mean curvature. Snakes are used for fitting a curve to an image when there is no underlying template, with the aim that the resulting estimated curve is smooth. Let the outline be $f(t) \in \mathbb{C}, t \in [0, 1], f(0) = f(1)$. The penalty in the snake can be written as:

$$- 2 \log P(t) \propto \int_0^1 \alpha(t) |f''(t)|^2 dt + \int_0^1 \beta(t) |f'(t)|^2 dt, \qquad (17.2)$$

where $\alpha(t)$ and $\beta(t)$ denote the degree of stiffness and elasticity at t, respectively. For $t_j = j/k, j = 0, 1, \dots, k$, denote $f(t_j) = u_j + iv_j$. We find that the right-hand side of Equation (17.2) can be written approximately for large k as (Mardia 1996a):

$$\left\{ \sum \alpha_j (u_{j+1} + u_{j-1} - 2u_j)^2 + \sum \beta_j (u_{j+1} - u_j)^2 \right\}$$
$$+ \left\{ \sum \alpha_j (v_{j+1} + v_{j-1} - 2v_j)^2 + \sum \beta_j (v_{j+1} - v_j)^2 \right\}.$$

Thus $\{u_j\}$ and $\{v_j\}$ also form a separable Gaussian MRF of order 2.

17.3.6   Inference

Inference about a scene in an image is made through the posterior distribution of S obtained from Equation (17.1). The full range of Bayesian statistical inference tools can be used and, as stated in Section 17.2, the maximum of the posterior density (the MAP estimate) is frequently used, as well as the posterior mean. There may be alternative occasions when particular template parameters are of great interest, in which case one would maximize the appropriate marginal posterior densities (Phillips and Smith 1994). One way to calculate the posterior mean is by a simulation method which does not depend on the complicated normalizing constant in $\pi(\theta|x)$. For example, a MCMC procedure using the Metropolis–Hastings algorithm (Metropolis *et al.* 1953; Hastings 1970) could be used. This procedure generates a Markov chain whose equilibrium distribution is the posterior distribution of $\theta|x$.

17.4   Warping and image averaging

17.4.1   Warping

Images can be deformed using deformations such as those described in Chapter 12 and this process is called **image warping**. Warping or morphing of images is used in a wide set of applications, including in the cinema industry and medical image registration.

Definition 17.1   *Consider an image $f(t)$ defined on a region $D_1 \in \mathbb{R}^2$ and deformed to $f(\Phi(t))$, where $\Phi(t) \in D_2 \in \mathbb{R}^2$. For example, a set of landmarks T_1 could be located in the original image and then deformed to new positions T_2. We call $f(\Phi(t))$ the* **warped** *image.*

A practical approach to warping is to use the inverse deformation from D_2 to D_1 to 'look up' the corresponding grey levels from the region D_1. Consider a pixel location $t \in D_2$. The deformation $\Phi(t)$ is computed from the deformed region D_2 to the original plane D_1. Then assign the new grey level at t as the grey level $f(\Phi(t))$ [in practice the closest pixel to location $\Phi(t)$].

The advantage of using the reverse deformation is that if the original image is defined on a regular lattice, then the warped image is still defined on a regular lattice. An alternative approach is to map from D_1 to D_2 resulting in a irregular non-square lattice for D_2, and then linear interpolation is used to obtain the final image on a regular lattice (Mardia and Hainsworth 1993).

Examples of warping include data fusion for medical imaging. For example, we may wish to combine information from an X-ray image (which has anatomical information) and a nuclear medicine image (which shows functional information). In Figure 17.2 we see the display tool of Mardia and Little (1994) which involves a

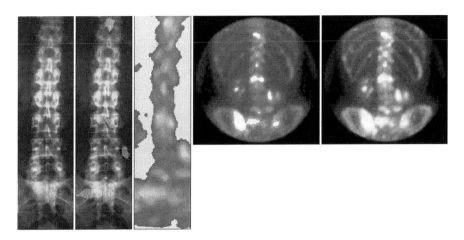

Figure 17.2   The display tool of Mardia and Little (1994) for combining information from an X-ray (far left) and a nuclear medicine image (far right). Thin-plate spline transformations are used to register the images, and perhaps the most useful image is the second from the left which displays the highest nuclear medicine grey levels on top of the X-ray image. These bright areas could indicate possible locations of tumours or other abnormalities. The precise location is easier to see in this new image compared with the pure nuclear medicine image. Source: Adapted from Mardia & Little 1994.

deformation from an X-ray image to a nuclear medicine image. For some other statistical work see Johnson *et al.* (1995) and Hurn *et al.* (1996). Also, Bookstein (2001) considers the relation between the use of warping with landmarks and image-based warping.

17.4.2 Image averaging

We can use the warping approach to construct an average from several images of objects (Mardia 1993).

Definition 17.2   *Consider a random sample of images $f_1, \ldots, f_n$ containing landmark configuration $X_1, \ldots, X_n$, from a population mean image f with a population mean configuration μ. We wish to estimate μ and f up to arbitrary Euclidean similarity transformations. The shape of μ can be estimated by the full Procrustes mean of the landmark configurations $X_1, \ldots, X_n$. Let Φ_i^* be the deformation obtained from the estimated mean shape $[\hat{\mu}]$ to the ith configuration. The* **average image** *has the grey level at pixel location t given by:*

$$\bar{f}(t) = \frac{1}{n} \sum_{i=1}^{n} f_i \{ \Phi_i^*(t) \}. \tag{17.3}$$

We consider $\bar{f}$ to be an estimate of the population mean image f.

Example 17.2   In Figure 17.3 we see five T1 (first thoracic) mouse vertebral images. Twelve landmarks were located on each of the vertebral outlines and the full Procrustes mean was computed. The average image is obtained using the above procedure in Equation (17.3) and is displayed in Figure 17.4. This approach leads to rather blurred averages in regions away from landmarks (here the central 'hole' in the centre of the bones which has no landmarks). □

Other examples of image averaging include average faces (Craw and Cameron 1992) and the average brain images obtained using thin-plate spline transformations (Bookstein 1991, Appendix A.1.3) and diffeomorphisms (Beg *et al.*, 2005).

Galton (1878, 1883) obtained averaged pictures of faces using photographic techniques in the 19th century, see Figure 17.5, and the technique was called composite photography. Galton was interested in the typical average face of groups of people, such as criminals and tuberculosis patients. He believed that faces contained information that could be used in many applications, for example the susceptibility of people to particular diseases. Another early example is the 'average' photograph of 12 American mathematicians obtained by Raphael Pumpelly taken in 1884 (Stigler 1984). Shapes of landmark configurations from photographs of the faces of individuals are used for forensic identification (e.g. see Mardia *et al.* 1996b; Evison and Vorder Bruegge 2010). A state of the art method of face recognition is DeepFace (Taigman *et al.*, 2014).

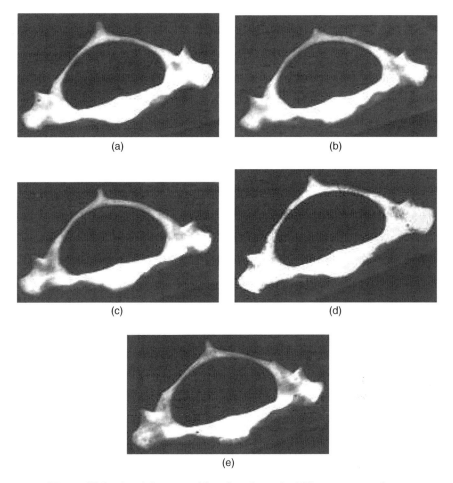

(a)   (b)

(c)   (d)

(e)

Figure 17.3   (a–e) Images of five first thoracic (T1) mouse vertebrae.

17.4.3   Merging images

The same technique of warping can be used to merge two images f_1 and f_2 by forming a weighted configuration $\bar{X} = \alpha X_1 + (1 - \alpha)X_2, 0 < \alpha < 1$. The merged image is then (Mardia and Hainsworth 1993):

$$\bar{f}_\alpha(t) = \alpha f_1\{\Phi_1^*(t)\} + (1 - \alpha)f_2\{\Phi_2^*(t)\},$$

where Φ_i^* is the deformation from $\bar{X}$ to X_i.

Example 17.3   In an illustrative example we see in Figure 17.6 a merged image composed from photographs of Einstein at ages 49 and 71. A set of 11 landmarks

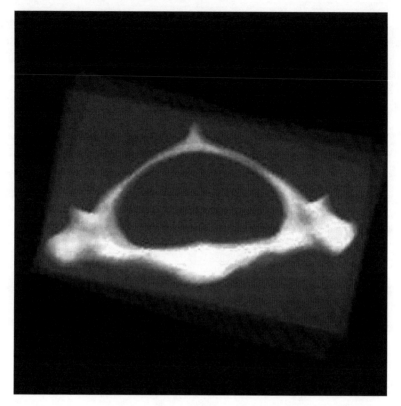

Figure 17.4   An average T1 vertebral image obtained from five vertebral images.

are chosen on each face – the eyes, mouth corners, tip of nose, bottom of ears, bottom of chin, forehead outline (three landmarks) – together with four landmarks in the extreme corners of each image and four landmarks at midpoints of the sides of each image. A thin-plate spline transformation of each image is calculated to a linear interpolation of the two sets of landmarks, and the grey levels are averaged (Mardia and Hainsworth 1993). The merged photograph has characteristics from both the younger and older images. An additional example merging images of Newton and Einstein is given in Figure 17.7.   □

Some other popular examples of image merging and warping include the merging of face images and age changes to the face, such as produced by Benson and Perrett (1993) and Burt and Perrett (1995), using a different technique based on PCA. Other applications include predicting a face image after a number of years from a photograph of a missing child. See Section 17.4.5 for further examples.

Glasbey and Mardia (2001) have proposed a novel approach to image warping using a penalized likelihood with the von Mises–Fourier similarity measure and the

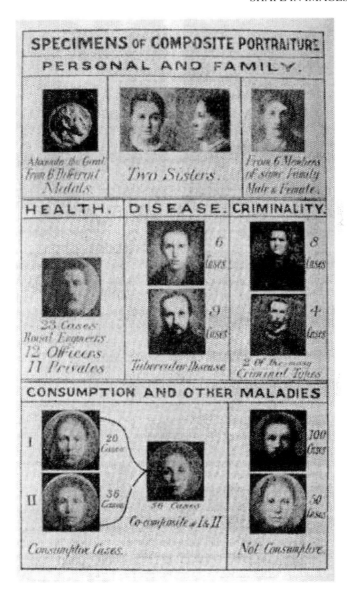

Figure 17.5   Composite photographs produced by Galton by taking multiple exposures of several portraits. Source: Galton 1883.

null distortion criterion. Various applications are given including registering images with a map, discrimination without landmarks and fusing images. Mardia *et al.* (2006c) have used this approach for image averaging and discrimination using a combination of landmark and normalized image information.

(a)   (b)

(c)

Figure 17.6   Photographs of Einstein at ages (a) 49 and (b) 71. Image (c) is a merged image between (a) and (b). Source (a): Photograph by Keystone/Getty Images. (b): Photograph by Doreen Spooner/Keystone Features/Getty Images.

17.4.4   Consistency of deformable models

Although much of the groundwork for Bayesian image analysis has been in place since the early 1990s, there have been many technical developments in recent years. In particular, statistical properties have been investigated in detail for deformable template estimation, including the study of consistency and stochastic approximation algorithms (Allassonnière *et al.* 2007, 2010a,b) and the consistency of Fréchet means in deformable template models (Bigot and Charlier 2011). In particular, Allassonnière

(a)   (b)

(c)

Figure 17.7   Photographs of (a) Newton and (b) Einstein. Image (c) is a merged image between (a) and (b) – 'Newstein'. Source (b): Photograph by Fred Stein Archive/Archive Photos/Getty Images.

et al. (2007) make use of the Bernstein–von Mises theorem (van der Vaart 1998, p. 141) in order to demonstrate Bayesian consistency, and Cheng *et al.* (2016) also use this result to show consistency for function and curve estimation.

Mardia *et al.* (2006b) fit a rigorous stochastic model for a deformation between landmarks and to assess the error of the fitted deformation whereas Mardia *et al.* (2006a) assess the effects of deformations through a Kullback–Leibler divergence measure. In particular, Mardia *et al.* (2006b) have also provided principal warps based on intrinsic random fields.

17.4.5   Discussion

Over the last few decades statistical algorithms have emerged for image analysis which are widely applicable and reliable. There is also an advantage in using explicit stochastic models so that we have a better understanding of the working behind the algorithms and we can make confident statements about conclusions. Another area of application of Bayesian image analysis is where the aim is not only object recognition but knowledge representation of objects, such as the creation of anatomical atlases of the brain. In such cases deformable templates and associated probability distributions can be used to describe normal subjects and patients (e.g. Grenander 1994; Beg *et al.* 2005; Villalon *et al.* 2011). Brain mapping methods are widely used for higher dimensional manifolds, that is landmarks (dimension 0) to sulci (lines of dimension 1), cortex and related surfaces (dimension 2), volumes and subvolumes (dimension 3) (e.g. see Grenander and Miller 1994; Joshi *et al.* 1995; Ashburner and Friston 2000; Durrleman *et al.* 2007). Another important area of image analysis is the use of scale space techniques (e.g. Witkin 1983) and multiscale methods for registration and template modelling (e.g. Fritsch *et al.* 1994; Ying-Lie *et al.* 1994; Wilson 1995; Wilson and Johnson 1995; Mardia *et al.* 1997b). For further discussion of the very broad area of shape and image analysis see, for example, Ying-Lie *et al.* (1994); Krim and Yezzi (2006); Grenander and Miller (2007); Davies *et al.* (2008b); Younes (2010); Neale and Russ (2012); Sotiras *et al.* (2013); and Turaga and Srivastava (2015).

18

Object data and manifolds

This final chapter contains some extensions to shape analysis, especially applications to object data and more general manifolds.

18.1 Object oriented data analysis

The techniques of shape analysis have natural extensions to other application areas. The very broad field of object oriented data analysis (OODA) is concerned with analysing different types of data objects compared with the conventional univariate or multivariate data. The concept of OODA was proposed by J.S. Marron. Examples of object data include functions, images, shapes, manifolds, dynamical systems, and trees. The main aims of shape analysis extend more generally to object data, for example defining a distance between objects, estimation of a mean, summarizing variability, reducing dimension to important components, specifying distributions of objects and carrying out hypothesis tests. From Marron and Alonso (2014) in any study an important consideration is to decide what are the atoms (most basic parts) of the data. A key question is 'what should be the data objects?', and the answer will then lead to appropriate methodology for statistical analysis. The subject is fast developing following initial definitions in Wang and Marron (2007), and a recent summary with discussion is given by Marron and Alonso (2014) with applications to Spanish human mortality functional data, shapes, trees and medical images.

One of the key aspects of object data analysis is that registration of the objects must be considered as part of the analysis. In addition the identifiability of models, choice of regularization and whether to marginalize or optimize as part of the inference (see Chapters 14 and 16) are important aspects of object data analysis, as they are in shape analysis (Dryden 2014). We have already described many applications of object oriented data, for example functional outline data analysis in Section 16.4;

Statistical Shape Analysis, with Applications in R, Second Edition. Ian L. Dryden and Kanti V. Mardia.
© 2016 John Wiley & Sons, Ltd. Published 2016 by John Wiley & Sons, Ltd.

averaging images in Section 17.4; and shape and size-and-shape analysis of landmark data.

18.2   Trees

A particularly challenging type of object data is that of trees, for example collections of phylogenetic trees (Holmes 2003) or geometric trees constructed from medical images of the brain (Skwerer *et al.* 2014).

Billera *et al.* (2001) introduced ground-breaking work on the study of the geometry of the space of phylogenetic trees, and in particular showed that the space has negative curvature. The space of trees can be thought of as an 'open book', where 'pages' of the book are Euclidean spaces which are all joined down a common spine. Each page of the open book represents trees with the same topological structure, but different lengths of branches. Trees on different pages have different topological structures. The geodesic path between two trees is just a straight line between two trees on the same page but between different pages the geodesic consists of straight lines that join at the spine. The space of trees is an example of a **stratified** space, where the Euclidean pages are the different strata. Phylogenetic trees have a fixed set of leaves. Kendall and Colijn (2015) provide a metric for comparing phylogenetic trees and fast computation of median trees.

Wang and Marron (2007) studied sets of trees and emphasized that different types of methods are required to analyse such object data, and they coined the phrase 'object oriented data analysis' from this work. However, working with trees is difficult. Owen (2011) provided a fast algorithm for computing geodesic distances in tree space which is required for statistical analysis, for example using the bootstrap (Holmes 2003), PCA (Aydin *et al.* 2009), a geodesic principal path (Nye 2011), and diffusion models (Nye and White 2014).

Feragen *et al.* (2010, 2013) introduce a distance for comparing geometric tree shapes, the quotient Euclidean distance, via quotienting out different representations of the same tree. The edges contain attributes, for example geometric points along a curve. Feragen *et al.* (2010, 2013) apply the methodology for averaging and comparing geometric trees to examples of trees from medical images of lung airways, and compare with the classical tree-edit distance.

Tree space contain strata that introduce curious effects, in particular the notion of 'stickyness' where a mean tree shape sticks to a particular lower dimensional stratum. For example, Hotz *et al.* (2013) develop sticky central limit theorems in tree space, and Barden *et al.* (2013) study central limit theorems for Fréchet means in the space of phylogenetic trees.

There are other types of stratified manifolds where it is of interest to develop statistical procedures. For example, the cone of symmetric positive semi-definite matrices contains different strata for the subspaces of different ranks [e.g. see Bhattacharya *et al.* (2013) and Hotz *et al.* (2013) for further discussion of stratified manifolds].

18.3   Topological data analysis

Topological data analysis is a relatively new area of study bringing together pure mathematical ideas and data analysis. Carlsson (2009) provides a summary and introduction to the area. The basic idea is that topological features (e.g. number of holes, Betti numbers) are recorded as a filtration of simplicial complexes is applied to the dataset. Particular types of filtration are the Čech complex, the Vietoris–Rips complex and Delaunay complex. Sets of geometrical shapes/patterns (simplicial complexes) are obtained by applying the filtration with different parameters, and observing how the features persist over different scales is key to the concept of persistent homology. Persistence diagrams are constructed and then compared using suitable distance metrics, such as the Wasserstein metric. Fréchet means can be constructed (Munch *et al.* 2015) and general statistical procedures can be applied. For example, Gamble and Heo (2010) explore the uses of persistent homology for statistical analysis of landmark-based shape data and Heo *et al.* (2012) introduce a topological analysis of variance with applications in dentistry. Other work includes that by Bubenik *et al.* (2010) who consider a non-parametric regression problem on a compact Riemannian manifold to estimate persistent homology consistently, Turner *et al.* (2014) who introduce a persistent homology transform for modelling shapes and surfaces, which is a sufficient statistic for representing shape and Kim *et al.* (2014) discuss the potential for using persistent homology to study phylogenetic trees in the microbiome. Nakamura *et al.* (2015) use a kernel-based distance between persistence diagrams in order to describe differences between different types of materials.

18.4   General shape spaces and generalized Procrustes methods

18.4.1   Definitions

We can also study more general shape spaces. The work follows Carne (1990); see also Dryden (1999) and Dryden and Mardia (1991a).

Definition 18.1   *Consider a point T in a differentiable manifold M. Define a group G which acts on M, and denote the transformation of T by $g \in G$ by $g(T)$. The **general shape space** $\Sigma(M, G) \equiv M/G$ is defined as the orbit space of M under the action of G. The equivalence class in $\Sigma(M, G)$ corresponding to $T \in M$ is the set $G(T)$, and we call this set the **general shape** of T.*

A suitable choice of distance between T and Y in $\Sigma(M, G)$ is given by:

$$d_\Sigma(T, Y) = \inf_{g \in G} \text{dist}(h(Y) - g(T)),$$

where $h \in G$ is chosen so that $d_\Sigma(T, Y) = d_\Sigma(Y, T)$, and $dist(\cdot)$ is a suitable choice of distance in M. For example $dist()$ could be the Euclidean metric if M was a Euclidean space.

The special case of the shape space of Chapter 3 that we have primarily considered is when $M = \mathbb{R}^{km} \setminus C$ with Euclidean metric (C is the set of coincident points) and G is the Euclidean similarity group of transformations. The size-and-shape space of Chapter 5 involves the isometry group of rotation and translation only. Ambartzumian (1990) considers the general affine shape, when $M = \mathbb{R}^{km} \setminus D$ (D is the set of deficient rank configurations) and G is the affine transformation group (see Section 12.2). Also, in Section 16.4 we had $M = L_2$, the Hilbert space of square integrable functions and G is the space of diffeomorphic warpings.

For some choices of manifold M a non-Euclidean metric will be the natural choice. For example, if $M = S^p$ (the unit sphere), then the great circle metric could be used. Le (1989a,b) has considered the shape space for points on a sphere, where the registration group is rotations only. Stoyan and Molchanov (1995) considered the manifold to be the system of non-empty compact sets in $\mathbb{R}^m$, and the registration group consisted of rotations and translations.

18.4.2   Two object matching

As we saw in the Section 4.1.1 calculation of distances in shape space can be regarded as finding the minimum Euclidean distances between configurations, by superimposing or registering them as close as possible using the similarity transformations. Consider two configurations of k points in $\mathbb{R}^m$: $\{t_j : j = 1, \ldots, k, \ t_j \in \mathbb{R}^m\}$ and $\{y_j : j = 1, \ldots, k, \ y_j \in \mathbb{R}^m\}$ which we write as $k \times m$ matrices:

$$T = [t_1, \ldots, t_k]^\mathrm{T}, \quad Y = [y_1, \ldots, y_k]^\mathrm{T}.$$

We shall assume throughout this section that $k \geq m + 1$. We wish to match T to Y, using

$$Y = h(T, B) + E, \tag{18.1}$$

where $h(\cdot)$ is a known smooth function, B are the parameters and E is the error matrix. We could consider estimating B by minimizing some objective function $s^2(E)$ – a function of the error matrix $E = Y - h(T, B)$.

Alternatively, we could formulate the situation in terms of a statistical model, where Y is perturbed from $h(T, B)$ by the random zero mean matrix E from some specified distribution. The unknown parameters B, and any parameters in the distribution of E (e.g. a variance parameter), could then be estimated say by maximum likelihood or some other inferential procedure.

Our primary concern throughout has been the least squares case. Hence, in this case the objective function used is $s^2(E) = \|E\|^2 = \mathrm{trace}(E^\mathrm{T} E)$. Equivalently, the implicit model for the errors $\mathrm{vec}(E)$ is independent multivariate normal. If the functions $h(T, B)$ are similarity transformations of T, then this special case reduces to the Procrustes analysis of Chapter 7. Some other choices for objective function are discussed in Section 13.6.

18.4.3   Generalized matching

If we have a random sample of n objects $T_1, \ldots, T_n$ ($T_j \in M$, $k \times m$ matrices) from a population, then it is of interest to obtain an estimate of the population mean shape μ ($k \times m$ matrix) and to explore the structure of variability, up to invariances in the set of transformations G.

An estimate of the population mean configuration μ up to invariances in G, denoted by $\hat{\mu}$, can be obtained by simultaneously matching each T_j to μ ($j = 1, \ldots, n$) and choosing $\hat{\mu}$ as a suitable M-estimator (subject to certain constraints on μ). In particular, $\hat{\mu}$ is obtained from the constrained minimization

$$\hat{\mu} = \arg\inf_{\mu} \sum_{j=1}^{n} \inf_{g_j} \phi(s(E_j)), \quad E_j = g_j(T_j) - \mu, \qquad (18.2)$$

where $\phi(x) \geq 0$ is a penalty function (an increasing function) on $\mathbb{R}^+$, $\phi(0) = 0$, $g_j(T_j)$ is a transformed version of T_j by $g_j \in G$, $s(E)$ is the objective function for matching two configurations and general restrictions need to be imposed on μ to avoid degeneracies. A common choice of estimator has $\phi(x) = x^2$, the least squares choice. In this special case, if g_j are known then the minimizing value of μ is:

$$\hat{\mu} = \frac{1}{n} \sum_{j=1}^{n} g_j(T_j).$$

Example 18.1   For $M = \mathbb{R}^{km} \setminus C$ and G are the Euclidean similarity transformations, the full Procrustes mean shape is obtained by solving Equation (18.2) with $\phi(x) = x^2, s(E)^2 = \|E\|^2 = \text{trace}(E^T E)$ subject to the constraint that $\|\mu\| = 1$, $\mu^T 1_k = 0$ and μ is rotated to a fixed axis.   □

18.5   Other types of shape

There are many studies into different aspects of shape in several disciplines. For example, the Gestalt theory of shape looks at shape from physical, physiological and behavioural points of view, for example Zusne (1970). Nagy (1992) gave some general observations about shape theory.

There is a vast amount of literature on shape in computer graphics, computer science and geometry, including the work of Koenderink (1990). Mumford (1991) gives some important observations about theories of shape and questions whether or not they model human perception. As well as discussing curvature he describes various metrics for shape. Following the work of Tversky (1977) he argues that human perception about the similarity of objects A and B is often different from the perception of similarity between objects B and A, and may also be affected by contextual information. Hence a metric, which is symmetric in its arguments, may not always be the best way to measure 'similarity'.

Further wide ranging reviews of notions of shape, particularly in image analysis, are given by Neale and Russ (2012).

18.6   Manifolds

The main ideas of shape analysis can be extended to many other types of manifolds too. Of fundamental importance is the need to define a distance, to define population and sample means, to describe variability, to define probability distributions and to carry out statistical inference.

Mardia and Jupp (2000) describe analysis of circles and spheres in detail, which are the simplest types of non-Euclidean manifold. There is a great deal of relevant distributional work given by Watson (1983) and Chikuse (2003) on spheres and other symmetric spaces. The topic of manifold data analysis is too large and broad to summarize here, and so instead we give a few select examples. Turaga et al. (2011) study Stiefel manifolds and Grassmannians for image and video-based recognition; Rosenthal et al. (2014) develop spherical regression using projective transformations; pure states in quantum physics can be represented as points in complex projective space (Kent 2003; Brody 2004) and Guta et al. (2012) develop maximum likelihood estimation methods for the extension to mixed quantum states; Kume and Le (2000, 2003) consider Fréchet means in simplex shape spaces, which are hyperbolic spaces which have negative curvature; Mardia et al. (1996d), Mardia and Patrangenaru (2005) and Kent and Mardia (2012) consider cross-ratios and data analysis with projective invariance; Dryden et al. (2009b) and Zhou et al. (2013, 2016) study symmetric positive-definite (SPD) matrices (covariance matrices), and use a Procrustes procedure applied to diffusion tensors in medical image analysis; Jayasumana et al. (2013a) use kernel methods for studying SPD matrices; Arsigny et al. (2006) and Fillard et al. (2007) give many practical examples on the same manifolds using the log-Euclidean metric with applications to clinical diffusion tensor MRI estimation, smoothing and white matter fibre tracking; Yuan et al. (2013) study varying coefficient models for diffusion tensors; and Pigoli et al. (2014) extend the Procrustes approach for distances in Hilbert spaces between covariance functions, with applications to comparing sounds in romance languages.

18.7   Reviews

A summary of such a broad and wide-ranging topic of shape analysis cannot be complete. We conclude by mentioning some other reviews and collections of material on shape analysis, which naturally have different emphases from our own.

Some books on the topic include those by Bookstein (1991), Lele and Richtsmeier (2001); Claude (2008), Weber and Bookstein (2011), Bookstein (2014) and Zelditch *et al.* (2012) with many biological examples; Kendall *et al.* (1999), Small (1996) and Younes (2010) from a mathematical perspective; Grenander and Miller (2007), Davies *et al.* (2008b) and Neale and Russ (2012) on deformation models and image analysis; da Fontoura Costa and Marcondes Cesar Jr (2009) on shape classification; Bhattacharya and Bhattacharya (2008), Brombin and Salmaso (2013) and Patrange-naru and Ellingson (2015) on non-parametric shape analysis; Stoyan and Stoyan (1994, Part II) on outline data; and Stoyan *et al.* (1995, Chapter 8) on unlabelled configurations.

The topic of shape analysis has entries in the *Encyclopaedia of Statistical Sciences* (Kotz and Johnson 1988) under the titles of 'Shape Statistics' (D.G. Kendall), 'Size and Shape Analysis' (Mosimann) and 'Landmark Data' (Mardia) and in the *Encyclopaedia of Biostatistics* (Dryden 2005a). Finally, some general reviews include Kendall (1989), Rohlf and Marcus (1993), Pavlidis (1995), Loncaric (1998), Adams *et al.* (2004, 2013), Heo and Small (2006), Slice (2007), Mitteroecker and Gunz (2009) and Kendall and Lee (2010); and some edited volumes involving shape analysis include those by Ying-Lie *et al.*, (1994), MacLeod and Forey (2002), Slice (2005), Krim and Yezzi (2006), Evison and Bruegge (2010), Hamelryck *et al.* (2012), Breuß *et al.* (2013), Dryden and Kent (2015) and Turaga and Srivastava (2015).

Exercises

1. Which of the triangles in Figure E.1 have the same size and the same shape (assume by 'shape' we mean labelled shape)?
   Which triangles have the same shape but different sizes?
   Which triangles have different shapes but the same 'reflection shape' (i.e. a translation, rotation, scale and a reflection will match them exactly)?

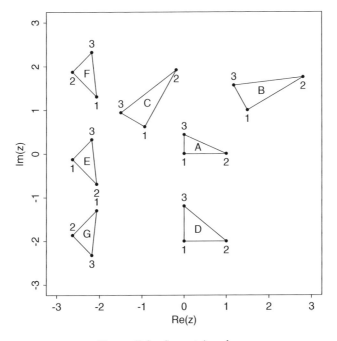

Figure E.1   Some triangles.

2. Rank the triangles in Figure E.1 in order of increasing centroid size. Repeat the rankings using baseline size (with baseline 1, 2) and square root of area.

Statistical Shape Analysis, with Applications in R, Second Edition. Ian L. Dryden and Kanti V. Mardia.
© 2016 John Wiley & Sons, Ltd. Published 2016 by John Wiley & Sons, Ltd.

3. Informally rank the triangles in Figure E.1 in order of 'closeness' to the anti-clockwise labelled equilateral triangle.

4. In a forensic study photographs of the front view of the faces of alleged criminals are taken. Some landmarks are to be located on the face. For each of the following landmarks decide whether it is a Bookstein's Type I, II or III landmark. Also, state whether they are anatomical, mathematical or pseudo-landmarks.
   a. Pupil of the eye
   b. Tip of the nose
   c. Corners of the mouth
   d. Point on the cheek halfway between the pupil of the eye and corner of the mouth.
   e. Centre of the forehead.
   f. Lowest point on the chin.
   Which landmarks are easy to locate and which are difficult?

5. Prove the result that

$$\sum_{j=1}^{m}\sum_{i=1}^{k}\sum_{l=1}^{k}(X_{ij} - X_{lj})^2 = 2k\sum_{i=1}^{k}\sum_{j=1}^{m}(X_{ij} - \bar{X}_j)^2$$
$$= 2kS(X)^2.$$

6. Consider the Helmert submatrix H. Write down H for $k = 3, 4, 5$ and verify that $H^T H = C$ and $HH^T = I$.
   Prove the general results for $k \geq 2$ that $H^T H = C$ and $HH^T = I$.

7. Rank the triangles in Figure E.1 in terms of Riemannian distance from the anti-clockwise labelled equilateral triangle.

8. **a.** For the male gorilla data (see Section 1.4.8) find the centroid sizes of the gorilla landmarks.
   b. Obtain Bookstein's shape variables for the specimens and provide a scatter plot of the data, registered on the baseline 1, 2.
   c. Give the Bookstein's shape variables for the largest specimen using landmarks 1, 2 (pr, l) as the baseline.
   d. Provide a scatter plot of the data using landmarks 1, 6 (pr, na) as the baseline, and comment on the scatter compared with that using baseline 1, 2.
   Hint: to permute the rows 1, 2, 3, 4 to 1, 4, 2, 3 for an array x with 4 rows, one would use the command x <- x[c(1, 4, 2, 3),,].
   e. Obtain the mean shape from Bookstein's shape variables (using the arithmetic mean) using both baselines. Are they the same shape?
   f. Find the specimen which is the furthest away from the mean in terms of Riemannian distance.

9. Prove the linear relationship between Bookstein coordinates U^B and Kendall coordinates U^K.

10. Consider a pre-shape $z = x + iy$ ($k - 1$−vector). Obtain the pre-shapes which are the furthest possible Riemannian distance from this pre-shape.

11. On Kendall's spherical shape space where are the right-angled triangles located (in terms of the spherical coordinates θ and ϕ)?

12. Obtain an expression for the full Procrustes distance d_F in terms of the partial Procrustes distance d_P. Obtain a Taylor series expansion of d_F in terms of d_P giving the first two non-zero terms explicitly.

13. Prove that the Riemannian distance ρ is a distance, that is show that: (i) $\rho(x, y) = 0$ if and only if $x = y$; (ii) $\rho(x, y) = \rho(y, x)$ for all x, y; and (iii) $\rho(x, y) \leq \rho(x, z) + \rho(z, y)$ for all x, y, z.

14. Plot the full and partial Procrustes and Riemannian distances versus ρ over the range of ρ. Discuss the relative values of the distances over the range of ρ.

15. For a triangle with Bookstein's shape variables (U^B, V^B) find an expression for the Riemannian distance ρ to the (a) closest equilateral shape and (b) closest collinear shape.

16. Consider two centred configurations z_1, z_2 of k points in two dimensions, which are not necessarily of unit size. Find the closest Euclidean distance between z_1 and z_2 by rotating (but not scaling) z_2 to be as close as possible to z_1 (complex notation should help). Express the distance in terms of the centroid sizes of z_1 and z_2 and the Riemannian shape distance $\rho(z_1, z_2)$.
    Find the same expression for k points in m dimensions.

17. Consider two configurations of k points in m dimensions with pre-shapes Z_1, Z_2. Derive an expression for the partial Procrustes distance d_R between pre-shapes Z_1, Z_2 where the minimization is over the orthogonal group $O(m)$ rather than $SO(m)$.

18. Consider the female gorilla data. Find the partial Procrustes tangent coordinates v for the largest female gorilla skull, using the full Procrustes mean as the pole of the projection.
    By using the inverse projection from the tangent plane to the figure space obtain a suitable icon for the largest gorilla, which is Procrustes rotated to be as close as possible to the full Procrustes mean female gorilla shape.
    Provide a plot of all the female gorilla icons, Procrustes rotated to the full Procrustes mean (i.e. for each skull find the tangent coordinates and project back to an icon, using the full Procrustes mean as the pole).
    Provide a pairwise plot of the coordinates of the icons and comment on any structure present in the plots.

19. Find the full Procrustes mean shapes of the the the male and female gorillas separately by using the routine procGPA with options eigen=TRUE and eigen=FALSE. Check whether they give the same result (up to an arbitrary scale and rotation). Compare the percentages of variability explained by the first three PCs using the routines with either option.

20. What percentage of variability is explained by the first three PCs for the three groups of mouse T2 vertebrae (Small, Large, Control). Describe the geometrical features highlighted by each PC.

21. Give an educated guess as to what the full Procrustes mean shape would be from a dataset of two equilateral triangles – one clockwise and the other anti-clockwise labelled. Find the full Procrustes mean shape using the routine procGPA with options eigen=TRUE and eigen=FALSE, and comment on any differences or similarities in the results.

22. Provide plots of the thin-plate spline transformation from the orangutan female to male mean shapes. Test whether the mean shapes are significantly different or not.

23. Carry out relative warps analysis of the orangutan data, with respect to bending energy and inverse bending energy, and comment on the differences.

24. Plot the partial warp scores for the female and male orangutan data, and comment.

25. Investigate allometry in the 3D macaque data.

Appendix

$$1_k \quad \text{Vector of ones } (k \times 1) \text{ vector}$$
$$A^{\mathrm{T}} \quad \text{Transpose of matrix } A$$
$$\|A\| = \sqrt{\text{trace}(A^*A)} \quad \text{Euclidean norm}$$
$$\overline{a} \quad \text{Complex conjugate of } a$$
$$a^* \quad \text{Transpose of the complex conjugate of } a$$
$$C = I_k - \tfrac{1}{k} 1_k 1_k^{\mathrm{T}} \quad \text{Centring matrix } (k \times k)$$
$$D(X) \quad \text{Baseline size (positive scalar)}$$
$$d_P(X_1, X_2) \quad \text{Partial Procrustes distance } (0 \le d_P \le \sqrt{2})$$
$$d_F(X_1, X_2) \quad \text{Full Procrustes distance } (0 \le d_F \le 1)$$
$$\rho(X_1, X_2) \quad \text{Riemannian distance } (0 \le \rho \le \pi/2)$$
$$d_S(X_1, X_2) \quad \text{Riemannian distance in size-and-shape space } (0 \le d_S < \infty)$$
$$H \quad \text{Helmert sub-matrix } ((k-1) \times k)$$
$$H_{hat} = X_D (X_D^{\mathrm{T}} X_D)^{-1} X_D \quad \text{Hat matrix}$$
$$H_j = X_{Dj} (X_{Dj}^{\mathrm{T}} X_{Dj})^{-1} X_{Dj} \quad j\text{th hat matrix}$$
$$I_k \quad \text{Identity matrix } (k \times k)$$
$$k \quad \text{Number of landmarks}$$
$$q = (k-1)m - \tfrac{m}{2}(m-1) - 1 \quad \text{Dimension of shape space}$$
$$m \quad \text{Real dimension of object}$$
$$R \quad \text{Rotation and reflection matrix } (\in O(m))\ (m \times m)$$
$$S(X) \quad \text{Centroid size (positive scalar)}$$
$$S^l \quad \text{Unit sphere in } l + 1 \text{ real dimensions}$$
$$S^l(r) \quad \text{Sphere in } l + 1 \text{ real dimensions with radius } r$$
$$T = [t_1, \dots, t_k]^{\mathrm{T}} \in \mathbb{R}^{km} \quad \text{Coordinates of landmarks to be matched } (k \times m \text{ matrix})$$
$$U^B = (u_3^B, \dots, u_k^B, v_3^B, \dots, v_k^B)^{\mathrm{T}} \quad \text{Bookstein coordinates for 2D data } ((2k-4) \times 1 \text{ vector})$$
$$U^K = (u_3^K, \dots, u_k^K, v_3^K, \dots, v_k^K)^{\mathrm{T}} \quad \text{Kendall coordinates for 2D data } ((2k-4) \times 1 \text{ vector})$$
$$v \quad \text{Tangent plane coordinates}$$

Statistical Shape Analysis, with Applications in R, Second Edition. Ian L. Dryden and Kanti V. Mardia.
© 2016 John Wiley & Sons, Ltd. Published 2016 by John Wiley & Sons, Ltd.

$W = X_H \Gamma$ Size-and-shape of X ($\Gamma \in SO(k)$)

X Configuration matrix of landmark coordinates ($k \times m$ matrix)

$[X]$ Shape of X

$[X]_I$ Icon (representative configuration)

$[X]_S$ Shape of X

$[X]_R$ Reflection shape of X

$[X]_{RS}$ Reflection size-and-shape of X

$X_H = HX$ Helmertized landmark coordinates ($(k-1) \times m$ matrix)

$X_D = I_m \otimes [1_k, T]$ Design matrix

X_{Dj} Design matrix for the jth configuration

X^P Full Procrustes fitted configuration

$Y = [y_1, \ldots, y_k]^T \in \mathbb{R}^{km}$ Coordinates of reference configuration ($k \times m$ matrix)

$Y = [1_k, T]B$ Affine transformation between configurations

$\text{vec}(Y) = X_D \beta$ Vectorized equation for affine/shape transformation

$y = At + c$ Affine transformation between points $y \in \mathbb{R}^2, t \in \mathbb{R}^2$

$Z = X_H / \|X_H\|$ Pre-shape ($(k-1) \times m$ matrix)

$Z_C = H^T Z_H$ Centred pre-shape

z Complex pre-shape ($(k-1) \times 1$ complex vector)

z^o Original complex landmark coordinates ($k \times 1$ complex vector)

z_H Helmertized complex landmarks ($(k-1) \times 1$ complex vector)

Γ Rotation matrix ($\in SO(m)$) ($m \times m$ matrix)

$\Phi(t) = [\Phi_1(t), \ldots, \Phi_m(t)]^T$ Deformation from $\mathbb{R}^m$ to $\mathbb{R}^m$

$\Theta^B = (\theta_3^B, \ldots, \theta_k^B, \phi_3^B, \ldots, \phi_k^B)^T$ Bookstein coordinates of population mean shape

$\Theta^K = (\theta_3^K, \ldots, \theta_k^K, \phi_3^K, \ldots, \phi_k^K)^T$ Kendall coordinates of population mean shape

μ Population average configuration (e.g., mean or mode)

Σ Covariance matrix of Helmertized landmarks

Σ_m^k Shape space for k points in $\mathbb{R}^m$

$S\Sigma_m^k$ Size-and-shape space

Ω Covariance matrix of original landmarks

C Set of coincident points

$\mathbb{C}^k$ k-dimensional complex space

$\mathbb{C}P^l$ l-dimensional complex projective space

$\mathbb{C}S^l$ Unit complex sphere in $l+1$ complex dimensions $= S^{2l+1}$

D Set of less than full-rank points

${}_1F_1(\cdot)$ Confluent hypergeometric function

$I_\nu(\cdot)$ Bessel function of the first kind

$\mathcal{L}_j$ Simple Laguerre polynomial of degree j

$\mathcal{P}_j$ Legendre polynomial of degree j

$\mathbb{R}^l$ l-dimensional real space

S_j j-dimensional simplex

arginf   value that gives the infemum

argsup   value that gives the supremum

Arg   argument of a complex number

References

Abramowitz, M. and Stegun, I. A. (1970). *Handbook of Mathematical Functions*. Dover, New York. page 228, 248, 256, 262, 263

Adams, D. C. and Otárola-Castillo, E. (2013). geomorph: an R package for the collection and analysis of geometric morphometric shape data. *Methods in Ecology and Evolution*, 4(4): 393–399. page 173

Adams, D. C., Rohlf, F. J., and Slice, D. E. (2004). Geometric morphometrics: Ten years of progress following the revolution. *Italian Journal of Zoology*, 71: 5–16. page 397

Adams, D. C., Rohlf, F. J., and Slice, D. E. (2013). A field comes of age: geometric morphometrics in the 21st century. *Hystrix, the Italian Journal of Mammalogy*, 24(1): 7–14. page 397

Afsari, B. (2011). Riemannian L^p center of mass: existence, uniqueness, and convexity. *Proceedings of the American Mathematical Society*, 139(2): 655–673. page 111, 112, 318

Afsari, B., Tron, R., and Vidal, R. (2013). On the convergence of gradient descent for finding the Riemannian center of mass. *SIAM Journal on Control and Optimization*, 51(3): 2230–2260. page 320

Airoldi, C. A., Bergonzi, S., and Davies, B. (2010). Single amino acid change alters the ability to specify male or female organ identity. *PNAS*, 107: 18898–18902. page 208, 209, 212

Aitchison, J. (1986). *The Statistical Analysis of Compositional Data*. Chapman and Hall, London. page 40

Albers, C. J. and Gower, J. C. (2010). A general approach to handling missing values in Procrustes analysis. *Advanced Data Analysis Classification*, 4(4): 223–237. page 339

Allassonnière, S., Amit, Y., and Trouvé, A. (2007). Towards a coherent statistical framework for dense deformable template estimation. *Journal of the Royal Statistical Society: Series B (Statistical Methodology)*, 69(1): 3–29. page 388

Allassonnière, S., Kuhn, E., and Trouvé, A. (2010a). Bayesian consistent estimation in deformable models using stochastic algorithms: applications to medical images. *Journal de la Société Français de Statistique*, 151(1): 1–16. page 388

Allassonnière, S., Kuhn, E., and Trouvé, A. (2010b). Construction of Bayesian deformable models via a stochastic approximation algorithm: a convergence study. *Bernoulli*, 16(3): 641–678. page 388

Alshabani, A. K. S., Dryden, I. L., and Litton, C. D. (2007a). Partial size-and-shape distributions. *Journal of Multivariate Analysis*, 98(10): 1988–2001. page 259

Alshabani, A. K. S., Dryden, I. L., Litton, C. D., and Richardson, J. (2007b). Bayesian analysis of human movement curves. *Journal of the Royal Statistical Society, Series C*, 56(4): 415–428. page 259, 325

Amaral, G. J. A., Dore, L. H., Lessa, R. P., and Stosic, B. (2010a). *k*-means algorithm in statistical shape analysis. *Communications in Statistics – Simulation and Computation*, 39(5): 1016–1026. page 323

Amaral, G. J. A., Dryden, I. L., Patrangenaru, V., and Wood, A. T. A. (2010b). Bootstrap confidence regions for the planar mean shape. *Journal of Statistical Planning and Inference*, 140(11): 3026–3034. page 322

Amaral, G. J. A., Dryden, I. L., and Wood, A. T. A. (2007). Pivotal bootstrap methods for *k*-sample problems in directional statistics and shape analysis. *Journal of the American Statistical Association*, 102(478): 695–707. page 193, 194, 203, 204, 322, 323

Amaral, G. J. A. and Wood, A. T. A. (2010). Empirical likelihood methods for two-dimensional shape analysis. *Biometrika*, 97(3): 757–764. page 323

Ambartzumian, R. V. (1982). Random shapes by factorisation. In: *Statistics in Theory and Practice* (ed. B. Ranneby). Swedish University of Agricultural Science, Umea. page 273

Ambartzumian, R. V. (1990). *Factorization, Calculus and Geometric Probability*, pp. 35–42. Cambridge University Press, Cambridge. page 273, 394

Amit, Y. (1997). Graphical shape templates for automatic anatomy detection with applications to MRI scans. *IEEE Transactions on Medical Imaging*, 16: 28–40. page 381

Amit, Y. and Geman, D. (1997). Shape quantization and recognition with randomized trees. *Neural Computation*, 9: 1545–1588. page 381

Amit, Y. and Kong, A. (1996). Graphical templates for model registration. *IEEE Transactions on Pattern Analysis and Machine Intelligence*, 18: 225–236. page 381

Andersen, H., Højbjerre, M., Sørensen, D., and Eriksen, P. S. (1995). *Linear and Graphical Models for the Multivariate Complex Normal Distribution*. Springer-Verlag, New York. page 220

Anderson, C. R. (1997). *Object recognition using statistical shape analysis*. PhD thesis, University of Leeds. page 11

Arad, N., Dyn, N., Reisfeld, D., and Yeshurun, Y. (1994). Image warping by radial basis functions: application to facial expressions. *Computer Vision and Graphical Image Processing: Graphical Models and Image Processing*, 56: 161–172. page 309

Arsigny, V., Fillard, P., Pennec, X., and Ayache, N. (2006). Geometric means in a novel vector space structure on symmetric positive-definite matrices. *SIAM Journal on Matrix Analysis and Applications*, 29(1): 328–347 (electronic). page 315, 396

Ashburner, J. (2007). A fast diffeomorphic image registration algorithm. *Neuroimage*, 38: 95–113. page 315

Ashburner, J. and Friston, K. J. (2000). Voxel-based morphometry – the methods. *Neuroimage*, 11: 805–821. page 390

Awate, S. P., Yu, Y., and Whitaker, R. T. (2014). Kernel principal geodesic analysis. In: *Machine Learning and Knowledge Discovery in Databases* (eds T. Calders, F. Esposito, E. Hüllermeier, and R. Meo). Proceedings of the European Conference, ECML PKDD 2014, Nancy, France, September 15–19, 2014. Part I, Vol. 8724 of *Lecture Notes in Computer Science*, pp. 82–98. Springer, Heidelberg. page 375

Aydin, B., Pataki, G., Wang, H., Bullitt, E., and Marron, J. S. (2009). A principal component analysis for trees. *Annals of Applied Statistics*, 3(4): 1597–1615. page 392

Baddeley, A. J. and van Lieshout, M. N. M. (1993). Stochastic geometry models in high-level vision. In: *Statistics and Images* (eds K. V. Mardia and G. K. Kanji), Vol. 1, pp. 231–256. Carfax, Oxford. page 380

Ball, F. G., Dryden, I. L., and Golalizadeh, M. (2006). Discussion to 'Exact and computationally efficient likelihood-based estimation for discretely observed diffusion processes' by A Beskos *et al*. *Journal of the Royal Statistical Society: Series B (Statistical Methodology)*, 68(3): 367–368. page 266

Ball, F. G., Dryden, I. L., and Golalizadeh, M. (2008). Brownian motion and Ornstein-Uhlenbeck processes in planar shape space. *Methodology and Computing in Applied Probability*, 10(1): 1–22. page 106, 266

Bandulasiri, A. and Patrangenaru, V. (2005). Algorithms for nonparametric inference on shape manifolds. In: *Proceedings of the Joint Statistical Meetings*, pp. 1617–1622. American Statistical Association, Alexandria, VA. page 355

Bär, C. (2010). *Elementary Differential Geometry*. Cambridge University Press, Cambridge. page 59

Barden, D., Le, H., and Owen, M. (2013). Central limit theorems for Fréchet means in the space of phylogenetic trees. *Electronic Journal of Probability*, 18(25): 25. page 392

Barry, S. J. E. and Bowman, A. W. (2008). Linear mixed models for longitudinal shape data with applications to facial modeling. *Biostatistics*, 9: 555–565. page 324

Bartlett, M. S. (1933). On the theory of statistical regression. *Proceedings of the Royal Society of Edinburgh*, 53: 260. page 264

Bates, D. (2005). Fitting linear mixed models in R. *R News*, 5(1): 27–30. page 324

Bauer, M., Bruveris, M., and Michor, P. W. (2014). Overview of the geometries of shape spaces and diffeomorphism groups. *Mathematical Imaging Vision*, 50: 60–97. page 315

Beg, M. F., Miller, M. I., Trouvé, A., and Younes, L. (2005). Computing large deformation metric mappings via geodesic flows of diffeomorphisms. *International Journal of Computer Vision*, 61(2): 139–157. page 314, 384

Bennani Dosse, M., Kiers, H. A. L., and Ten Berge, J. M. F. (2011). Anisotropic generalized Procrustes analysis. *Computer Statistics Data Analysis*, 55(5): 1961–1968. page 148

Benson, P. J. and Perrett, D. I. (1993). Extracting prototypical facial images from exemplars. *Perception*, 22: 257–261. page 386

Berman, H. M., Westbrook, J., Feng, Z., *et al*. (2000). The protein data bank. *Nucleic Acids Research*, 28: 235–242. page 21

Besag, J. E. (1986). On the statistical analysis of dirty pictures (with discussion). *Journal of the Royal Statistical Society, Series B*, 48: 259–302. page 379

Besag, J. E., Green, P. J., Higdon, D., and Mengersen, K. L. (1995). Bayesian computation and stochastic systems. *Statistical Science*, 10(1): 3–66. page 234, 379

Besl, P. J. and McKay, N. D. (1992). A method for registration of 3D shapes. *IEEE Transactions on Pattern Analysis and Machine Intelligence*, 14: 239–256. page 375

Bhattacharya, A. (2008). Statistical analysis on manifolds: a nonparametric approach for inference on shape spaces. *Sankhyā*, 70(2, Ser. A): 223–266. page 355

Bhattacharya, A. and Bhattacharya, R. (2008). Statistics on Riemannian manifolds: asymptotic distribution and curvature. *Proceedings of the American Mathematical Society*, 136(8): 2959–2967. page 322, 397

Bhattacharya, A. and Bhattacharya, R. (2012). *Nonparametric Inference on Manifolds*, Vol. 2 of *Institute of Mathematical Statistics (IMS) Monographs*. Cambridge University Press, Cambridge. page 322

Bhattacharya, A. and Dunson, D. (2012a). Nonparametric Bayes classification and hypothesis testing on manifolds. *Journal of Multivariate Analysis*, 111: 1–19. page 323

Bhattacharya, A. and Dunson, D. B. (2010). Nonparametric Bayesian density estimation on manifolds with applications to planar shapes. *Biometrika*, 97(4): 851–865. page 323

Bhattacharya, A. and Dunson, D. B. (2012b). Strong consistency of nonparametric Bayes density estimation on compact metric spaces with applications to specific manifolds. *Annals of the Institute of Statistical Mathematics*, 64(4): 687–714. page 323

Bhattacharya, R. and Patrangenaru, V. (2003). Large sample theory of intrinsic and extrinsic sample means on manifolds. I. *Annals of Statistics*, 31(1): 1–29. page 112, 120, 318, 322

Bhattacharya, R. and Patrangenaru, V. (2005). Large sample theory of intrinsic and extrinsic sample means on manifolds. II. *Annals of Statistics*, 33(3): 1225–1259. page 112, 120, 318, 322

Bhattacharya, R. N., Buibas, M., Dryden, I. L., *et al.* (2013). Extrinsic data analysis on sample spaces with a manifold stratification. In: *Advances in Mathematics* (eds L. Beznea, V. Brzaneseu, M. Iosifeseu, G. Marinosehi, R. Purice and D. Timotin), pp. 227–240. The Publishing House of the Romanian Academy, Bucharest. page 392

Bhavnagri, B. (1995a). Construction of a Markov process to model a process arising in vision. In: *Current Issues in Statistical Shape Analysis* (eds K. V. Mardia and C. A. Gill), pp. 76–81, University of Leeds Press, Leeds. page 350

Bhavnagri, B. (1995b). Connected components of the space of simple, closed non-degenerate polygons. In: *Current Issues in Statistical Shape Analysis* (eds K. V. Mardia and C. A. Gill), pp. 187–188. University of Leeds Press, Leeds. page 273

Biasotti, S., Cerri, A., Bronstein, A. M., and Bronstein, M. M. (2014). Quantifying 3D shape similarity using maps: Recent trends, applications and perspectives. In: *Eurographics 2014* (eds S. Lefebvre and M. Spagnuolo), pp. 135–159. Eurographics Association, Geneva. page 375

Bigot, J. and Charlier, B. (2011). On the consistency of Fréchet means in deformable models for curve and image analysis. *Electronic Journal of Statistics*, 5: 1054–1089. page 388

Billera, L. J., Holmes, S. P., and Vogtmann, K. (2001). Geometry of the space of phylogenetic trees. *Advances in Applied Mathematics*, 27(4): 733–767. page 392

Bingham, C., Chang, T., and Richards, D. (1992). Approximating the matrix Fisher and Bingham distributions: applications to spherical regression and Procrustes analysis. *Journal of Multivariate Analysis*, 41: 314–337. page 222

Boas, F. (1905). The horizontal plane of the skull and the general problem of the comparison of variable forms. *Science*, 21(544): 862–863. page 125, 132

Bock, M. T. and Bowman, A. W. (2006). On the measurement and analysis of asymmetry with applications to facial modelling. *Journal of the Royal Statistical Society, Series C*, 55(1): 77–91. page 171, 324

Bookstein, F. L. (1978). *The Measurement of Biological Shape and Shape Change*. Lecture Notes on Biomathematics, Vol. 24. Springer-Verlag, New York. page 32, 33, 305, 306, 307

Bookstein, F. L. (1984). A statistical method for biological shape comparisons. *Journal of Theoretical Biology*, 107: 475–520. page 41, 249

Bookstein, F. L. (1986). Size and shape spaces for landmark data in two dimensions (with discussion). *Statistical Science*, 1: 181–242. page 28, 33, 35, 41, 42, 53, 108, 249, 259, 260, 261, 262, 307

Bookstein, F. L. (1989). Principal warps: thin-plate splines and the decomposition of deformations. *IEEE Transactions on Pattern Analysis and Machine Intelligence*, 11: 567–585. page xxii, 270, 279, 282, 284, 288, 292, 293

Bookstein, F. L. (1991). *Morphometric Tools for Landmark Data: Geometry and Biology*. Cambridge University Press, Cambridge. page xx, 4, 22, 152, 239, 270, 279, 291, 384, 396

Bookstein, F. L. (1994a). The morphometric synthesis: a brief intellectual history. In: *Commemorative 100th Volume of Lecture Notes in Biomathematics* (ed. S. A. Levin), pp. 212–237. Springer, New York. page 32

Bookstein, F. L. (1994b). Can biometrical shape be a homologous character? In: *Homology: the Hierarchical Basis of Comparative Biology* (ed. B. K. Hall), pp. 197–227. Academic Press, New York. page 272

Bookstein, F. L. (1995). Metrics and symmetries of the morphometric synthesis. In: *Proceedings in Current Issues in Statistical Shape Analysis* (eds K. V. Mardia and C. A. Gill), pp. 139–153. University of Leeds Press, Leeds. page 297

Bookstein, F. L. (1996a). Applying landmark methods to biological outline data. In: *Proceedings in Image Fusion and Shape Variability Techniques* (eds K. V. Mardia, C. A. Gill, and I. L. Dryden), pp. 59–70. University of Leeds Press, Leeds. page 5, 368, 369

Bookstein, F. L. (1996b). Biometrics, biomathematics and the morphometric synthesis. *Bulletin of Mathematical Biology*, 58: 313–365. page 15, 16, 81, 289, 297, 299, 303, 304

Bookstein, F. L. (1996c). Landmark methods for forms without landmarks: morphometrics of group differences in outline shape. *Medical Image Analysis*, 1: 225–243. page 368, 369

Bookstein, F. L. (1997). Shape and the information in medical images: a decade of the morphometric synthesis. *Computer Vision and Image Understanding*, 66: 97–118. page 193, 201

Bookstein, F. L. (2000). Creases as local features of deformation grids. *Medical Image Analysis*, 4: 93–110. page 203, 285

Bookstein, F. L. (2001). "Voxel-based morphometry" should not be used with imperfectly registered images. *Neuroimage*, 14: 1454–1462. page 384

Bookstein, F. L. (2002). Creases as morphometric characters. In: *Morphology, Shape and Phylogeny* (eds N. MacLeod and P. L. Forey), Vol. 64, pp. 139–174. Taylor and Francis, London. page 285

Bookstein, F. L. (2013a). Allometry for the twenty-first century. *Biological Theory*, 7(1): 10–25. page 109, 305

Bookstein, F. L. (2013b). It is not our data that are non-Euclidean, but only our models. In: *Proceedings of LASR 2013 – Statistical Models and Methods for non-Euclidean Data with Current Scientific Applications* (eds K. V. Mardia, A. Gusnanto, A. D. Riley and J. Voss), pp. 33–37. University of Leeds Press, Leeds. page 259

Bookstein, F. L. (2014). *Measuring and Reasoning: Numerical Inference in the Sciences*. Cambridge University Press, Cambridge. page xx, 197, 239, 285, 299, 326, 396

Bookstein, F. L. (2015a). Integration, disintegration, and self-similarity: Characterizing the scales of shape variation in landmark data. *Evolutionary Biology*, 42(4): 395–426. page 147, 150, 289, 295, 298, 311, 362

Bookstein, F. L. (2015b). The relation between geometric morphometrics and functional morphology, as explored by Procrustes interpretation of individual shape measures pertinent to junction. *The Anatomical Record* 298: 314–327. page 163

Bookstein, F. L. (2016). The inappropriate symmetries of multivariate statistical analysis in geometric morphometrics. *Evolutionary Biology*. DOI: 10.1007/s11692-016-9382-7 page 111, 152, 165

Bookstein, F. L. and Green, W. D. K. (1993). A feature space for edgels in images with landmarks. *Journal of Mathematical Imaging and Vision*, 3: 231–261. page 313

Bookstein, F. L. and Mardia, K. V. (2001). EM-type algorithms for missing morphometric data. In: *Proceedings of the International Conference on Recent Devlopments in Statistics and Probability and their Applications*, pp. 66–68. page 339

Bookstein, F. L. and Sampson, P. D. (1990). Statistical models for geometric components of shape change. *Communications in Statistics –Theory and Methods*, 19: 1939–1972. page 212, 307, 333, 334, 335

Bookstein, F. L., Schäfer, K., Prossinger, H., *et al.* (1999). Comparing frontal cranial profiles in archaic and modern homo by morphometric analysis. *The Anatomical Record*, 257(6): 217–224. page 196

Boomsma, W., Mardia, K. V., Taylor, C. C., Ferkinghoff-Borg, J., Krogh, A., and Hamelryck, T. (2008). A generative, probabilistic model of local protein structure. *Proceedings of the National Academy of Sciences of the United States of America*, 105(26): 8932–8937. page 208, 209

Borg, I. and Groenen, P. (1997). *Modern Multidimensional Scaling: Theory and Applications*. Springer, New York. page 126

Bowman, A. (2008). Statistics with a human face. *Significance*, 5(2): 74–77. page 324

Bowman, A. W. and Bock, M. T. (2006). Exploring variation in three-dimensional shape data. *Journal of Computational and Graphical Statistics*, 15(3): 524–541. page 324

Breuß, M., Bruckstein, A. M., and Maragos, P. (eds) (2013). *Innovations for Shape Analysis, Models and Algorithms*. Springer, Berlin. page 397

Brignell, C. J. (2007). *Shape analysis and statistical modelling in brain imaging*. PhD thesis, University of Nottingham. page 148, 236

Brignell, C. J., Browne, W. J., and Dryden, I. L. (2005). Covariance weighted Procrustes analysis. In: *LASR 2005 – Quantitative Biology, Shape Analysis, and Wavelets* (eds S. Barber, P. D. Baxter, K. V. Mardia and R. E. Walls), pp. 107–110. University of Leeds, Leeds. page 148

Brignell, C. J., Dryden, I. L., and Browne, W. J. (2015). Covariance weighted Procrustes analysis. In: *Riemannian Computing in Computer Vision* (eds P. K. Turaga and A. Srivastava), pp. 189–209. Springer, New York. page 148, 149

Brignell, C. J., Dryden, I. L., Gattone, S. A., *et al.* (2010). Surface shape analysis, with an application to brain surface asymmetry in schizophrenia. *Biostatistics*, 11: 609–630. page 4, 25, 171, 367

Broadbent, S. (1980). Simulating the ley hunter. *Journal of the Royal Statistical Society, Series A*, 143: 109–140. page 30, 244

Brody, D. C. (2004). Shapes of quantum states. *Journal of Physics A*, 37(1): 251–257. page 396

Brombin, C., Pesarin, F., and Salmaso, L. (2011). Dealing with more variables than the sample size: an application to shape analysis. In: *Nonparametric Statistics and Mixture Models*, pp. 28–44. World Scientific Publishing, Hackensack, NJ. page 323

Brombin, C. and Salmaso, L. (2009). Multi-aspect permutation tests in shape analysis with small sample size. *Computer Statistics Data Analysis*, 53(12): 3921–3931. page 323

Brombin, C. and Salmaso, L. (2013). *Permutation Tests in Shape Analysis*. Springer Briefs in Statistics. Springer, New York. page xx, 323, 397

Brombin, C., Salmaso, L., Fontanella, L., and Ippoliti, L. (2015). Non-parametric combination-based tests in dynamic shape analysis. *Journal of Nonparametric Statistics*, 27: 460–484. page 323

Bronstein, A. M., Bronstein, M. M., Kimmel, R., Mahmoudi, M., and Sapiro, G. (2010). A Gromov-Hausdorff framework with diffusion geometry for topologically-robust non-rigid shape matching. *International Journal of Computer Vision*, 89(2-3): 266–286. page 375

Bubenik, P., Carlsson, G., Kim, P. T., and Luo, Z.-M. (2010). Statistical topology via Morse theory persistence and nonparametric estimation. In: *Algebraic Methods in Statistics and Probability II*, Vol. 516 of *Contemporary Mathematics*, pp. 75–92. American Mathematical Society, Providence, RI. page 393

Burl, M. and Perona, P. (1996). Recognition of planar object classes. In: *IEEE Conference on Computer Vision and Pattern Recognition (CVPR)*, pp. 223–230. IEEE Computer Society Press, Los Alamitos, CA. page 258

Burt, D. M. and Perrett, D. I. (1995). Perception of age in facial aging in adult caucasian male faces: computer graphic manipulation of shape and colour information. *Proceedings of the Royal Society of London, Series B*, 259: 137–143. page 386

Buser, P. and Karcher, H. (1981). *Gromov's Almost Flat Manifolds*, Vol. 81 of *Astérisque*. Société Mathématique de France, Paris. page 318

Cao, Y., Zhang, Z., Czogiel, I., Dryden, I. L., and Wang, S. (2011). 2D non-rigid partial shape matching using MCMC and contour subdivision. In: *IEEE Conference on Computer Vision and Pattern Recognition (CVPR)*, pp. 2345–2352. IEEE Computer Society Press, Los Alamitos, CA. page 349

Cardini, A. (2014). Missing the third dimension in geometric morphometrics: how to assess if 2D images really are a good proxy for 3D structures? *Hystrix, the Italian Journal of Mammalogy*, 25(2): 73–81. page 173

Cardini, A., Polly, D., Dawson, R., and Milne, N. (2015). Why the long face? Kangaroos and wallabies follow the same 'rule' of cranial evolutionary allometry (CREA) as placentals. *Evolutionary Biology*, 42(2): 169–176. page 109

Carlsson, G. (2009). Topology and data. *Bulletin of the American Mathematical Society (NS)*, 46(2): 255–308. page 393

Carne, T. K. (1990). The geometry of shape spaces. *Proceedings of the London Mathematical Society*, 61: 407–432. page 393

Charpiat, G., Faugeras, O., and Keriven, R. (2005). Approximations of shape metrics and application to shape warping and empirical shape statistics. *Foundations of Computational Mathematics*, 5(1): 1–58. page 364

Cheng, W., Dryden, I. L., Hitchcock, D. B., and Le, H. (2014). Analysis of AneuRisk65 data: internal carotid artery shape analysis. *Electronic Journal of Statistics*, 8: 1905–1913. page 4, 374

Cheng, W., Dryden, I. L., and Huang, X. (2016). Bayesian registration of functions and curves. *Bayesian Analysis*, 11: 447–475. page 4, 325, 371, 372, 373, 374, 389

Cheverud, J. M., Lewis, J. L., Bachrach, W., and Lew, W. D. (1983). The measurement of form and variation in form: and application of three dimensional quantitive morphology by finite element methods. *American Journal of Physical Anthropology*, 62: 151–165. page 305

Cheverud, J. M. and Richtsmeier, J. T. (1986). Finite element scaling applied to sexual dimorphism in rhesus macaque (Macaca mulatta) facial growth. *Systematic Zoology*, 35: 381–399. page 303, 305, 307

Chikuse, Y. (2003). *Statistics on Special Manifolds*, Vol. 174 of *Lecture Notes in Statistics*. Springer, New York. page 396

Christaller, W. (1933). *Die Zentralen Orte in Suddeutschland*. Prentice Hall, Upper Saddle River, NJ. page 28

Christensen, G., Rabbitt, R. D., and Miller, M. I. (1996). Deformable templates using large deformation kinematics. *IEEE Transactions on Image Processing*, 5: 1435–1447. page 314

Chui, H. and Rangarajan, A. (2000). A feature registration framework using mixture models. In: *IEEE Workshop on Mathematical Methods in Biomedical Image Analysis (MMBIA)*, pp. 190–197. page 341, 348

Chui, H. and Rangarajan, A. (2003). A new point matching algorithm for non-rigid registration. *Computer Vision and Image Understanding*, 89(2–3): 114–141. page 341, 348

Chung, M. K., Dalton, K. M., Shen, L., Evans, A. C., and Davidson, R. J. (2007). Weighted Fourier series representation and its application to quantifying the amount of gray matter. *IEEE Transactions on Medical Imaging*, 26(4): 566–581. page 365

Claeskens, G., Silverman, B. W., and Slaets, L. (2010). A multiresolution approach to time warping achieved by a Bayesian prior-posterior transfer fitting strategy. *Journal of the Royal Statistical Society: Series B (Statistical Methodology)*, 72(5): 673–694. page 325

Claude, J. (2008). *Morphometrics with R*. Springer, New York. page xx, 396

Cliff, N. (1966). Orthogonal rotation to congruence. *Psychometrika*, 31: 33–42. page 125

Cohen, F. and Sternberg, J. (1980). On the prediction of protein structure: The significance of the root-mean-square deviation. *Journal of Molecular Biology*, 38: 321–333. page 100

Coombes, A. M., Moss, J. P., Linney, A. D., Richards, R., and James, D. R. (1991). A mathematical method for the comparison of three dimensional changes in the facial surface. *European Journal of Orthodontics*, 13: 95–110. page 375

Cooper, W., Goodall, C. R., Suryawanshi, S., and Tan, H. (1995). Euclidean shape tensor analysis. In: *Proceedings in Current Issues in Statistical Shape Analysis* (eds K. V. Mardia and C. A. Gill), pp. 179–180. University of Leeds Press, Leeds. page 361

Cootes, T. F., Edwards, G. J., and Taylor, C. J. (2001). Active appearance models. *IEEE Transactions on Pattern Analysis and Machine Intelligence*, 23(6): 681–685. page 381

Cootes, T. F. and Taylor, C. J. (1995). Active shape models: A review of recent work. In: *Proceedings in Current Issues in Statistical Shape Analysis*, pp. 108–114. University of Leeds Press, Leeds. page 381

Cootes, T. F., Taylor, C. J., Cooper, D. H., and Graham, J. (1992). Training models of shape from sets of examples. In: *Proceedings of the British Machine Vision Conference* (eds D. C. Hogg and R. D. Boyle), pp. 9–18. Springer-Verlag, Berlin. page 4, 95, 150, 163, 380

Cootes, T. F., Taylor, C. J., Cooper, D. H., and Graham, J. (1994). Image search using flexible shape models generated from sets of examples. In: *Statistics and Images*, (ed. K. V. Mardia), Vol. 2, pp. 111–139. Carfax, Oxford. page 4, 163, 164, 381

Cox, T. F. and Cox, M. A. A. (1994). *Multidimensional Scaling*. Chapman and Hall, London. page 126, 133

Craw, I. and Cameron, P. (1992). Face recognition by computer. In: *Proceedings of the British Machine Vision Conference* (eds D. Hogg and R. D. Boyle), pp. 498–507. Springer-Verlag, Berlin. page 384

Crawley, M. (2007). *The R Book*. John Wiley & Sons, Ltd, Chichester. page 7

Cressie, N. A. C. (1993). *Statistics for Spatial Data*, Revised Edition. John Wiley & Sons, Inc., New York. page 287, 308, 311

Czogiel, I. (2010). *Statistical inference for molecular shapes*. PhD thesis, The University of Nottingham. page 195

Czogiel, I., Dryden, I. L., and Brignell, C. J. (2011). Bayesian matching of unlabeled marked point sets using random fields, with an application to molecular alignment. *Annals of Applied Statistics*, 5: 2603–2629. page 13, 349, 375

da Fontoura Costa, L. and Marcondes Cesar Jr, R. (2009). *Shape Classification and Analysis: Theory and Practice*, 2nd edn. CRC Press, Boca Raton. page xx, 397

Darroch, J. N. and Mosimann, J. E. (1985). Canonical and principal components of shape. *Biometrika*, 72: 241–252. page 40

Das, S. and Vaswani, N. (2010). Nonstationary shape activities: Dynamic models for landmark shape change and applications. *IEEE Transactions on Pattern Analysis and Machine Intelligence*, 32(4): 579–592. page 325

Davies, R. H., Twining, C. J., Cootes, T. F., Waterton, J. C., and Taylor, C. J. (2002). A minimum description length approach to statistical shape modeling. *IEEE Transactions on Medical Imaging*, 21(5): 525–537. page 349

Davies, R. H., Twining, C. J., and Taylor, C. J. (2008a). Groupwise surface correspondence by optimization: Representation and regularization. *Medical Image Analysis*, 12(6): 787–796. page 349

Davies, R. H., Twining, C. J., and Taylor, C. J. (2008b). *Statistical Models of Shape: Optimisation and Evaluation*. Springer, New York. page xx, 349, 377, 381, 390, 397

de Souza, K. M. A., Jackson, A. L., Kent, J. T., Mardia, K. V., and Soames, R. W. (2001a). An assessment of the accuracy of stereolithographic skull models. *Clinical Anatomy*, 14: 296. page 36

de Souza, K. M. A., Jackson, A. L., Kent, J. T., Mardia, K. V., and Soames, R. W. (2001b). A technique for assessing the accuracy of stereolithographic models. *Clinical Anatomy*, 14: 296. page 36

Deane, C. M., Dunbar, J., Fuchs, A., Mardia, K. V., Shi, J., and Wilman, H. R. (2013). Describing protein structure geometry to aid in functional understanding. In: *Proceedings of LASR 2013 – Statistical Models and Methods for non-Euclidean Data with Current Scientific Applications* (eds K. V. Mardia, A. Gusnanto, A. D. Riley and J. Voss), pp. 49–51. University of Leeds Press, Leeds. page 374

Dette, H. and Wiens, D. P. (2009). Robust designs for 3D shape analysis with spherical harmonic descriptors. *Statistica Sinica*, 19(1): 83–102. page 365

Dryden, I. L. (1989). *The statistical analysis of shape data*. PhD thesis, University of Leeds. page 9, 121

Dryden, I. L. (1991). Discussion to 'Procrustes methods in the statistical analysis of shape' by C.R. Goodall. *Journal of the Royal Statistical Society, Series B*, 53: 327–328. page 136, 146, 228, 232

Dryden, I. L. (1999). General shape and registration analysis. In: *Stochastic Geometry: Likelihood and Computation* (eds O. Barndorff-Nielsen, W. S. Kendall and M. N. M. van Lieshout), pp. 333–364. Chapman and Hall, London. page 393

Dryden, I. L. (2003). Statistical shape analysis in high-level vision. In: *Mathematical Methods in Computer Vision* (eds P. J. Olver and A. Tannenbaum), pp. 37–55. Springer-Verlag, New York. page 163

Dryden, I. L. (2005a). Shape analysis. In: *Encyclopaedia of Biostatistics*. (eds P. Armitage and T. Colton), 2nd edn, Vol. 7, pp. 4919–4928. John Wiley & Sons, Ltd, Chichester. page 397

Dryden, I. L. (2005b). Statistical analysis on high-dimensional spheres and shape spaces. *Annals of Statistics*, 33(4): 1643–1665. page 223, 367

Dryden, I. L. (2014). Shape and object data analysis [discussion of the paper by Marron and Alonso (2014)]. *Biometrical Journal*, 56(5): 758–760. page 391

Dryden, I. L. (2015). *shapes: Statistical shape analysis*. R package version 1.1-11. http://cran.r-project.org/package=shapes (accessed 20 March 2016). page xix, 7

Dryden, I. L., Bai, L., Brignell, C. J., and Shen, L. (2009a). Factored principal components analysis, with applications to face recognition. *Statistics and Computing*, 19(3): 229–238. page 147

Dryden, I. L., Faghihi, M. R., and Taylor, C. C. (1995). Investigating regularity in spatial point patterns using shape analysis. In: *Proceedings in Current Issues in Statistical Shape Analysis* (eds K. V. Mardia and C. A. Gill), pp. 40–48. University of Leeds Press, Leeds. page 351

Dryden, I. L., Faghihi, M. R., and Taylor, C. C. (1997). Procrustes shape analysis of spatial point patterns. *Journal of the Royal Statistical Society, Series B*, 59: 353–374. page 351, 362

Dryden, I. L., Hirst, J. D., and Melville, J. L. (2007). Statistical analysis of unlabeled point sets: comparing molecules in chemoinformatics. *Biometrics*, 63(1): 237–251, 315. page 13, 234, 341, 345, 346, 347, 348, 349

Dryden, I. L. and Kent, J. T. (eds) (2015). *Geometry Driven Statistics*. John Wiley & Sons, Ltd, Chichester. page 397

Dryden, I. L., Koloydenko, A., and Zhou, D. (2009b). Non-Euclidean statistics for covariance matrices, with applications to diffusion tensor imaging. *Annals of Applied Statistics*, 3(3): 1102–1123. page 396

Dryden, I. L., Kume, A., Le, H., and Wood, A. T. A. (2008a). A multi-dimensional scaling approach to shape analysis. *Biometrika*, 95(4): 779–798. page 355, 356

Dryden, I. L., Kume, A., Le, H., and Wood, A. T. A. (2010). Statistical inference for functions of the covariance matrix in the stationary Gaussian time-orthogonal principal components model. *Annals of the Institute of Statistical Mathematics*, 62(5): 967–994. page 325

Dryden, I. L., Kume, A., Le, H., Wood, A. T. A., and Laughton, C. (2002). Size-and-shape analysis of DNA molecular dynamics simulations. In: *Proceedings of LASR 2002* (eds K. V. Mardia, R. G. Aykroyd and P. McDonnell), pp.23–26. University of Leeds Press, Leeds. page 325

Dryden, I. L., Le, H., Preston, S. P., and Wood, A. T. (2014). Mean shapes, projections and intrinsic limiting distributions. *Journal of Statistical Planning and Inference*, 145: 25–32. page 115, 117, 197, 322, 354, 355

Dryden, I. L. and Mardia, K. V. (1991a). Distributional and theoretical aspects of shape analysis. In: *Probability Measures on Groups X* (ed. H. Heyer), pp. 95–116. Plenum, New York. page 393

Dryden, I. L. and Mardia, K. V. (1991b). General shape distributions in a plane. *Advances in Applied Probability*, 23: 259–276. page 121, 147, 239, 252, 253

Dryden, I. L. and Mardia, K. V. (1992). Size and shape analysis of landmark data. *Biometrika*, 79: 57–68. page 35, 36, 259, 260, 263

Dryden, I. L. and Mardia, K. V. (1993). Multivariate shape analysis. *Sankhya Series A*, 55: 460–480. page 12, 93, 95, 193

Dryden, I. L. and Mardia, K. V. (1998). *Statistical Shape Analysis*. John Wiley & Sons, Ltd, Chichester. page xix, 5

Dryden, I. L., Oxborrow, N., and Dickson, R. (2008b). Familial relationships of normal spine shape. *Statistics in Medicine*, 27(11): 1993–2003. page 236

Dryden, I. L., Taylor, C. C., and Faghihi, M. R. (1999). Size analysis of nearly regular delaunay triangulations. *Methodology and Computing in Applied Probability*, 1: 97–117. page 362

Dryden, I. L. and Walker, G. (1998). Shape analysis using highly resistant regression. Technical Report STAT98/03, Department of Statistics, University of Leeds. page 338

Dryden, I. L. and Walker, G. (1999). Highly resistant regression and object matching. *Biometrics*, 55: 820–825. page 336, 337

Dryden, I. L. and Zempléni, A. (2006). Extreme shape analysis. *Journal of the Royal Statistical Society, Series C*, 55: 103–121. page 351

Du, J., Dryden, I. L., and Huang, X. (2015). Size and shape analysis of error-prone shape data. *Journal of the American Statistical Association*, 110(509): 368–379. page 63, 146

Dubuisson, M.-P. and Jain, A. K. (1994). A modified Hausdorff distance for object matching. In: *Proceedings of the International Conference on Pattern Recognition*, pp. 566–568. IEEE Computer Society Press, Los Alamitos, CA. page 364

Duchon, J. (1976). Interpolation des fonctions de deux variables suivant la principe de la flexion des plaques minces. *RAIRO Analyse Numérique*, 10: 5–12. page 280, 287

Durbin, R., Eddy, S., Krogh, A., and Mitchison, G. (1998). *Biological Sequence Analysis: Probabilistic Models of Proteins and Nucleic Acids*. Cambridge University Press, Cambridge. page 208

Dürer, A. (1528). *Vier Bücher von Menschlicher Proportion*. Hieronymus Formschneider, Nuremberg. page 303, 304

Durrleman, S., Pennec, X., Trouvé, A., and Ayache, N. (2007). Measuring brain variability via sulcal lines registration: a diffeomorphic approach. In: *Proceedings of Medical Image Computing and Computer Assisted Intervention (MICCAI)* (eds N. Ayache, S. Ourselin and A. Maeder), Vol. 4791 of *Lecture Notes in Computer Science*. Brisbane. page 390

Durrleman, S., Pennec, X., Trouvé, A., and Ayache, N. (2009). Statistical models on sets of curves and surfaces based on currents. *Medical Image Analysis*, 13(5): 793–808. page 315

Durrleman, S., Pennec, X., Trouvé, A., Thompson, P. M., and Ayache, N. (2008). Inferring brain variability from diffeomorphic deformations of currents: An integrative approach. *Medical Image Analysis*, 12(5): 626–637. page 315

Dutilleul, P. (1999). The MLE algorithm for the matrix normal distribution. *Journal of Statistical Computation and Simulation*, 64: 105–123. page 147

Elad, A. and Kimmel, R. (2003). On bending invariant signatures for surfaces. *IEEE Transactions on Pattern Analysis and Machine Intelligence*, 25(10): 1285–1295. page 375

Everitt, B. S. and Rabe-Hesketh, S. (1997). *The Analysis of Proximity Data*. Kendall's Library of Statistics 4. Arnold, London. page 126

Evison, M. P., Morecroft, L., Fieller, N. R. J., and Dryden, I. L. (2010). A large database sample of 3D facial images and measurements. In: *Computer-Aided Forensic Facial Comparison* (eds M. P. Evison and R. W. V. Bruegge), pp. 53–69. CRC Press, Boca Raton. page 163

Evison, M. P. and Vorder Bruegge, R. W. (eds) (2010). *Computer-aided Forensic Facial Comparison*. CRC Press, Boca Raton. page 384, 397

Exner, H. E. and Hougardy, H. P. (1988). *Quantitative Image Analysis of Microstructures*. DGM Informationsgesellschaft mbH, Oberursel. page 363

Faghihi, M. R. (1996). *Shape analysis of spatial point patterns*. PhD thesis, University of Leeds. page 351

Faghihi, M. R., Taylor, C. C., and Dryden, I. L. (1999). Procrustes shape analysis of triangulations of a two coloured point pattern. *Statistics and Computing*, 9: 43–53. page 362

Falconer, D. S. (1973). Replicated selection for body weight in mice. *Genetical Research Cambridge*, 22: 291–321. page 8

Faraway, J. (2004a). Human animation using nonparametric regression. *Journal of Computational and Graphical Statistics*, 13: 537–553. page 324, 325

Faraway, J. J. (2004b). Modeling continuous shape change for facial animation. *Statistics and Computing*, 14(4): 357–363. page 324

Faraway, J. J. and Trotman, C.-A. (2011). Shape change along geodesics with application to cleft lip surgery. *Journal of the Royal Statistical Society: Series C (Applied Statistics)*, 60(5): 743–755. page 327

Fawcett, C. D. and Lee, A. (1902). A second study of the variation and correlation of the human skull, with special reference to the Naqada crania. *Biometrika*, 1: 408–467. page 31

Feragen, A., Lauze, F., and Nielsen, M. (2010). Fundamental geodesic deformations in spaces of tree-like shapes. In: *International Conference for Pattern Recognition*, pp. 2089–2093. IEEE Computer Society Press, Los Alamitos, CA. page 392

Feragen, A., Lo, P., de Bruijne, M., Nielsen, M., and Lauze, F. (2013). Toward a theory of statistical tree-shape analysis. *IEEE Transactions on Pattern Analysis and Machine Intelligence*, 35(8): 2008–2021. page 392

Fillard, P., Arsigny, V., Pennec, X., and Ayache, N. (2007). Clinical DT-MRI estimation, smoothing and fiber tracking with log-Euclidean metrics. *IEEE Transactions on Medical Imaging*, 26(11): 1472–1482. page 396

Fisher, N. I., Lewis, T., and Embleton, B. J. (1987). *Statistical Analysis of Spherical Data*. Cambridge University Press, Cambridge. page 325

Fletcher, P. T. (2013). Geodesic regression and the theory of least squares on Riemannian manifolds. *International Journal of Computer Vision*, 105(2): 171–185. page 327

Fletcher, P. T., Lu, C., Pizer, S. M., and Joshi, S. C. (2004). Principal geodesic analysis for the study of nonlinear statistics of shape. *IEEE Transactions on Medical Imaging*, 23(8): 995–1005. page 326, 327

Fletcher, P. T., Venkatasubramanian, S., and Joshi, S. (2009). The geometric median on Riemannian manifolds with application to robust atlas estimation. *Neuroimage*, 45(1): S143–S152. page 323

Fotouhi, H. and Golalizadeh, M. (2012). Exploring the variability of DNA molecules via principal geodesic analysis on the shape space. *Journal of Applied Statistics*, 39(10): 2199–2207. page 327

Fréchet, M. (1948). Les éléments aléatoires de nature quelconque dans un espace distancié. *Annales de l'Institut Henri Poincaré*, 10: 215–310. page 103, 111

Free, S. L., O'Higgins, P., Maudgil, D. D., *et al.* (2001). Landmark-based morphometrics of the normal adult brain using MRI. *Neuroimage*, 13: 801–813. page 163

Friedman, J., Hastie, T., and Tibshirani, R. (2008). Sparse inverse covariance estimation with the graphical Lasso. *Biostatistics*, pp. 432–441. page 197

Fright, W. R. and Linney, A. D. (1993). Registration of 3-D head surfaces using multiple landmarks. *IEEE Transactions on Medical Imaging*, 12: 515–520. page 139

Fritsch, D. S., Pizer, S. M., Chaney, E. L., Lui, A., Raghavan, S., and Shah, T. (1994). Cores for image registration. In: *Proceedings of SPIE Medical Imaging '94* (ed. M. H. Loew), Vol. 2167, pp. 128–142. SPIE Press, Bellingham, WA. page 390

Galileo (1638). *Discorsi e dimostrazioni matematiche, informo a due nuoue scienze attenti alla mecanica i movimenti locali*. appresso gli Elsevirii; Opere VIII. page 2, 3

Galton, F. (1878). Composite portraits. *Journal of the Anthropological Institute of Great Britain and Ireland*, 8: 132–142. page 384

Galton, F. (1883). *Enquiries into Human Faculty and Development*. Dent, London. page 384, 387

Galton, F. (1907). Classification of portraits. *Nature*, 76: 617–618. page 36, 42

Gamble, J. and Heo, G. (2010). Exploring uses of persistent homology for statistical analysis of landmark-based shape data. *Journal of Multivariate Analysis*, 101(9): 2184–2199. page 393

Gardner, R. J., Hobolth, A., Jensen, E. B. V., and Sørensen, F. B. (2005). Shape discrimination by total curvature, with a view to cancer diagnostics. *Journal of Microscopy*, 217(1): 49–59. page 368

Gates, J. (1994). Shape distributions for planar triangles by dual construction. *Advances in Applied Probability*, 26: 324–333. page 351

Gelman, A., Carlin, J. B., Stern, H. S., Dunson, D. B., Vehtari, A., and Rubin, D. B. (2014). *Bayesian Data Analysis*, 3rd edn. CRC Press, Boca Raton. page 379

Geman, S. and Geman, D. (1984). Stochastic relaxation, Gibbs distributions and the Bayesian restoration of images. *IEEE Transactions of Pattern Analysis and Machine Intelligence*, 6: 721–741. page 379

Gilks, W. R., Richardson, S., and Spiegelhalter, D. J. (eds) (1996). *Markov Chain Monte Carlo in Practice*. Chapman and Hall, London. page 234, 379

Glasbey, C. A. and Horgan, G. W. (1995). *Image Analysis for the Biological Sciences*. John Wiley & Sons, Ltd, Chichester. page 363

Glasbey, C. A., Horgan, G. W., Gibson, G. J., and Hitchcock, D. (1995). Fish shape analysis using landmarks. *Biometrical Journal*, 37: 481–495. page 147

Glasbey, C. A. and Mardia, K. V. (2001). A penalized likelihood approach to image warping. *Journal of the Royal Statistical Society: Series B (Statistical Methodology)*, 63(3): 465–514. page 375, 386

Glaunès, J., Vaillant, M., and Miller, M. I. (2004). Landmark matching via large deformation diffeomorphisms on the sphere. *Journal of Mathematical Imaging and Vision*, 20(1–2): 179–200. page 315

Good, P. (1994). *Permutation Tests*. Springer-Verlag, New York. page 193

Goodall, C. R. (1991). Procrustes methods in the statistical analysis of shape (with discussion). *Journal of the Royal Statistical Society, Series B*, 53: 285–339. page 33, 35, 95, 125, 133, 135, 138, 146, 147, 197, 198, 200, 205, 277

Goodall, C. R. (1995). Procrustes methods in the statistical analysis of shape revisited. In: *Current Issues in Statistical Shape Analysis* (eds K. V. Mardia and C. A. Gill), pp. 18–33. University of Leeds Press, Leeds. page 66, 147, 361

Goodall, C. R. and Bose, A. (1987). Models and Procrustes methods for the analysis of shape differences. In: *Proceedings of the 19th INTERFACE Symposium* (eds R. M. Heiberger), pp. 86–92. Interface Foundation, Fairfax Station. page 138

Goodall, C. R. and Lange, N. (1989). Growth curve models for correlated triangular shapes. In: *Proceedings of the 21st Symposium on the Interface between Computing Science and Statistics* (eds K. Berk and L. Malone), pp. 445–454. Interface Foundation, Fairfax Station. page 22, 325

Goodall, C. R. and Mardia, K. V. (1991). A geometrical derivation of the shape density. *Advances in Applied Probability*, 23: 496–514. page 243, 260, 264, 265, 266

Goodall, C. R. and Mardia, K. V. (1992). The noncentral Bartlett decompositions and shape densities. *Journal of Multivariate Analysis*, 40: 94–108. page 68, 106, 121, 264, 267

Goodall, C. R. and Mardia, K. V. (1993). Multivariate aspects of shape theory. *Annals of Statistics*, 21: 848–866. page 106, 121, 264, 267, 273

Goodman, N. R. (1963). Statistical analysis based on a certain multivariate complex Gaussian distribution (an introduction). *Annals of Mathematical Statistics*, 34: 152–177. page 220

Gordon, A. D. (1995). Local transformations of facial features. *Journal of Applied Statistics*, 22: 179–184. page 276

Gower, J. C. (1966). Some distance properties of latent root and vector methods used in multivariate analysis. *Biometrika*, 53: 325–338. page 79

Gower, J. C. (1971). Statistical methods of comparing different multivariate analyses of the same data. In: *Mathematics in the Archaeological and Historical Sciences* (eds F. R. Hodson, D. G. Kendall and P. Tautu), pp. 138–149. Edinburgh University Press, Edinburgh. page 125

Gower, J. C. (1975). Generalized Procrustes analysis. *Psychometrika*, 40: 33–50. page 125, 133, 136, 138

Green, B. F. (1952). The orthogonal approximation of an oblique structure in factor analysis. *Psychometrika*, 17: 429–440. page 125

Green, P. J., Latuszyński, K., Pereyra, M., and Robert, C. P. (2015). Bayesian computation: a summary of the current state, and samples backwards and forwards. *Statistics and Computing*, 25(4): 835–862. page 379

Green, P. J. and Mardia, K. V. (2004). Bayesian alignment using hierarchical models, with applications in protein bioinformatics. Technical Report, University of Bristol. arXiv:math/0503712v1. page 21

Green, P. J. and Mardia, K. V. (2006). Bayesian alignment using hierarchical models, with applications in protein bioinformatics. *Biometrika*, 93: 235–254. page 21, 234, 341, 342, 343, 344, 345, 349

Green, P. J., Mardia, K. V., Nyirongo, V. B., and Ruffieux, Y. (2010). Bayesian modelling for matching and alignment of biomolecules. In: *The Oxford Handbook of Applied Bayesian Analysis* (eds A. O'Hagan and M. West), pp. 27–50. Oxford University Press, Oxford. page 344

Green, P. J. and Sibson, R. (1978). Computing Dirichlet tessalations in the plane. *Computer Journal*, 21: 168–173. page 29

Green, P. J. and Silverman, B. W. (1994). *Nonparametric Regression and Generalized Linear Models: A Roughness Penalty Approach*. Chapman and Hall, London. page 280, 287, 288, 313

Green, W. D. K. (1995). A simple construction of triangle shape space. In: *Current Issues in*

Statistical Shape Analysis (eds K. V. Mardia and C. A. Gill), pp. 160–166. University of Leeds Press, Leeds. page 86

Green, W. D. K. (1996). The thin-plate spline and images with curving features. In: *Image Fusion and Shape Variability* (eds K. V. Mardia, C. A. Gill and I. L. Dryden), pp. 79–87. University of Leeds Press, Leeds. page 5, 369

Grenander, U. (1994). *General Pattern Theory*. Clarendon Press, Oxford. page 365, 379, 390

Grenander, U., Chow, Y., and Keenan, D. M. (1991). *Hands: A Pattern Theoretic Study of Biological Shapes*. Research Notes in Neural Computing, Vol. 2. Springer-Verlag, New York. page 4

Grenander, U. and Keenan, D. M. (1993). Towards automated image understanding. In: *Statistics and Images* (eds K. V. Mardia and G. K. Kanji), Vol. 1, pp. 89–103. Carfax, Oxford. page 379

Grenander, U. and Manbeck, K. M. (1993). A stochastic shape and colour model for defect detection in potatoes. *Journal of Computational and Statistical Graphics*, 2: 131–151. page 365, 366, 368

Grenander, U. and Miller, M. I. (1994). Representations of knowledge in complex systems (with discussion). *Journal of the Royal Statistical Society, Series B*, 56: 549–603. page 4, 367, 368, 390

Grenander, U. and Miller, M. I. (2007). *Pattern Theory: from Representation to Inference*. Oxford University Press, Oxford. page xx, 314, 365, 377, 379, 390, 397

Groisser, D. (2005). On the convergence of some Procrustean averaging algorithms. *Stochastics*, 77(1): 31–60. page 138

Grove, K. and Karcher, H. (1973). How to conjugate C^1-close group actions. *Mathematische Zeitschrift*, 132: 11–20. page 111, 318

Grove, K., Karcher, H., and Ruh, E. A. (1974). Jacobi fields and Finsler metrics on compact Lie groups with an application to differentiable pinching problems. *Mathematische Annalen*, 211: 7–21. page 318

Grove, K., Karcher, H., and Ruh, E. A. (1975). Group actions and curvature. *Bulletin of the American Mathematical Society*, 81: 89–92. page 318

Gruvaeus, G. T. (1970). A general approach to Procrustes pattern rotation. *Psychometrika*, 35: 493–505. page 125

Gunz, P., Mitteroecker, P., and Bookstein, F. L. (2005). Semi-landmarks in three dimensions. In: *Modern Morphometrics in Physical Anthropology* (ed. D. E. Slice), pp. 73–98. Kluwer Academic/Plenum Publishers, New York. page 5

Gunz, P., Mitteroecker, P., Neubauer, S., Weber, G. W., and Bookstein, F. L. (2009). Principles for the virtual reconstruction of hominin crania. *Journal of Human Evolution*, 57: 48–62. page 339

Guta, M., Kypraios, T., and Dryden, I. (2012). Rank-based model selection for multiple ions quantum tomography. *New Journal of Physics*, 14(10): 105002. page 396

Hainsworth, T. J. and Mardia, K. V. (1993). A Markov random field restoration of image sequences. In: *Markov Random Fields* (eds R. Chellappa and A. Jain), pp. 409–445. Academic Press, Boston. page 379

Hall, P., Marron, J. S., and Neeman, A. (2005). Geometric representation of high dimension, low sample size data. *Journal of the Royal Statistical Society: Series B (Statistical Methodology)*, 67(3): 427–444. page 8

Hamelryck, T., Mardia, K. V., and Ferkinghoff-Borg, J. (eds) (2012). *Bayesian Methods in Structural Bioinformatics*. Springer, New York. page 349, 397

Hammer, Ø., Harper, D. A. T., and Ryan, P. D. (2001). PAST: Paleontological statistics software package for education and data analysis. *Palaeontologia Electronica*, 4: 9. page 173

Harris, J. (1992). *Algebraic Geometry: A First Course*. Springer-Verlag, Berlin. page 273

Harris, S. A., Gavathiotis, E., Searle, M. S., Orozco, M., and Laughton, C. A. (2001). Co-operativity in drug-DNA recognition: a molecular dynamics study. *Journal of the American Chemical Society*, 123: 12658–12663. page 139

Hastie, T. and Kishon, E. (1991). Discussion to Goodall (1991). *Journal of the Royal Statistical Society, Series B*, 53: 330–331. page 277, 314

Hastie, T. and Simard, P. Y. (1998). Metrics and models for handwritten character recognition. *Statistical Science*, 13(1): 54–65. page 11

Hastie, T. and Tibshirani, R. (1994). Handwritten digit recognition via deformable prototypes. Technical Report, AT&T Bell Laboratories. page 11

Hastings, W. K. (1970). Monte Carlo sampling methods using Markov chains and their applications. *Biometrika*, 57: 97–109. page 382

Hendriks, H. and Landsman, Z. (1996a). Asymptotic behavior of sample mean location for manifolds. *Statistics & Probability Letters*, 26(2): 169–178. page 318

Hendriks, H. and Landsman, Z. (1996b). Asymptotic tests for mean location on manifolds. *Comptes Rendus de l'Academiedes Sciences – Series I – Mathematics*, 322(8): 773–778. page 318

Hendriks, H. and Landsman, Z. (1998). Mean location and sample mean location on manifolds: asymptotics, tests, confidence regions. *Journal of Multivariate Analysis*, 67(2): 227–243. page 318, 322

Hendriks, H. and Landsman, Z. (2007). Asymptotic data analysis on manifolds. *Annals of Statistics*, 35(1): 109–131. page 318

Hendriks, H., Landsman, Z., and Ruymgaart, F. (1996). Asymptotic behavior of sample mean direction for spheres. *Journal of Multivariate Analysis*, 59(2): 141–152. page 318

Henry, G. and Rodriguez, D. (2009). Robust nonparametric regression on Riemannian manifolds. *Journal of Nonparametric Statistics*, 21(5): 611–628. page 323

Heo, G., Gamble, J., and Kim, P. T. (2012). Topological analysis of variance and the maxillary complex. *Journal of the American Statistical Association*, 107. page 393

Heo, G. and Small, C. G. (2006). Form representations and means for landmarks: a survey and comparative study. *Computer Vision and Image Understanding*, 102(2): 188–203. page 358, 397

Hinkle, J., Fletcher, P. T., and Joshi, S. (2014). Intrinsic polynomials for regression on Riemannian manifolds. *Journal of Mathematical Imaging and Vision*, 50(1–2): 32–52. page 328

Hobolth, A., Kent, J. T., and Dryden, I. L. (2002). On the relation between edge and vertex modelling in shape analysis. *Scandinavian Journal of Statistics*, 29: 355–374. page 368

Hobolth, A., Pedersen, J., and Jensen, E. B. V. (2003). A continuous parametric shape model. *Annals of the Institute of Statistical Mathematics*, 55(2): 227–242. page 368

Hobolth, A. and Vedel Jensen, E. B. (2000). Modelling stochastic changes in curve shape, with an application to cancer diagnostics. *Advances in Applied Probability*, 32(2): 344–362. page 368

Hodgman, T. C., Ugartechea-Chirino, Y., Tansley, G., and Dryden, I. L. (2006). The implications for bioinformatics of integration across the scales. *Journal of Integrative Bioinformatics*, 3: Article 39. page 163

Hoerl, A. E. and Kennard, R. W. (1970). Ridge regression: Biased estimation for nonorthogonal problems. *Technometrics*, 12: 55–67. page 214

Holmes, S. (2003). Bootstrapping phylogenetic trees: theory and methods. *Statistical Science*, 18(2): 241–255. page 392

Hopkins, J. W. (1966). Some considerations in multivariate allometry. *Biometrics*, 22: 747–760. page 31

Horgan, G. W., Creasey, A., and Fenton, B. (1992). Superimposing two-dimensional gels to study genetic variation in malaria parasites. *Electrophoresis*, 13: 871–875. page 24, 25, 274

Horn, A. (1954). Doubly stochastic matrices and the diagonal of a rotation matrix. *American Journal of Mathematics*, 76: 620–630. page 70

Horn, B. K. P. (1987). Closed-form solution of absolute orientation using unit quaternions. *Journal of the Optical Society of America*, 4: 629–642. page 63

Hotz, T. and Huckemann, S. (2015). Intrinsic means on the circle: uniqueness, locus and asymptotics. *Annals of the Institute of Statistical Mathematics*, 67(1): 177–193. page 115

Hotz, T., Huckemann, S., Le, H., *et al.* (2013). Sticky central limit theorems on open books. *Annals of Applied Probability*, 23(6): 2238–2258. page 392

Hotz, T., Huckemann, S., Munk, A., Gaffrey, D., and Sloboda, B. (2010). Shape spaces for pre-aligned star-shaped objects—studying the growth of plants by principal components analysis. *Journal of the Royal Statistical Society: Series C (Applied Statistics)*, 59(1): 127–143. page 326, 327

Huang, C., Styner, M., and Zhu, H. (2015). Clustering high-dimensional landmark-based two-dimensional shape data. *Journal of the American Statistical Association*, 110: 946–961. page 258

Huber, P. J. (1964). Robust estimation of a location parameter. *Annals of Mathematical Statistics*, 35: 73–101. page 337

Huckemann, S. (2011a). Inference on 3D Procrustes means: tree bole growth, rank deficient diffusion tensors and perturbation models. *Scandinavian Journal of Statistics*, 38(3): 424–446. page 322

Huckemann, S. (2012). On the meaning of mean shape: Manifold stability, locus and the two sample test. *Annals of the Institute of Statistical Mathematics*, 64: 1227–1259. page 112, 113, 115, 117, 120

Huckemann, S. and Hotz, T. (2009). Principal component geodesics for planar shape spaces. *Journal of Multivariate Analysis*, 100(4): 699–714. page 327

Huckemann, S. and Hotz, T. (2014). On means and their asymptotics: circles and shape spaces. *Journal of Mathematical Imaging and Vision*, 50(1-2): 98–106. page 120

Huckemann, S., Hotz, T., and Munk, A. (2010). Intrinsic shape analysis: geodesic PCA for Riemannian manifolds modulo isometric Lie group actions. *Statistica Sinica*, 20(1): 1–58. page 122, 327

Huckemann, S. and Ziezold, H. (2006). Principal component analysis for Riemannian manifolds, with an application to triangular shape spaces. *Advances in Applied Probability*, 38(2): 299–319. page 122, 327

Huckemann, S. F. (2011b). Intrinsic inference on the mean geodesic of planar shapes and tree discrimination by leaf growth. *Annals of Statistics*, 39(2): 1098–1124. page 326, 327

Hull, J. J. (1990). Character recognition: the reading of text by computer. In: *Encyclopaedia of Artificial Intelligence*, (ed. S. C. Shapiro), Vols 1 and 2, pp. 82–88. John Wiley & Sons, Inc., New York. page 11

Hurley, J. R. and Cattell, R. B. (1962). The Procrustes program: producing direct rotation to test a hypothesised factor structure. *Behavioural Science*, 7: 258–262. page 72

Hurn, M. (1998). Confocal flourescence microscopy of leaf cells: an application of Bayesian image analysis. *Applied Statistics*, 47: 361–377. page 380

Hurn, M., Steinsland, I., and Rue, H. (2001). Parameter estimation for a deformable template model. *Statistics and Computing*, 11(4): 337–346. page 368

Hurn, M. A., Mardia, K. V., Hainsworth, T. J., Kirkbride, J., and Berry, E. (1996). Bayesian fused classification of medical images. *IEEE Transactions on Medical Imaging*, 15: 850–858. page 384

Huxley, J. S. (1924). Constant differential growth ratios and their significance. *Nature*, 114: 895–896. page 31, 107

Huxley, J. S. (1932). *Problems of Relative Growth*. Methuen, London. page 31, 107

Hyvärinen, A., Karhunen, J., and Oja, E. (2001). *Independent Components Analysis*. John Wiley & Sons, Inc., New York. page 169

Iaci, R., Yin, X., Sriram, T. N., and Klingenberg, C. P. (2008). An informational measure of association and dimension reduction for multiple sets and groups with applications in morphometric analysis. *Journal of the American Statistical Association*, 103(483): 1166–1176. page 168

Ihaka, R. and Gentleman, R. (1996). R: A language for data analysis and graphics. *Journal of Computational and Graphical Statistics*, 5(3): 299–314. page 7

Jackson, C. M. (1915). *Morris's Human Anatomy*. Churchill, London. page 303

James, A. T. (1964). Distributions of matrix variates and latent roots derived from normal samples. *Annals of Mathematical Statistics*, 35: 475–501. page 267

James, G. S. (1954). Tests of linear hypotheses in univariate and multivariate analysis when the ratios of the population variances are unknown. *Biometrika*, 41: 19–43. page 203

Jardine, N. (1969). The observational and theoretical components of homology: a study based on the morphology of the derma-roofs of rhipidistan fishes. *Biological Journal of the Linnean Society*, 1: 327–361. page 3

Jayasumana, S., Hartley, R., Salzmann, M., Li, H., and Harandi, M. (2013a). Kernel methods on the Riemannian manifold of symmetric positive definite matrices. In: *IEEE Conference on Computer Vision and Pattern Recognition (CVPR)*, pp. 73–80. IEEE Computer Society Press, Los Alamitos, CA. page 396

Jayasumana, S., Hartley, R., Salzmann, M., Li, H., and Harandi, M. (2015). Kernel methods on Riemannian manifolds with Gaussian RBF kernels. *IEEE Transactions on Pattern Analysis and Machine Intelligence*, 37(12): 2464–2477. page 375

Jayasumana, S., Salzmann, M., Li, H., and Harandi, M. (2013b). A framework for shape analysis via Hilbert space embedding. In: *IEEE International Conference on Computer Vision (ICCV)*, pp. 1249–1256. IEEE Computer Society Press, Los Alamitos, CA. page 375

Jermyn, I. H., Kurtek, S., Klassen, E., and Srivastava, A. (2012). Elastic shape matching of parameterized surfaces using square root normal fields. In: *Proceedings of Computer*

Vision – ECCV 2012 – 12th European Conference on Computer Vision, (eds A. W. Fitzgibbon, S. Lazebnik, P. Perona, Y. Sato and C. Schmid) Part V, Vol. 7576 of *Lecture Notes in Computer Science*, pp. 804–817. Springer, Heidelberg. page 374

Johnson, D. R., O'Higgins, P., and McAndrew, T. J. (1988). The effect of replicated selection for body weight in mice on vertaebral shape. *Genetical Research Cambridge*, 51: 129–135. page 8

Johnson, D. R., O'Higgins, P., McAndrew, T. J., Adams, L. M., and Flinn, R. M. (1985). Measurement of biological shape: a general method applied to mouse vertebrae. *Journal of Embryology and Experimental Morphology*, 90: 363–377. page 8, 9, 365

Johnson, V. E., Bowsher, J. E., Jaszczak, R. J., and Turking, T. G. (1995). Analysis and reconstruction of medical images using prior information. In: *Case Studies in Bayesian Statistics* (eds C. Gastonis, J. S. Hodges, R. E. Kass and N. D. Singpurwalla), Vol. II, pp. 149–218. Springer-Verlag, New York. page 384

Jolicoeur, J. and Mosimann, J. E. (1960). Size and shape variation in the painted turtle. *Growth*, 24: 339–354. page 40

Jones, M. C. (1987). On moments of ratios of quadratic forms in normal variables. *Statistics and Probability Letters*, 6: 129–136, 369. page 251

Joshi, S., Klassen, E., Srivastava, A., and Jermyn, I. (2007). A novel representation for Riemannian analysis of elastic curves in $\mathbb{R}^n$. In: *IEEE Conference on Computer Vision and Pattern Recognition (CVPR)*, pp. 1–7. IEEE Computer Society Press, Los Alamitos, CA. page 371

Joshi, S. C. and Miller, M. I. (2000). Landmark matching via large deformation diffeomorphisms. *IEEE Transactions on Image Processing*, 9(8): 1357–1370. page 315

Joshi, S. C., Wang, J., Miller, M. I., Van Essen, D. C., and Grenander, U. (1995). On the differential geometry of the cortical surface. *Proceedings of SPIE*, 2573: 304–311. page 390

Jung, S., Dryden, I. L., and Marron, J. S. (2012). Analysis of principal nested spheres. *Biometrika*, 99(3): 551–568. page 122, 328

Jung, S., Foskey, M., and Marron, J. S. (2011). Principal arc analysis on direct product manifolds. *Annals of Applied Statistics*, 5(1): 578–603. page 328

Jupp, P. and Mardia, K. V. (1989). A unified view of the theory of directional statistics, 1975–1988. *International Statistical Review*, 57: 261–294. page 264

Jupp, P. E. and Kent, J. T. (1987). Fitting smooth paths to spherical data. *Journal of the Royal Statistical Society, Series C*, 36(1): 34–46. page 328

Karcher, H. (1977). Riemannian center of mass and mollifier smoothing. *Communications on Pure and Applied Mathematics*, 30(5): 509–541. page 111, 112, 318, 319

Kass, M., Witkin, A., and Terzopoulos, D. (1988). Snakes: active contour models. *International Journal of Computer Vision*, 1: 321–331. page 382

Kelley, L. A., Mezulis, S., Yates, C. M., Wass, M. N., and Sternberg, M. J. E. (2015). The Phyre2 web portal for protein modeling, prediction and analysis. *Nature Protocols*, 10: 845–858. page 211

Kendall, D. G. (1977). The diffusion of shape. *Advances in Applied Probability*, 9: 428–430. page 1, 33, 351

Kendall, D. G. (1983). The shape of Poisson-Delaunay triangles. In: *Studies in Probability and Related Topics* (eds M. C. Demetreseu and M. Iosifescu), pp. 321–330. Nagard, Montreal. page 29, 56, 86, 325, 351

Kendall, D. G. (1984). Shape manifolds, Procrustean metrics and complex projective spaces. *Bulletin of the London Mathematical Society*, 16: 81–121. page xxi, 7, 33, 34, 49, 61, 63, 64, 68, 70, 83, 84, 90, 125, 218, 239, 244, 321, 350

Kendall, D. G. (1985). Exact distributions for shapes of random triangles in convex sets. *Advances in Applied Probability*, 17: 308–329. page 350

Kendall, D. G. (1989). A survey of the statistical theory of shape (with discussion). *Statistical Science*, 4: 87–120. page 29, 32, 63, 74, 397

Kendall, D. G. (1991a). Discussion to 'Procrustes methods in the statistical analysis of shape' by C. R. Goodall. *Journal of the Royal Statistical Society, Series B*, 53: 321–324. page 350

Kendall, D. G. (1991b). The Mardia–Dryden distribution for triangles – a stochastic calculus approach. *Journal of Applied Probability*, 28: 225–230. page 239, 247

Kendall, D. G. (1995). Looking at geodesics in the shape space for 4 points in 3 dimensions. In: *Current Issues in Statistical Shape Analysis* (eds K. V. Mardia and C. A. Gill), pp. 6–8. University of Leeds Press, Leeds. page 32

Kendall, D. G., Barden, D., Carne, T. K., and Le, H. (1999). *Shape and Shape Theory*. John Wiley & Sons, Ltd, Chichester. page xx, 74, 75, 120, 397

Kendall, D. G. and Kendall, W. S. (1980). Alignments in two dimensional random sets of points. *Advances in Applied Probability*, 12: 380–424. page 30, 350

Kendall, D. G. and Le, H.-L. (1987). The structure and explicit determination of convex-polygonally generated shape densities. *Advances in Applied Probability*, 19: 896–916. page 350

Kendall, M. and Colijn, C. (2015). A tree metric using structure and length to capture distinct phylogenetic signals. Technical Report, Imperial College London. http://arxiv.org/abs/1507.05211. page 392

Kendall, W. S. (1988). Symbolic computation and the diffusion of triads. *Advances in Applied Probability*, 20: 775–797. page 266

Kendall, W. S. (1990a). The diffusion of Euclidean shape. In: *Disorder in Physical Systems* (eds G. R. Grimmett and D. J. A. Welch), pp. 203–217. Oxford University Press, Oxford. page 319, 351

Kendall, W. S. (1990b). Probability, convexity, and harmonic maps with small image. I. Uniqueness and fine existence. *Proceedings of the London Mathematical Society*, 61(2): 371–406. page 112, 318, 319

Kendall, W. S. (1998). A diffusion model for Bookstein triangle shape. *Advances in Applied Probability*, 30(2): 317–334. page 279

Kendall, W. S. (2015). Barycentres and hurricane trajectories. In: *Geometry Driven Statistics* (eds I. L. Dryden and J. T. Kent), pp. 146–160. John Wiley & Sons, Ltd, Chichester. page 122

Kendall, W. S. and Le, H. (2010). Statistical shape theory. In: *New Perspectives in Stochastic Geometry* (eds W. S. Kendall and I. Molchanov), pp. 348–373. Oxford University Press, Oxford. page 397

Kendall, W. S. and Le, H. (2011). Limit theorems for empirical Fréchet means of independent and non-identically distributed manifold-valued random variables. *Brazilian Journal of Probability and Statistics*, 25(3): 323–352. page 322

Kenobi, K. and Dryden, I. L. (2012). Bayesian matching of unlabeled point sets using Procrustes and configuration models. *Bayesian Analysis*, 7(3): 547–565. page 234, 343, 347

Kenobi, K., Dryden, I. L., and Le, H. (2010). Shape curves and geodesic modelling. *Biometrika*, 97(3): 567–584. page 22, 122, 327

Kent, J. T. (1991). Discussion to 'Procrustes methods in the statistical analysis of shape' by C. R. Goodall. *Journal of the Royal Statistical Society, Series B*, 53: 324–325. page 179

Kent, J. T. (1992). New directions in shape analysis. In: *The Art of Statistical Science* (ed. K. V. Mardia), pp. 115–127. John Wiley & Sons, Ltd, Chichester. page 95, 113, 115, 143, 146, 152, 179, 232, 323

Kent, J. T. (1994). The complex Bingham distribution and shape analysis. *Journal of the Royal Statistical Society, Series B*, 56: 285–299. page 32, 33, 47, 85, 150, 178, 179, 182, 212, 220, 221, 224, 228, 231, 250, 251, 277, 335, 354, 362

Kent, J. T. (1995). Current issues for statistical inference in shape analysis. In: *Proceedings in Current Issues in Statistical Shape Analysis* (eds K. V. Mardia and C. A. Gill), pp. 167–175. University of Leeds Press, Leeds. page 90

Kent, J. T. (1997). Data analysis for shapes and images. *Journal of Statistical Inference and Planning*, 57: 181–193. page 192, 231

Kent, J. T. (2003). Discussion on the paper by Barndorff-Nielsen, Gill and Jupp. *Journal of the Royal Statistical Society: Series B (Statistical Methodology)*, 65(4): 809. page 396

Kent, J. T., Constable, P. D. L., and Er, F. (2004). Simulation for the complex Bingham distribution. *Statistics and Computing*, 14(1): 53–57. page 226

Kent, J. T., Dryden, I. L., and Anderson, C. R. (2000). Using circulant symmetry to model featureless objects. *Biometrika*, 87(3): 527–544. page 22, 367

Kent, J. T., Ganeiber, A. M., and Mardia, K. V. (2013). A new method to simulate the Bingham and related distributions in directional data analysis with applications. arXiv:1310.8110. page 226

Kent, J. T., Lee, D., Mardia, K. V., and Linney, A. D. (1996a). Using curvature information in shape analysis. In: *Image Fusion and Shape Variability* (eds K. V. Mardia, C. A. Gill and I. L. Dryden), pp. 88–99. University of Leeds Press, Leeds. page 375

Kent, J. T. and Mardia, K. V. (1994a). The link between kriging and thin-plate splines. In: *Probability, Statistics and Optimization: a Tribute to Peter Whittle* (ed. F. P. Kelly), pp 325–339. John Wiley & Sons, Ltd, Chichester. page 282, 287, 311

Kent, J. T. and Mardia, K. V. (1994b). Statistical shape methodology. In: *NATO Conference on Shape in Pictures* (eds O. Ying-Lie, A. Toet, D. Foster, H. J. A. M. Heijmans and P. Meer), pp. 443–452. Springer-Verlag, Berlin. page 305

Kent, J. T. and Mardia, K. V. (1997). Consistency of Procrustes estimators. *Journal of the Royal Statistical Society, Series B*, 59: 281–290. page 121, 318

Kent, J. T. and Mardia, K. V. (2001). Shape, Procrustes tangent projections and bilateral symmetry. *Biometrika*, 88(2): 469–485. page 70, 71, 74, 90, 171

Kent, J. T. and Mardia, K. V. (2012). A geometric approach to projective shape and the cross ratio. *Biometrika*, 99(4): 833–849. page 396

Kent, J. T. and Mardia, K. V. (2013). Discrimination for spherical data. In: *Proceedings of LASR 2013* (eds K. V. Mardia, A. Gusnanto, A. D. Riley and J. Voss), pp. 71–74. University of Leeds Press, Leeds. page 169

Kent, J. T., Mardia, K. V., and McDonnell, P. (2006). The complex Bingham quartic distribution and shape analysis. *Journal of the Royal Statistical Society: Series B (Statistical Methodology)*, 68(5): 747–765. page 231

Kent, J. T., Mardia, K. V., Morris, R. J., and Aykroyd, R. G. (2001). Functional models of growth for landmark data. In: *Proceedings of LASR 2001* (eds K. V. Mardia and R. G. Aykroyd), pp. 109–115. University of Leeds Press, Leeds. page 22, 324

Kent, J. T., Mardia, K. V., and Taylor, C. C. (2010). An EM interpretation of the softassign algorithm for alignment problems. In: *Proceedings of LASR 2010* (eds A. Gusnanto, K. V. Mardia, C. J. Fallaize and J. Voss), pp. 29–32. University of Leeds Press, Leeds. page 348

Kent, J. T., Mardia, K. V., and Walder, A. N. (1996b). Conditional cyclic Markov random fields. *Advances in Applied Probability*, 28: 1–12. page 365

Khatri, C. G. and Mardia, K. V. (1977). The von Mises–Fisher matrix distribution in orientation statistics. *Journal of the Royal Statistical Society, Series B*, 39(1): 95–106. page 343

Killian, B. J., Kravitz, J. Y., and Gilson, M. K. (2007). Extraction of configurational entropy from molecular simulations via an expansion approximation. *The Journal of Chemical Physics*, 127: 024107. page 104, 105

Kim, K., Kim, P. T., Koo, J., and Pierrynowski, M. R. (2013). Frenet–Serret and the estimation of curvature and torsion. *IEEE Journal of Selected Topics in Signal Processing*, 7(4): 646–654. page 374

Kim, P. T., Pinder, S., and Rush, S. T. A. (2014). Fréchet analysis and the microbiome. *Journal of Statistical Planning and Inference*, 145: 37–41. page 393

Kimeldorf, G. and Wahba, G. (1971). Some results on Tchebycheffian spline functions. *Journal of Mathematical Analysis and Applications*, 33: 82–95. page 311

Klassen, E., Srivastava, A., Mio, W., and Joshi, S. H. (2003). Analysis of planar shapes using geodesic paths on shape spaces. *IEEE Transactions on Pattern Analysis and Machine Intelligence*, 26(3): 372–383. page 369

Klingenberg, C. P. (1996). Multivariate allometry. In: *Advances in Morphometrics* (eds L. Marcus, M. Corti, A. Loy, G. Naylor and D. Slice), Vol. 284 of *NATO ASI Series*, pp. 23–49. Springer, New York. page 109

Klingenberg, C. P. (2011). MorphoJ: an integrated software package for geometric morphometrics. *Molecular Ecology Resources*, 11(2): 353–357. page 173

Klingenberg, C. P. and McIntyre, G. S. (1998). Geometric morphometrics of developmental instability: analyzing patterns of fluctuating asymmetry with Procrustes methods. *Evolution*, 52: 1363–1375. page 173

Kneip, A. and Gasser, T. (1992). Statistical tools to analyze data representing a sample of curves. *Annals of Statistics*, 20(3): 1266–1305. page 325

Kneip, A., Li, X., MacGibbon, K. B., and Ramsay, J. O. (2000). Curve registration by local regression. *Canadian Journal of Statistics*, 28(1): 19–29. page 325

Kobayashi, S. and Nomizu, K. (1969). Foundations of Differential Geometry, Vol. 2, Wiley, New York. page 83

Koch, I. (2014). Analysis of Multivariate and High-Dimensional Data. Cambridge University Press, Cambridge. page 126

Koenderink, J. J. (1990). *Solid Shape*. MIT Press Series in Artificial Intelligence. MIT Press, Cambridge, MA. page 330, 375, 376, 395

Koenderink, J. J. and van Doorn, A. J. (1992). Surface shape and curvature scales. *Image and Vision Computing*, 10(8): 557–564. page 375

Koenker, R. (2006). The median is the message: toward the Fréchet median. *Journal de la Société Française de Statistique*, 147(2): 61–64. page 323

Kotz, S. and Johnson, N. L. (eds) (1988). *Encyclopedia of Statistical Sciences*, Vol. 8. John Wiley & Sons, Inc., New York. page 397

Krim, H. and Yezzi, A. J. (eds) (2006). *Statistics and Analysis of Shapes* Birkhäuser, Boston. page 390, 397

Kristof, W. and Wingersky, B. (1971). Generalization of the orthogonal Procrustes rotation procedure to more than two matrices. In: *Proceedings of the 79th Annual Convention of the American Psychological Association*, pp. 89–90. American Psychological Association, Washington, DC. page 137

Krzanowski, W. J. and Marriott, F. H. C. (1994). *Multivariate Analysis, Part 1: Distributions, Ordination and Inference*. Edward Arnold, London. page 126, 133

Kshirsagar, A. M. (1963). Effect of non-centrality on Bartlett decompositions of a Wishart matrix. *Annals of the Institute of Statistical Mathematics*, 14: 217–228. page 265

Kume, A., Dryden, I. L., and Le, H. (2007). Shape-space smoothing splines for planar landmark data. *Biometrika*, 94(3): 513–528. page 24, 26, 328, 329, 330, 331

Kume, A., Dryden, I. L., and Wood, A. T. A. (2015). Shape inference based on multivariate normal matrix distributions. Technical Report, University of Kent. page 258

Kume, A. and Le, H. (2000). Estimating Fréchet means in Bookstein's shape space. *Advances in Applied Probability*, 32(3): 663–674. page 396

Kume, A. and Le, H. (2003). On Fréchet means in simplex shape spaces. *Advances in Applied Probability*, 35(4): 885–897. page 396

Kume, A. and Walker, S. G. (2009). On the Fisher–Bingham distribution. *Statistical Computing*, 19(2): 167–172. page 226

Kume, A. and Welling, M. (2010). Maximum likelihood estimation for the offset-normal shape distribution using EM. *Journal of Computational and Graphical Statistics*, 19(3): 702–723. page 258

Kume, A. and Wood, A. T. A. (2005). Saddlepoint approximations for the Bingham and Fisher–Bingham normalising constants. *Biometrika*, 92(2): 465–476. page 225

Kume, A. and Wood, A. T. A. (2007). On the derivatives of the normalising constant of the Bingham distribution. *Statistics & Probability Letters*, 77(8): 832–837. page 225

Kurtek, S., Klassen, E., Ding, Z., and Srivastava, A. (2010). A novel Riemannian framework for shape analysis of 3D objects. In: *IEEE Conference on Computer Vision and Pattern Recognition (CVPR)*, pp. 1625–1632. IEEE Computer Society Press, Los Alamitos, CA. page 374

Kurtek, S., Klassen, E., Gore, J. C., Ding, Z., and Srivastava, A. (2011). Classification of mathematics deficiency using shape and scale analysis of 3d brain structures. *Proceedings of SPIE*, 7962: 796244. page 4

Langron, S. P. and Collins, A. J. (1985). Perturbation theory for generalized Procrustes analysis. *Journal of the Royal Statistical Society, Series B*, 47: 277–284. page 125, 138, 197

Le, H. (2001). Locating Fréchet means with application to shape spaces. *Advances in Applied Probability*, 33(2): 324–338. page 112

Le, H. (2003). Unrolling shape curves. *Journal of the London Mathematical Society*, 68(2): 511–526. page 328, 330

Le, H. and Barden, D. (2014). On the measure of the cut locus of a Fréchet mean. *Bulletin of the London Mathematical Society*, 46(4): 698–708. page 320

Le, H. and Bhavnagri, B. (1997). On simplifying shapes by subjecting them to collinearity constraints. *Mathematical Proceedings of the Cambridge Philosophical Society*, 122(2): 315–323. page 350

Le, H. and Kume, A. (2000a). Detection of shape changes in biological features. *Journal of Microscopy*, 200: 140–147. page 22, 326

Le, H. and Kume, A. (2000b). The Fréchet mean shape and the shape of the means. *Advances in Applied Probability*, 32(1): 101–113. page 120

Le, H.-L. (1987). Explicit formulae for polygonally generated shape densities in the basic tile. *Mathematical Proceedings of the Cambridge Philosophical Society*, 101: 313–332. page 350

Le, H.-L. (1988). *Shape theory in flat and curved spaces, and shape densities with uniform generators*. PhD thesis, University of Cambridge. page 101, 103

Le, H.-L. (1989a). Random spherical triangles I: geometrical background. *Advances in Applied Probability*, 21: 570–580. page 394

Le, H.-L. (1989b). Random spherical triangles II: shape densities. *Advances in Applied Probability*, 21: 581–594. page 394

Le, H.-L. (1991a). On geodesics in Euclidean shape spaces. *Journal of the London Mathematical Society*, 44: 360–372. page 75, 88, 90, 350

Le, H.-L. (1991b). A stochastic calculus approach to the shape distribution induced by a complex normal model. *Mathematical Proceedings of the Cambridge Philosophical Society*, 109: 221–228. page 71, 243, 247, 253, 319, 351

Le, H.-L. (1992). The shapes of non-generic figures, and applications to collinearity testing. *Proceedings of the Royal Society of London, Series A*, 439: 197–210. page 350

Le, H.-L. (1994). Brownian motions on shape and size-and-shape spaces. *Journal of Applied Probability*, 31: 101–113. page 351

Le, H.-L. (1995). Mean size-and-shapes and mean shapes: a geometric point of view. *Advances in Applied Probability*, 27: 44–55. page 101, 125, 146, 318

Le, H. L. and Kendall, D. G. (1993). The Riemannian structure of Euclidean shape spaces: a novel environment for statistics. *Annals of Statistics*, 21(3): 1225–1271. page 33, 74, 77, 319

Lele, S. (1991). Some comments on a coordinate free and scale invariant method in morphometrics. *American Journal of Physical Anthropology*, 85: 407–418. page 357

Lele, S. (1993). Euclidean distance matrix analysis (EDMA): estimation of mean form and mean form difference. *Mathematical Geology*, 25(5): 573–602. page 146, 147, 317, 357, 358

Lele, S. and Cole, T. M. (1995). Euclidean distance matrix analysis: a statistical review. In: *Current Issues in Statistical Shape Analysis* (eds K. V. Mardia and C. A. Gill), pp. 49–53. University of Leeds Press, Leeds. page 359

Lele, S. and Richtsmeier, J. T. (1991). Euclidean distance matrix analysis: a coordinate-free approach for comparing biological shapes using landmark data. *American Journal of Physical Anthropology*, 86: 415–427. page 239, 359

Lele, S. R. and McCulloch, C. E. (2002). Invariance, identifiability, and morphometrics. *Journal of the American Statistical Association*, 97(459): 796–806. page 318

Lele, S. R. and Richtsmeier, J. T. (2001). *An Invariant Approach to the Statistical Analysis of Shapes*. Chapman and Hall/CRC, Boca Raton. page xx, 32, 357, 358, 396

Lestrel, P. (ed) (1997). *Fourier Descriptors and their Application in Biological Science*. Cambridge University Press, Cambridge. page 365

Leu, R. and Damien, P. (2014). Bayesian shape analysis of the complex Bingham distribution. *Journal of Statistical Planning and Inference*, 149: 183–200. page 236

Levitt, M. (1976). A simpified representation of protein conformations for rapid simulation of protein folding. *Journal of Molecular Biology*, 104: 59–107. page 100

Lew, W. D. and Lewis, J. L. (1977). An anthropometric scaling method with application to the knee joint. *Journal of Biomechanics*, 10: 171–184. page 305

Lipman, Y. and Daubechies, I. (2010). Conformal Wasserstein distances: comparing surfaces in polynomial time. *Advances in Mathematics*, 227: 1047–1077. page 375

Lipman, Y., Puente, J., and Daubechies, I. (2013). Conformal Wasserstein distance: II. Computational aspects and extensions. *Mathematics of Computation*, 82(281): 331–381. page 375

Liu, W., Srivastava, A., and Zhang, J. (2011). A mathematical framework for protein structure comparison. *PLoS Computational Biology*, 7(2): e1001075, 10. page 349

Lohmann, G. P. (1983). Eigenshape analysis of microfossils: a general morphometric procedure for describing changes in shape. *Mathematical Geology*, 15: 659–672. page 4, 28

Loncaric, S. (1998). A survey of shape analysis techniques. *Pattern Recognition*, 31: 983–1001. page 397

Loubes, J.-M. and Pelletier, B. (2008). A kernel-based classifier on a Riemannian manifold. *Statistical Decisions*, 26(1): 35–51. page 323

MacLeod, N. and Forey, P. L. (eds) (2002). *Morphology, Shape and Phylogeny*, Vol. 64 of *Systematics Association Special Volume Series*. Taylor and Francis, London. page 397

Mallet, X. D. G., Dryden, I. L., Vorder Bruegge, R., and Evison, M. (2010). An exploration of sample representativeness in anthropometric facial comparison. *Journal of Forensic Sciences*, 55: 1025–1031. page 163

Mancham, A. and Molchanov, I. S. (1996). Stochastic model of randomly perturbed images and related estimation problems. In: *Image Fusion and Shape Variability* (eds K. V. Mardia, C. A. Gill and I. L. Dryden), pp. 43–49. University of Leeds Press, Leeds. page 367

Mannion, D. (1988). A Markov chain of triangle shapes. *Advances in Applied Probability*, 20: 348–370. page 351

Mannion, D. (1990a). Convergence to collinearity of a sequence of random triangle shapes. *Advances in Applied Probability*, 22: 831–844. page 351

Mannion, D. (1990b). The invariant distribution of a sequence of random collinear triangle shapes. *Advances in Applied Probability*, 22: 845–865. page 351

Marchini, J., Heaton, C., and Ripley, B. D. (2013). *fastICA package in R*. R package version 1.2-0. http://cran.r-project.org/package=fastICA (accessed 20 March 2016). page 169

Mardia, K. V. (1972). *Statistics of Directional Data*. Academic Press, London. page 63

Mardia, K. V. (1975). Statistics of directional data (with discussion). *Journal of the Royal Statistical Society, Series B*, 37: 349–393. page 366

Mardia, K. V. (1977). Mahalanobis distance and angles. In: *Multivariate Analysis IV* (ed. P. R. Krishnaiah), pp. 495–511. North-Holland, Amsterdam. page xxi, 297

Mardia, K. V. (1980). Discussion to 'Simulating the ley hunter' by S. R. Broadbent. *Journal of the Royal Statistical Society, Series A*, 143: 147. page 219, 241, 248

Mardia, K. V. (1984). Spatial discrimination and classification maps. *Communications in Statistics – Theory and Methods*, 13: 2181–2197. page 147

Mardia, K. V. (1989a). Discussion to 'A survey of the statistical theory of shape' by D. G. Kendall. *Statistical Science*, 4: 108–111. page 9

Mardia, K. V. (1989b). Shape analysis of triangles through directional techniques. *Journal of the Royal Statistical Society, Series B*, 51: 449–458. page 54, 56, 86, 226, 248, 351, 352

Mardia, K. V. (1991). New advances in shape analysis with applications to image processing. Lectures under the late Professor M. C. Chakrabarti memorial lectureship endowment, University of Bombay. Technical Report, University of Leeds. page 43

Mardia, K. V. (1993). Discussion to papers on 'Gibbs sampler and other MCMC methods'. *Journal of the Royal Statistical Society, Series B*, 55: 83–84. page 384

Mardia, K. V. (ed.) (1994). *Statistics and Images*, Vol. 2, Carfax, Oxford. page 377

Mardia, K. V. (1995). Shape advances and future perspectives. In: *Proceedings in Current Issues in Statistical Shape Analysis* (eds K. V. Mardia and C. A. Gill), pp. 57–75. University of Leeds Press, Leeds. page 251, 297

Mardia, K. V. (1996a). The art and science of Bayesian object recognition. In: *Proceedings in Image Fusion and Shape Variability Techniques* (eds K. V. Mardia, C. A. Gill and I. L. Dryden), pp. 21–35. University of Leeds Press, Leeds. page 365, 382

Mardia, K. V. (1996b). Shape analysis. Technical Report, Department of Statistics, University of Leeds. page 86, 226

Mardia, K. V. (1997). Bayesian image analysis. *Journal of Theoretical Medicine*, 1: 63–77. page 165

Mardia, K. V. (2009). Statistical complexity in protein bioinformatics. In: *Statistical Tools for Challenges in Bioinformatics* (eds A. Gusnanto, K. V. Mardia and C. J. Fallaize), pp. 9–20. Leeds University Press, Leeds. page 104

Mardia, K. V. (2010). Statistical complexity in protein bioinformatics II. In: *Proceedings of LASR 2010* (eds A. Gusnanto, K. V. Mardai, C. J. Fallaize and J. Voss), pp. 9–16. Leeds University Press, Leeds. page 100

Mardia, K. V. (2013a). Some aspects of geometry driven statistical models. In: *Proceedings of LASR 2013* (eds K. V. Mardia, A. Gusnanto, A. D. Riley and J. Voss), pp. 21–29. University of Leeds Press, Leeds. page 313

Mardia, K. V. (2013b). Statistical approaches to three key challenges in protein structural bioinformatics. *Journal of the Royal Statistical Society: Series C (Applied Statistics)*, 62(3): 487–514. page 21, 208, 209, 344

Mardia, K. V., Angulo, J. M., and Goitia, A. (2006a). Synthesis of image deformation strategies. *Image and Vision Computing*, 24: 1–12. page 389

Mardia, K. V., Baczkowski, A. J., Feng, X., and Millner, P. A. (1996a). A study of three dimensional curves. *Journal of Applied Statistics*, 23: 139–148. page 236

Mardia, K. V., Bookstein, F. L., and Kent, J. T. (2013a). Alcohol, babies and the death penalty: Saving lives by analysing the shape of the brain. *Significance*, 10(3): 12–16. page 16, 17, 168

Mardia, K. V., Bookstein, F. L., Kent, J. T., and Meyer, C. R. (2006b). Intrinsic random fields and image deformations. *Journal of Mathematical Imaging and Vision*, 26: 59–71. page 389, 390

Mardia, K. V., Bookstein, F. L., and Moreton, I. J. (2000). Statistical assessment of bilateral symmetry of shapes. *Biometrika*, 87(2): 285–300. page 171

Mardia, K. V., Coombes, A., Kirkbride, J., Linney, A., and Bowie, J. L. (1996b). On statistical problems with face identification from photographs. *Journal of Applied Statistics*, 23: 655–675. page 40, 360, 384

Mardia, K. V. and Dryden, I. L. (1989a). Shape distributions for landmark data. *Advances in Applied Probability*, 21: 742–755. page 33, 120, 239, 247

Mardia, K. V. and Dryden, I. L. (1989b). The statistical analysis of shape data. *Biometrika*, 76: 271–282. page 8, 120, 239, 247, 335

Mardia, K. V. and Dryden, I. L. (1994). Shape averages and their bias. *Advances in Applied Probability*, 26: 334–340. page 136, 146, 251

Mardia, K. V. and Dryden, I. L. (1997). Bookstein's shape coordinates for three dimensions. Technical Report STAT97/03, Department of Statistics, University of Leeds. page 47

Mardia, K. V. and Dryden, I. L. (1999). The complex Watson distribution and shape analysis. *Journal of the Royal Statistical Society, Series B*, 61(4): 913–926. page 227, 228, 232

Mardia, K. V., Dryden, I. L., Hurn, M. A., Li, Q., Millner, P. A., and Dickson, R. A. (1994). Familial spinal shape. *Journal of Applied Statistics*, 21: 623–641. page 236

Mardia, K. V., Edwards, R., and Puri, M. L. (1977). Analysis of Central Place Theory. *Bulletin of the International Statistical Institute*, 47: 93–110. page 28, 29, 39, 40, 351

Mardia, K. V., Fallaize, C. J., Barber, S., Jackson, R. M., and Theobald, D. L. (2013b). Bayesian alignment of similarity shapes. *Annals of Applied Statistics*, 7: 989–1009. page 236, 344, 346, 349

Mardia, K. V., Ghali, N. M., Hainsworth, T. J., Howes, M., and Sheehy, N. (1993). Techniques for online gesture recognition on workstations. *Image and Vision Computing*, 11: 283–294. page 11

Mardia, K. V. and Gill, C. A. (eds) (1995). *Proceedings in Current Issues in Statistical Shape Analysis*. University of Leeds Press, Leeds. page xx, xxiii

Mardia, K. V., Gill, C. A., and Aykroyd, R. G. (eds) (1997a). *Proceedings of the Leeds Annual Statistics Research Workshop*. University of Leeds Press, Leeds. page xxiii

Mardia, K. V., Gill, C. A., and Dryden, I. L. (eds) (1996c). *Image Fusion and Shape Variability*. University of Leeds Press, Leeds. page xxiii

Mardia, K. V. and Goodall, C. R. (1993). Spatial-temporal analysis of multivariate environmental montioring data. In: *Multivariate Environmental Statistics* (eds G. P. Patil and C. R. Rao), pp. 347–386. North Holland, Amsterdam. page 282

Mardia, K. V., Goodall, C. R., and Walder, A. N. (1996d). Distributions of projective invariants and model-based vision. *Advances in Applied Probability*, 28: 641–661. page 396

Mardia, K. V., Gusnanto, A., Nooney, C., and Voss, J. (eds) (2015). *LASR 2015: Geometry-Driven Statistics and its Cutting Edge Applications: Celebrating Four Decades of Leeds Statistics Workshops*. University of Leeds Press, Leeds. page xx

Mardia, K. V. and Hainsworth, T. J. (1993). Image warping and Bayesian reconstruction with grey-level templates. In: *Statistics and Images* (eds K. V. Mardia and G. K. Kanji), Vol. 1, pp. 257–280. Carfax, Oxford. page 382, 383, 385, 386

Mardia, K. V. and Jupp, P. E. (2000). *Directional Statistics*. Wiley Series in Probability and Statistics. John Wiley & Sons Ltd, Chichester. page 40, 63, 85, 396

Mardia, K. V. and Kanji, G. K. (eds) (1993). *Statistics and Images*, Vol. 1. Carfax, Oxford. page 377

Mardia, K. V., Kent, J. T., and Bibby, J. M. (1979). *Multivariate Analysis*. Academic Press, London. page 72, 79, 81, 116, 125, 133, 166, 179, 186, 188, 196, 201, 220, 243, 274, 276, 297, 298, 309, 353

Mardia, K. V., Kent, J. T., Goodall, C. R., and Little, J. L. (1996e). Kriging and splines with derivative informations. *Biometrika*, 83: 207–221. page 286, 309, 311, 313

Mardia, K. V., Kent, J. T., and Walder, A. N. (1991). Statistical shape models in image analysis. In: *Computer Science and Statistics: Proceedings of the 23rd INTERFACE Symposium* (ed. E. M. Keramidas), pp. 550–557. Interface Foundation, Fairfax Station. page 4, 281, 310, 379, 382

Mardia, K. V., Kirkbride, J., and Bookstein, F. L. (2004). Statistics of shape, direction and cylindrical variables. *Journal of Applied Statistics*, 31(4): 465–479. page 313

Mardia, K. V. and Little, J. L. (1994). Image warping using derivative information. In: *Proceedings of Mathematical Methods in Medical Imaging III* (eds F. L. Bookstein, J. S. Duncan, N. Lange, and D. C. Wilson), Vol. 2299, pp. 16–31. SPIE, Washington. page 308, 313, 383

Mardia, K. V. and Marshall, R. J. (1984). Maximum likelihood estimation of models for residual covariance in spatial regression. *Biometrika*, 71: 135–146. page 308, 309

Mardia, K. V., McCulloch, C., Dryden, I. L., and Johnson, V. (1997b). Automatic scale-space method of landmark detection. In: *Proceedings of the Leeds Annual Statistics Research Workshop* (eds K. V. Mardia, C. A. Gill, and R. G. Aykroyd), pp. 17–29, University of Leeds Press, Leeds. page 390

Mardia, K. V., McDonnell, P., and Linney, A. D. (2006c). Penalized image averaging and discrimination with facial and fishery applications. *Journal of Applied Statistics*, 33(3): 339–371. page 387

Mardia, K. V., Morris, R. J., Walder, A. N., and Koenderink, J. J. (1999). Estimation of torsion. *Journal of Applied Statistics*, 26: 373–381. page 374

Mardia, K. V., Nyirongo, V. B., Fallaize, C. J., Barber, S., and Jackson, R. M. (2011). Hierarchical Bayesian modelling of pharmacophores in bioinformatics. *Biometrics*, 67(2): 611–619. page 344, 349

Mardia, K. V., Nyirongo, V. B., Green, P. J., Gold, N. D., and Westhead, D. R. (2007). Bayesian refinement of protein functional site matching. *BMC Bioinformatics*, 8: 257. page 344

Mardia, K. V. and Patrangenaru, V. (2005). Directions and projective shapes. *Annals of Statistics*, 33(4): 1666–1699. page 396

Mardia, K. V., Petty, E. M., and Taylor, C. C. (2012). Matching markers and unlabeled configurations in protein gels. *Annals of Applied Statistics*, 6(3): 853–869. page 274

Mardia, K. V. and Qian, W. (1995). Bayesian method for compact object recognition from noisy images. In: *Complex Stochastic Systems in Science and Engineering* (ed. D. M. Titterington), pp. 155–165. Clarendon Press, Oxford. page 366

Mardia, K. V., Qian, W., Shah, D., and De Souza, K. (1996f). Deformable template recognition of multiple occluded objects. *IEEE Transactions on Pattern Analysis and Machine Intelligence*, 19: 1036–1042. page 368

Mardia, K. V., Rabe, S., and Kent, J. T. (1995). Statistics, shape and images. In: *IMA Proceedings on Complex Stochastic Systems and Engineering Applications*, pp. 85–103. Clarendon Press, Oxford. page 376, 379

Mardia, K. V. and Walder, A. N. (1994a). Shape analysis of paired landmark data. *Biometrika*, 81: 185–196. page 333, 335

Mardia, K. V. and Walder, A. N. (1994b). Size-and-shape distributions for paired landmark data. *Advances in Applied Probability*, 26: 893–905. page 335

Marron, J. S. and Alonso, A. M. (2014). Overview of object oriented data analysis. *Biometrical Journal*, 56(5): 732–753. page 391

Marron, J. S., Jung, S., and Dryden, I. L. (2010). Speculation on the generality of the backward stepwise view of PCA. In: *Proceedings of MIR 2010: 11th ACM SIGMM International Conference on Multimedia Information Retrieval*, pp. 227–230. Association for Computing Machinery, Inc., Danvers, MA. page 122

Matheron, G. (1973). The intrinsic random functions and their applications. *Advances in Applied Probability*, 5: 439–468. page 310

Matheron, G. (1981). Splines and kriging: their formal equivalence. In: *Down-to-earth Statistics: Solutions Looking for Geological Problems* (ed. D. F. Merriam), pp. 77–95. Syracuse University Geology Contributions, Syracuse. page 311

Mazza, C. (1997). *Human type I 17beta-hydroxysteroid dehydrogenase: site directed mutagenesis and X-ray crystallography structure-function analysis*. PhD thesis, Université Joseph Fourier. page 21

McLachlan, A. D. (1972). A mathematical procedure for superimposing atomic coordinates of proteins. *Acta Crystallography*, A28. page 125

Medawar, P. B. (1944). The shape of a human being as a function of time. *Proceedings of the Royal Society of London, Series B*, 132: 133–141. page 32, 303, 305

Medawar, P. B. (1945). Size, shape and age. In: *Essays on Growth and Form presented to D'Arcy Wentworth Thompson* (eds W. E. Le Gros Clark and P. B. Medawar), pp. 157–187. Clarendon Press, Oxford. page 303

Meinguet, J. (1979). Multivariate interpolation at arbitrary points made simple. *Zeitschrift Angewandte Mathematik und Physik*, 30: 292–304. page 280, 287

Meinshausen, N. and Bühlmann, P. (2006). High-dimensional graphs and variable selection with the Lasso. *Annals of Statistics*, 34(3): 1436–1462. page 197

Metropolis, N., Rosenbluth, A. W., Rosenbluth, M. N., Teller, A. H., and Teller, E. (1953). Equations of state calculations by fast computing machines. *Journal of Chemical Physics*, 21: 1087–1092. page 382

Micheas, A. C. and Dey, D. K. (2005). Modeling shape distributions and inferences for assessing differences in shapes. *Journal of Multivariate Analysis*, 92(2): 257–280. page 233

Micheas, A. C., Dey, D. K., and Mardia, K. V. (2006). Complex elliptical distributions with applications to shape analysis. *Journal of Statistical Planning and Inference*, 136(9): 2961–2982. page 233

Micheas, A. C. and Peng, Y. (2010). Bayesian Procrustes analysis with applications to hydrology. *Journal of Applied Statistics*, 37(1–2): 41–55. page 236

Michor, P. W. and Mumford, D. (2006). Riemannian geometries on spaces of plane curves. *Journal of the European Mathematical Society (JEMS)*, 8(1): 1–48. page 374

Michor, P. W. and Mumford, D. (2007). An overview of the Riemannian metrics on spaces of curves using the Hamiltonian approach. *Applied and Computational Harmonic Analysis*, 23(1): 74–113. page 374

Miles, R. E. (1970). On the homogeneous planar Poisson point process. *Mathematical Biosciences*, 6: 85–127. page 36, 351

Miller, K. S. (1964). *Multidimensional Gaussian Distributions*. John Wiley & Sons, Inc., New York. page 262

Miller, M. I., Joshi, S., Maffit, D. R., McNally, J. G., and Grenander, U. (1994). Membranes, mitochondria and amoebae: shape models. In: *Statistics and Images*. (ed. K. V. Mardia), Vol. 2, pp. 141–163. Carfax, Oxford. page 367

Miller, M. I., Trouvé, A., and Younes, L. (2002). On the metrics and Euler–Lagrange equations of computational anatomy. *Annual Review of Biomedical Engineering*, 4(1): 375–405. page 315

Miller, M. I., Trouvé, A., and Younes, L. (2006). Geodesic shooting for computational anatomy. *Journal of Mathematical Imaging and Vision*, 24(2): 209–228. page 315

Mitchelson, J. R. (2013). MOSHFIT: Algorithms for occlusion-tolerant mean shape and rigid motion from 3D movement data. *Journal of Biomechanics*, 46: 2326–2329. page 320, 339

Mitteroecker, P. and Gunz, P. (2009). Advances in geometric morphometrics. *Evolutionary Biology*, 36(2): 235–247. page 397

Mitteroecker, P., Gunz, P., Bernhard, M., Schaefer, K., and Bookstein, F. L. (2004). Comparison of cranial ontogenetic trajectories among great apes and humans. *Journal of Human Evolution*, 46(6): 679–698. page 104, 109, 196

Monteiro, L. R. (1999). Multivariate regresion models and geometric morphometrics: The search for causal factors in the analysis of shape. *Systematic Biology*, 48(1): 192–199. page 22

Morecroft, L., Fieller, N. R., Dryden, I. L., and Evison, M. P. (2010). Shape variation in anthropometric landmarks in 3D. In: *Computer-Aided Forensic Facial Comparison* (eds M. P. Evison and R. W. V. Bruegge), pp. 35–52. CRC Press, Boca Raton. page 163

Morris, R. J., Kent, J. T., Mardia, K. V., Fidrich, M., Aykroyd, R. G., and Linney, A. (1999). Analysing growth in faces. In: *Proceedings of Conference on Imaging Science, Systems and Technology*, (ed. H. R. Arabnia), pp. 404–409. Computer Science Research, Education and Applications (CSREA) Press, Bogart, GA. page 324

Mosier, C. I. (1939). Determining a simple structure when loadings for certain tests are known. *Psychometrika*, 4: 149–162. page 125

Mosimann, J. E. (1970). Size allometry: Size and shape variables with characterizations of the lognormal and generalized gamma distributions. *Journal of the American Statistical Association*, 65: 930–948. page 38, 40

Mosimann, J. E. (1975a). Statistical problems of size and shape. I. Biological applications and basic theorems. In: *Statistical Distributions in Scientific Work*, (ed. G. P. Patil), Vol. 2, pp. 187–217. D. Reidel, Dordrecht. page 40

Mosimann, J. E. (1975b). Statistical problems of size and shape. II. Characterizations of the lognormal, gamma and Dirichlet distributions. In: *Statistical Distributions in Scientific Work* (ed. G. P. patil), Vol. 2, pp. 219–239. D. Reidel, Dordrecht. page 40

Mosimann, J. E. (1988). Size and shape analysis. In: *Encyclopedia of Statistical Sciences* (eds S. Kotz and N. L. Johnson), Vol. 8, pp. 497–507. John Wiley & Sons, Inc., New York. page 40

Moss, M. L., Vilman, H., Moss-Salentijn, L., Sen, K., Pucciarelli, H. M., and Skalak, R. (1987). Studies on orthocephalization: Growth behaviour of the rat skull in the period 13-19 days as described by the finite element method. *American Journal of Physical Anthropology*, 72: 323–342. page 305, 333, 335

Moss, S. and Hancock, E. R. (1996). Registering incomplete radar images using the EM algorithm. In: *Proceedings of the Seventh British Machine Vision Conference* (eds R. B. Fisher and E. Trucco), pp. 685–694. BMVA Press, Manchester. page 341, 348

Mumford, D. (1991). Mathematical theories of shape: do they model perception? In: *Proceedings of the SPIE*, 1570: 2–210. page 395

Mumford, D. and Michor, P. W. (2013). On Euler's equation and 'EPDiff'. *Journal of Geometric Mechanics*, 5(3): 319–344. page 315

Munch, E., Turner, K., Bendich, P., Mukherjee, S., Mattingly, J., and Harer, J. (2015). Probabilistic Fréchet means for time varying persistence diagrams. *Electronic Journal of Statistics*, 9(1): 1173–1204. page 393

Muralidharan, P. and Fletcher, P. T. (2012). Sasaki metrics for analysis of longitudinal data on manifolds. In: *2012 IEEE Conference on Computer Vision and Pattern Recognition*,

Providence, RI, USA, June 16-21, 2012, pp. 1027–1034. IEEE Computer Society Press, Los Alamitos, CA. page 333

Nagy, G. (1992). The dimensions of shape and form. In: *Visual Form: Analysis and Recognition* (eds C. Arcelli, L. P. Cordella, and G. Sanniti di Baja), pp. 409–419. Plenum, New York. page 395

Nakamura, T., Hiraoka, Y., Hirata, A., Escolar, E. G., and Nishiura, Y. (2015). Persistent homology and many-body atomic structure for medium-range order in the glass. *Nanotechnology*, 26(30): 304001. page 393

Neale, F. B. and Russ, J. C. (2012). *Measuring Shape*. CRC Press, Boca Raton. page 377, 390, 396

Niethammer, M., Huang, Y., and Vialard, F. (2011). Geodesic regression for image time-series. In: *Medical Image Computing and Computer-Assisted Intervention (MICCAI 2011)*, (eds G. Fichtinger, A. L. Martel, and F. Vialard), Part II, Vol. 6892 of *Lecture Notes in Computer Science*, pp. 655–662. Springer, New York. page 333

Niethammer, M. and Vialard, F.-X. (2013). Riemannian metrics for statistics on shapes: Parallel transport and scale invariance. In: *MICCAI Workshop Mathematical Foundations of Computational Anatomy* (eds X. Pennec, S. Joshi, M. Nielsen, P. T. Fletcher, S. Durrelmand and S. Sommer), pp. 113. http://www-sop.inria.fr/asclepios/events/MFCA13/ (accessed 20 March 2016). page 333

Nye, T. M. W. (2011). Principal components analysis in the space of phylogenetic trees. *Annals of Statistics*, 39(5): 2716–2739. page 392

Nye, T. M. W. and White, M. C. (2014). Diffusion on some simple stratified spaces. *Journal of Mathematical Imaging and Vision*, 50(1–2): 115–125. page 392

O'Higgins, P. (1989). *A morphometric study of cranial shape in the Hominoidea*. PhD thesis, University of Leeds. page 19

O'Higgins, P. (2000). The study of morphological variation in the hominid fossil record: biology, landmarks and geometry. *Journal of Anatomy*, 197(01): 103–120. page 196

O'Higgins, P. and Dryden, I. L. (1992). Studies of craniofacial development and evolution. *Archaeology and Physical Anthropology in Oceania*, 27: 105–112. page 23

O'Higgins, P. and Dryden, I. L. (1993). Sexual dimorphism in hominoids: further studies of craniofacial shape differences in *pan, gorilla, pongo. Journal of Human Evolution*, 24: 183–205. page 19, 212, 305, 308

O'Higgins, P. and Jones, N. (1998). Facial growth in *Cercocebus torquatus*: an application of three-dimensional geometric morphometric techniques to the study of morphological variation. *Journal of Anatomy*, 193(02): 251–272. page 196

Okabe, A., Boots, B., Sugihara, K., and Chiu, S. N. (2000). *Spatial Tessellations: Concepts and Applications of Voronoi Diagrams*, 2nd edn. Wiley Series in Probability and Statistics. John Wiley & Sons, Ltd., Chichester. page 29, 305

Owen, M. (2011). Computing geodesic distances in tree space. *SIAM Journal on Discrete Mathematics*, 25(4): 1506–1529. page 392

Panaretos, V. M. (2006). The diffusion of Radon shape. *Advances in Applied Probability*, 38(2): 320–335. page 266

Panaretos, V. M. (2008). Representation of Radon shape diffusions via hyperspherical Brownian motion. *Mathematical Proceedings of the Cambridge Philosophical Society*, 145(2): 457–470. page 266

Panaretos, V. M. (2009). On random tomography with unobservable projection angles. *Annals of Statistics*, 37(6A): 3272–3306. page 266

Panaretos, V. M. and Konis, K. (2011). Sparse approximations of protein structure from noisy random projections. *Annals of Applied Statistics*, 5(4): 2572–2602. page 266, 349

Panaretos, V. M., Pham, T., and Yao, Z. (2014). Principal flows. *Journal of the American Statistical Association*, 109(505): 424–436. page 332

Parui, S. and Majumder, D. D. (1983). Shape similarity measures for open curves. *Pattern Recognition Letters*, 1(3): 129–134. page 365

Patrangenaru, V. and Ellingson, L. (2015). *Nonparametric Statistics on Manifolds and Their Applications to Object Data Analysis*. CRC Press, Boca Raton. page xx, 397

Patrangenaru, V. and Mardia, K. V. (2003). Affine shape analysis and image analysis. In: *Proceedings of LASR 2003* (eds R. G. Aykroyd, K. V. Mardia, and M. J. Langdon), pp. 57–62. University of Leeds Press, Leeds. page 273

Pavlidis, T. (1995). A review of algorithms for shape analysis. In: *Document Image Analysis* (eds L. O'Gorman and R. Kasturi), pp. 145–160. IEEE Computer Society Press, Los Alamitos, CA. page 397

Pearson, K. (1926). On the coefficient of racial likeness. *Biometrika*, 18: 105–117. page 31, 41

Pelletier, B. (2005). Kernel density estimation on Riemannian manifolds. *Statistics & Probability Letters*, 73(3): 297–304. page 323

Pelletier, B. (2006). Non-parametric regression estimation on closed Riemannian manifolds. *Journal of Nonparametric Statistics*, 18(1): 57–67. page 323

Pennec, X. (1996). Multiple registration and mean rigid shapes: application to the 3D case. In: *Proceedings in Image Fusion and Shape Variability Techniques* (eds K. V. Mardia, C. A. Gill, and I. L. Dryden), pp. 178–185. University of Leeds Press, Leeds. page 233

Pennec, X. (1999). Probabilities and statistics on Riemannian manifolds: Basic tools for geometric measurements. In: *Proceedings of IEEE Workshop on Nonlinear Signal and Image Processing (NSIP99)* (eds A. Cetin, L. Akarun, A. Ertuzun, M. Gurcan, and Y. Yardimci), Vol. 1, pp. 194–198. IEEE, Antalya. page 233

Pennec, X. (2006). Intrinsic statistics on Riemannian manifolds: Basic tools for geometric measurements. *Journal of Mathematical Imaging and Vision*, 25(1): 127–154. page 233

Penrose, L. S. (1952). Distance, size and shape. *Annals of Eugenics*, 17: 337–343. page 41

Phillips, D. B. and Smith, A. F. M. (1993). Dynamic image analysis using Bayesian shape and texture models. In: *Statistics and Images*. (eds K. V. Mardi and G. K. Kanji), Vol. 1, pp. 299–322. Carfax, Oxford. page 5, 380

Phillips, D. B. and Smith, A. F. M. (1994). Bayesian faces via hierarchical template modeling. *Journal of the American Statistical Association*, 89: 1151–1163. page 5, 380, 382

Phillips, R., O'Higgins, P., Bookstein, F., *et al.* (2010). EVAN (European Virtual Anthropology Network) toolbox. page 173

Pigoli, D., Aston, J. A. D., Dryden, I. L., and Secchi, P. (2014). Distances and inference for covariance operators. *Biometrika*, 101(2): 409–422. page 396

Piras, P., Evangelista, A., Gabriele, S., *et al.* (2014). 4D-analysis of left ventricular heart cycle using Procrustes motion analysis. *PLOS One*, 9(4): e94673. page 328

Pizer, S. M., Fletcher, P. T., Joshi, S. C., *et al.* (2003). Deformable M-reps for 3D medical image segmentation. *International Journal of Computer Vision*, 55(2–3): 85–106. page 328

Pizer, S. M., Jung, S., Goswami, D., *et al.* (2013). Nested sphere statistics of skeletal models. In: *Innovations for Shape Analysis*, pp. 93–115. Springer, Heidelberg. page 328

Plamondon, R. and Srihari, S. (2000). Online and off-line handwriting recognition: a comprehensive survey. *IEEE Transactions on Pattern Analysis and Machine Intelligence*, 22(1): 63–84. page 11

Polly, P. D., Lawing, A. M., Fabre, A.-C., and Goswami, A. (2013). Phylogenetic principal components analysis and geometric morphometrics. *Hystrix, the Italian Journal of Mammalogy*, 24: 1–9. page 298

Powell, M. J. D. (1987). Radial basis functions for multivariate interpolation: a review. In: *Algorithms for Approximations* eds J. C. Mason and M. G. Cox), pp. 143–167. Clarendon Press, Oxford. page 309

Prentice, M. J. and Mardia, K. V. (1995). Shape changes in the plane for landmark data. *Annals of Statistics*, 23(6): 1960–1974. page 228, 335

Preston, S. P. and Wood, A. T. A. (2009). On definitions of mean reflection shape. In: 57th Session of the International Statistical Institute, Durban. pp. 356. page 355

Preston, S. P. and Wood, A. T. A. (2010). Two-sample bootstrap hypothesis tests for three-dimensional labelled landmark data. *Scandinavian Journal of Statistics*, 37(4): 568–587. page 356

Preston, S. P. and Wood, A. T. A. (2011). Bootstrap inference for mean reflection shape and size-and-shape with three-dimensional landmark data. *Biometrika*, 98(1): 49–63. page 356

Pukkila, T. M. and Rao, C. R. (1988). Pattern recognition based on scale invariant discriminant functions. *Information Sciences*, 45: 379–389. page 40

R Development Core Team (2015). *R: A Language and Environment for Statistical Computing*. R Foundation for Statistical Computing, Vienna. page xix, 7

Rainville, E. D. (1960). *Special Functions*. Macmillan, New York. page 244

Ramachandran, G., Ramakrishnan, C., and Sasisekharan, V. (1963). Stereochemistry of polypeptide chain configurations. *Journal of Molecular Biology*, 7(1): 95–99. page 40

Ramsay, J. O. and Li, X. (1998). Curve registration. *Journal of the Royal Statistical Society: Series B (Statistical Methodology)*, 60(2): 351–363. page 325

Rangarajan, A., Chui, H., and Bookstein, F. L. (1997). The Softassign Procrustes matching algorithm. In: *Information Processing in Medical Imaging* (eds J. Duncan and G. Gindi), pp. 29–42. Springer, New York. page 341, 348

Rao, C. R. (1948). Tests of significance in multivariate analysis. *Biometrika*, 35: 58–79. page 31

Rao, C. R. (1973). *Linear Statistical Inference and Its Applications, 2nd edn*. John Wiley & Sons, Inc., New York. page 281

Rao, C. R. and Suryawanshi, S. (1996). Statistical analysis of shape of objects based on landmark data. *Proceedings of the National Academy of Sciences of the United States of America*, 93: 12132–12136. page 360

Renaud, S. (1995). Shape analysis of outlines through Fourier methods using curvature measures. In: *Current Issues in Statistical Shape Analysis* (eds K. V. Mardia and C. A. Gill), pp. 181–182. University of Leeds Press, Leeds. page 365

Reyment, R. A., Blackith, R. E., and Campbell, N. A. (1984). *Multivariate Morphometrics*, 2nd. edn. Academic Press, New York. page 31, 40

Rice, J. A. and Silverman, B. W. (1991). Estimating the mean and covariance structure nonparametrically when the data are curves. *Journal of the Royal Statistical Society, Series B*, 53: 233–243. page 236, 313

Richstmeier, J. T. (1986). Finite element scaling analysis of human craniofacial growth. *Journal of Craniofacial Genetics and Devopmental Biology*, 6: 289–323. page 305

Richstmeier, J. T. (1989). Application of finite element scaling in primatology. *Folia Primatologica*, 53: 50–64. page 305

Ripley, B. D. (1988). *Statistical Inference for Spatial Processes*. Cambridge University Press, Cambridge. page 377

Ripley, B. D. and Sutherland, A. I. (1990). Finding spiral structures in galaxies. *Philosophical Transactions of the Royal Society of London, Series A*, 332: 477–485. page 380

Rodriguez, A. and Schmidler, S. C. (2014). Bayesian protein structure alignment. *Annals of Applied Statistics*, 8(4): 2068–2095. page 349

Rohlf, F. (2010). tpsRelw, relative warps analysis. Department of Ecology and Evolution, State University of New York, Stony Brook, NY. page 173

Rohlf, F. J. (1990). Fitting curves to outlines. In: *Proceedings of the Michigan Morphometrics Workshop*, Special publication No. 2, pp. 167–177. University of Michigan Museum of Zoology, Ann Arbor. page 365

Rohlf, F. J. (2000). Statistical power comparisons among alternative morphometric methods. *American Journal of Physical Anthropology*, 111: 463–478. page 360

Rohlf, F. J. and Archie, J. (1984). A comparison of Fourier methods for the description of wing shape in mosquitoes. *Systematic Zoology*, 33: 302–317. page 365

Rohlf, F. J. and Marcus, L. F. (1993). A revolution in morphometrics. *Trends in Ecology and Evolution*, 8(4): 129–132. page 397

Rohlf, F. J. and Slice, D. (1990). Extensions of the Procrustes method for the optimal superimposition of landmarks. *Systematic Zoology*, 39: 40–59. page 277, 336, 338

Rosenthal, M., Wu, W., Klassen, E., and Srivastava, A. (2014). Spherical regression models using projective linear transformations. *Journal of the American Statistical Association*, 109(508): 1615–1624. page 396

Rosman, G., Bronstein, M. M., Bronstein, A. M., and Kimmel, R. (2010). Nonlinear dimensionality reduction by topologically constrained isometric embedding. *International Journal of Computer Vision*, 89(1): 56–68. page 375

Rousseeuw, P. J. (1984). Least median of squares regression. *Journal of the American Statistical Association*, 79: 871–880. page 337, 338

Rousseeuw, P. J. and Yohai, V. J. (1984). Robust regression by means of S-estimators. In: *Robust and Non-linear Time Series Analysis* (eds J. Franke, W. Härdle, and R. D. Martin), pp. 256–272. Springer-Verlag, New York. page 336

Rubin, D. B. (1976). Inference and missing data. *Biometrika*, 63: 581–592. page 339

Rue, H. and Syversveen, A. R. (1998). Bayesian object recognition with Baddeley's delta loss. *Advances in Applied Probability*, 30(1): 64–84. page 368

Ruffieux, Y. and Green, P. J. (2009). Alignment of multiple configurations using hierarchical models. *Journal of Computational and Graphical Statistics*, 18(3): 756–773. page 344, 348

Salomon-Ferrer, R., Case, D. A., and Walker, R. C. (2013). An overview of the Amber biomolecular simulation package. *Wiley Interdisciplinary Reviews: Computational Molecular Science*, 3(2): 198–210. page 18

Samir, C., Absil, P.-A., Srivastava, A., and Klassen, E. (2012). A gradient-descent method for curve fitting on Riemannian manifolds. *Foundations of Computational Mathematics*, 12(1): 49–73. page 330

Sampson, P. D., Lewis, S., Guttorp, P., Bookstein, F. L., and Hurley, C. B. (1991). Computation and interpretation of deformations for landmark data in morphometrics and environmetrics. In: *Proceedings of the INTERFACE '91 Symposium* (ed. E. M. Keramidas), pp. 534–541. Interface Foundation, Fairfax Station. page 307

Sangalli, L. M., Secchi, P., and Vantini, S. (2014). AneuRisk65: a dataset of three-dimensional cerebral vascular geometries. *Electronic Journal of Statistics*, 8(2): 1879–1890. page 4

Schmidler, S. C. (2007). Fast Bayesian shape matching using geometric algorithms (with discussion). In: *Proceedings of the Valencia/ISBA 8th World Meeting on Bayesian Statistics* (eds J. M. Bernado, M. J. Bayarri, J. O. Berger *et al.*), pp. 471–490. Oxford University Press, Oxford. page 341, 345, 346, 349

Schmidler, S. C., Liu, J. S., and Brutlag, D. L. (2002). Bayesian protein structure prediction. In: *Case Studies in Bayesian Statistics*, Vol. V, Vol. 162 of *Lecture Notes in Statistics*, pp. 363–378. Springer, New York. page 349

Schmidler, S. C., Lucas, J. E., and Oas, T. G. (2007). Statistical estimation of statistical mechanical models: helix-coil theory and peptide helicity prediction. *Journal of Computational Biology*, 14(10): 1287–1310. page 349

Schönemann, P. H. (1966). A generalized solution to the orthogonal Procrustes problem. *Psychometrika*, 31: 1–10. page 125

Schönemann, P. H. (1968). On two sided orthogonal Procrustes problems. *Psychometrika*, 33: 19–33. page 125

Schönemann, P. H. and Carroll, R. M. (1970). Fitting one matrix to another under choice of central dilation and rigid motion. *Psychometrika*, 35: 245–255. page 125

Schwarz, G. (1978). Estimating the dimension of a model. *The Annals of Statistics*, 6: 461–464. page 349

Seber, G. A. F. (1984). *Multivariate Observations*. John Wiley & Sons, Inc., New York. page 203

Shelupsky, D. (1962). An introduction to spherical coordinates. *American Mathematical Monthly*, 69: 644–646. page 85

Sherer, E., Harris, S. A., Soliva, R., Orozco, M., and Laughton, C. A. (1999). Molecular dynamics studies of DNA A-tract structure and flexibility. *Journal of the American Chemical Society*, 121: 5981–5991. page 139

Sibson, R. (1978). Studies in the robustness of multidimensional scaling: Procrustes statistics. *Journal of the Royal Statistical Society, Series B*, 40: 234–238. page 125, 133, 197

Sibson, R. (1979). Studies in the robustness of multidimensional scaling: perturbation analysis of classical scaling. *Journal of the Royal Statistical Society, Series B*, 41: 217–229. page 125, 133, 197

Siegel, A. F. and Benson, R. H. (1982). A robust comparison of biological shapes. *Biometrics*, 38: 341–350. page 336, 338

Silverman, B. W. (1995). Incorporating parametric effects into functional principal components analysis. *Journal of the Royal Statistical Society, Series B*, 57(4): 673–689. page 325

Simard, P., Le Cun, Y., and Denker, J. (1993). Efficient pattern recognition using a new transformation distance. In: *Advances in Neural Information Processing Systems* (eds S. Hanson, J. Cowan, and C. Giles), Vol. 5, pp. 50–58. Morgan Kaufmann, San Mateo. page 11

Skwerer, S., Bullitt, E., Huckemann, S., *et al.* (2014). Tree-oriented analysis of brain artery structure. *Journal of Mathematical Imaging and Vision*, 50(1–2): 126–143. page 392

Slice, D. E., (ed.) (2005). *Modern Morphometrics in Physical Anthropology*. Springer, New York. page 397

Slice, D. E. (2007). Geometric morphometrics. *Annual Review of Anthropology*, 36: 261–281. page 397

Slice, D. E., Bookstein, F. L., Marcus, L. F., and Rohlf, F. J. (1996). A glossary for geometric morphometrics. In: *Advances in Morphometrics* (eds L. F. Marcus, A. Corti, A. Loy, G. J. P. Naylor, and D. Slice), Vol. 284 of *NATO ASI Series A*, pp. 531–551, Plenum, New York. page 32

Small, C. G. (1982). Random uniform triangles and the alignment problem. *Mathematical Proceedings of the Cambridge Philosophical Society*, 91: 315–322. page 351

Small, C. G. (1988). Techniques of shape analysis on sets of points. *International Statistical Review*, 56: 243–257. page 30, 279, 350

Small, C. G. (1996). *The Statistical Theory of Shape*. Springer, New York. page xx, xxiii, 74, 350, 397

Small, C. G. and Lewis, M. E. (1995). Shape metrics and Frobenius norms. In: *Proceedings in Current Issues in Statistical Shape Analysis* (eds K. V. Mardia and C. A. Gill), pp. 88–95. University of Leeds Press, Leeds. page 279

Smith, A. F. M. and Roberts, G. O. (1993). Bayesian computation via the Gibbs sampler and related Markov chain Monte Carlo methods (with discussion). *Journal of the Royal Statistical Society, Series B*, 55: 3–24. page 234

Smith, S. M. (2002). Fast robust automated brain extraction. *Human Brain Mapping*, 17: 143–155. page 26

Sneath, P. H. A. (1967). Trend surface analysis of transformation grids. *Journal of Zoology, London*, 151: 65–122. page 32, 125, 303, 306

Sommer, S., Lauze, F., Nielsen, M., and Pennec, X. (2013). Sparse multi-scale diffeomorphic registration: the kernel bundle framework. *Journal of Mathematical Imaging and Vision*, 46: 292–308. page 315

Sotiras, A., Davatzikos, C., and Paragios, N. (2013). Deformable medical image registration: A survey. *IEEE Transactions on Medical Imaging*, 32(7): 1153–1190. page 315, 377, 390

Southworth, R., Mardia, K. V., and Taylor, C. C. (2000). Transformation- and label-invariant neural network for the classification of landmark data. *Journal of Applied Statistics*, 27: 205–215. page 169

Sozou, P. D., Cootes, T. F., Taylor, C. J., and Mauro, E. C. D. (1995). Non-linear generalization of point distribution models using polynomial regression. *Image Vision Computing*, 13(5): 451–457. page 327

Sparr, G. (1992). Depth computations from polyhedral images. In: *Computer Vision – ECCV '92, Lecture Notes in Computer Science* (ed. G. Sandini), Vol. 588, pp. 378–386. Springer-Verlag, Berlin. page 273

Sprent, P. (1972). The mathematics of size and shape. *Biometrics*, 28: 23–37. page 31, 107

Srivastava, A., Klassen, E., Joshi, S. H., and Jermyn, I. H. (2011a). Shape analysis of elastic curves in Euclidean spaces. *IEEE Transactions on Pattern Analysis and Machine Intelligence*, 33(7): 1415–1428. page 369, 370, 371

Srivastava, A., Turaga, P. K., and Kurtek, S. (2012). On advances in differential-geometric approaches for 2D and 3D shape analyses and activity recognition. *Image and Vision Computing*, 30(6–7): 398–416. page 373

Srivastava, A., Wu, W., Kurtek, S., Klassen, E., and Marron, J. S. (2011b). Registration of functional data using the Fisher-Rao metric. Technical Report, Florida State University. arXiv:1103.3817v2 [math.ST]. page 325, 371

Stigler, S. M. (1984). Can you identify these mathematicians? *Mathematical Intelligencer*, 6: 72. page 384

Stone, G. (1988). *Bivariate splines*. PhD thesis, University of Bath. page 287

Stoyan, D. (1990). Estimation of distances and variances in Bookstein's landmark model. *Biometrical Journal*, 32: 843–849. page 357

Stoyan, D. (1997). Geometrical means, medians and variances for samples of particles. *Particle and Particle Systems Characterization*, 14: 30–34. page 22, 364, 368

Stoyan, D. and Frenz, M. (1993). Estimating mean landmark triangles. *Biometrical Journal*, 35: 643–647. page 146

Stoyan, D., Kendall, W. S., and Mecke, J. (1995). *Stochastic Geometry and its Applications*, 2nd edn, John Wiley & Sons, Ltd, Chichester. page xx, 30, 84, 350, 397

Stoyan, D. and Molchanov, I. S. (1995). Set-valued means of random particles. Technical Report BS-R9511, CWI, Amsterdam. page 363, 364, 394

Stoyan, D. and Stoyan, H. (1994). *Fractals, Random Shapes and Point Fields: Methods of Geometric Statistics*. John Wiley & Sons, Ltd, Chichester. page xx, xxiii, 22, 358, 363, 364, 365, 397

Stuart, A. and Ord, K. (1994). *Kendall's Advanced Theory of Statistics, Vol. 1: Distribution Theory*. Arnold, London. page 239, 247

Stuelpnagel, J. (1964). On the parametrization of the three-dimensional rotation group. *SIAM Review*, 6: 422–430. page 62

Su, J., Dryden, I., Klassen, E., Le, H., and Srivastava, A. (2012). Fitting smoothing splines to time-indexed, noisy points on nonlinear manifolds. *Image and Vision Computing*, 30(67): 428–442. page 77, 330, 331

Subsol, G., Thirion, J.-P., and Ayache, N. (1996). A general scheme for automatically building 3D morphometric anatomical atlases: application to a skull and brain atlas. In: *Image Fusion and Shape Variability* (eds K. V. Mardia, C. A. Gill, and I. L. Dryden), pp. 115–122. University of Leeds Press, Leeds. page 375

Tanaka, N., Nonaka, T., Nakanishi, M., Deyashiki, Y., Hara, A., and Mitsui, Y. (1996). Crystal structure of the ternary complex of mouse lung carbonyl reductase at 1.8 A resolution: the structural origin of coenzyme specificity in the short-chain dehydrogenase/reductase family. *Structure*, 4: 33–45. page 21

Taigman, Y., Yang, M., Ranzato, M., and Wolf, L. (2014). DeepFace: Closing the gap to human-level performance in face verification. *The IEEE Conference on Computer Vision and Pattern Recognition (CVPR)*, 1701–1708, IEEE Computer Society Press, Los Alamitos, CA. page 384

Taylor, C. C., Faghihi, M. R., and Dryden, I. L. (1995). An understanding of muscle fibre images. In: *Image Analysis and Processing, ICIAP95 Conference Proceedings* (eds C. Braccini, L. DeFloriani, and G. Vernazza), Vol. 974 of Lecture Notes in Computer Science, pp. 223–228. Springer Verlag, Berlin. page 351

Taylor, C. C., Mardia, K. V., and Kent, J. T. (2003). Matching unlabelled configurations using the EM algorithm. In: *Proceedings of LASR 2003* (eds R. G. Aykroyd, K. V. Mardia, and M. J. Langdon), pp. 19–21. page 341, 348

Taylor, S. (1996). Euclidean distance matrix analysis and MBU-20/P oxygen mask fit. In: *Proceedings in Image Fusion and Shape Variability Techniques* (eds K. V. Mardia, C. A. Gill, and I. L. dryden), pp. 220–221. University of Leeds Press, Leeds. page 359

Ten Berge, J. M. F. (1977). Orthogonal Procrustes rotation for two or more matrices. *Psychometrika*, 42: 267–276. page 125, 136, 138

Theobald, C. M., Glasbey, C. A., Horgan, G. W., and Robinson, C. D. (2004). Principal component analysis of landmarks from reversible images. *Journal of the Royal Statistical Society, Series C*, 53(1): 163–175. page 171

Theobald, D. L. (2005). Rapid calculation of RMSDs using a quaternion-based characteristic polynomial. *Acta Crystallographica Section A*, 61(4): 478–480. page 63

Theobald, D. L. and Mardia, K. V. (2011). Full Bayesian analysis of the generalized non-isotropic Procrustes problem with scaling. In: *Proceedings of LASR 2011* (eds A. Gusnanto, K. V. Mardia, C. J. Fallaize, and J. Voss), pp. 41–44. University of Leeds Press, Leeds. page 236

Theobald, D. L. and Wuttke, D. S. (2006). Empirical Bayes hierarchical models for regularizing maximum likelihood estimation in the matrix Gaussian Procrustes problem. *Proceedings of the National Academy of Sciences of the United States of America*, 103(49): 18521–18527. page 147

Theobald, D. L. and Wuttke, D. S. (2008). Accurate structural correlations from maximum likelihood superpositions. *PLoS Computational Biology*, 4(2): e43. page 147

Thompson, D. W. (1917). *On Growth and Form*. Cambridge University Press, Cambridge. page xxiii, 2, 32, 33, 270, 271, 282

Thompson, S. and Rosenfeld, A. (1994). Discrete stochastic growth for two-dimensional shapes. In: *NATO Conference on Shape in Pictures* (eds O. Ying-Lie, A. Toet, D. Foster, H. J. A. M. Heijmans, and P. Meer), pp. 443–452. Springer-Verlag, Berlin. page 333

Tibshirani, R. (1996). Regression shrinkage and selection via the Lasso. *Journal of the Royal Statistical Society, Series B*, 58(1): 267–288. page 214

Tjelmeland, H. and Eidsvik, J. (2004). On the use of local optimizations within Metropolis–Hastings updates. *Journal of the Royal Statistical Society: Series B (Statistical Methodology)*, 66(2): 411–427. page 347

Tjelmeland, H. and Hegstad, B. K. (2001). Mode jumping proposals in MCMC. *Scandinavian Journal of Statistics*, 28(1): 205–223. page 347

Tramontano, A. (2006). *Protein Structure Prediction*. John Wiley & Sons, Ltd, Chichester. page 209

Trevor, J. C. T. (1950). *Anthropometry*, pp. 458–462. Chambers Encyclopedia. George Newnes, London. page 31

Truslove, G. M. (1976). The effect of selection for body weight on the skeletal variation of the mouse. *Genetical Research Cambridge*, 28: 1–10. page 8

Turaga, P., Veeraraghavan, A., Srivastava, A., and Chellappa, R. (2011). Statistical computations on Grassmann and Stiefel manifolds for image and video-based recognition. *IEEE Transactions on Pattern Analysis and Machine Intelligence*, 33(11): 2273–2286. page 396

Turaga, P. K. and Srivastava, A. (eds) (2015). *Riemannian Computing in Computer Vision*. Springer, New York. page 390

Turner, K., Mukherjee, S., and Boyer, D. M. (2014). Persistent homology transform for modeling shapes and surfaces. *Information and Inference*, 3(4): 310–344. page 393

Tversky, A. (1977). Features of similarity. *Psychological Review*, 84: 327–352. page 395

Twining, C. J., Marsland, S., and Taylor, C. J. (2011). Metrics, connections, and correspondence: The setting for groupwise shape analysis. In: *Proceedings of Energy Minimazation Methods in Computer Vision and Pattern Recognition – 8th International Conference, EMMCVPR 2011* (eds Y. Boykov, F. Kahl, V. S. Lempitsky, and F. R. Schmidt), Vol., 6819 of *Lecture Notes in Computer Science*, pp. 399–412. Springer, New York. page 349

van der Vaart, A. W. (1998). *Asymptotic Statistics*, Vol. 3 of *Cambridge Series in Statistical and Probabilistic Mathematics*. Cambridge University Press, Cambridge. page 372, 389

Vaswani, N., Roy-Chowdhury, A., and Chellapa, R. (2005). Shape activity: a continuous-state HMM for moving/deforming shapes with application to abnormal activity detection. *IEEE Transactions on Image Processing*, 14: 1603–1616. page 325

Venables, W. N. and Ripley, B. D. (2002). *Modern Applied Statistics with S*, 4th edn. Springer, New York. page 7

Verboon, P. and Heiser, W. J. (1992). Resistant orthogonal Procrustes analysis. *Journal of Classification*, 9: 237–256. page 337

Vercauteren, T., Pennec, X., Perchant, A., and Ayache, N. (2009). Diffeomorphic demons: Efficient non-parametric image registration. *NeuroImage*, 45(1, Suppl. 1): S61–S72. page 315

Villalon, J., Joshi, A. A., Lepore, N., Brun, C. C., Toga, A. W., and Thompson, P. M. (2011). Comparison of volumetric registration algorithms for tensor-based morphometry. In: *ISBI* (eds S. Wright, X. Pan and M. Liebling), pp. 1536–1541. IEEE, Piscataway, NJ. page 315, 390

Wagener, M., Sadowski, J., and Gasteiger, J. (1995). Autocorrelation of molecular surface properties for modeling Corticosteroid Binding Globulin and Cytosolic Ah receptor activity by neural networks. *Journal of the American Chemical Society*, 117: 7769–7775. page 13

Wahba, G. (1990). *Spline Models for Observational Data*. SIAM, Philadelphia. page 280, 287

Walker, G. (1999). *Robust, non-parametric and automatic methods for matching spatial point patterns*. PhD thesis, University of Leeds. page 348

Wang, H. and Marron, J. S. (2007). Object oriented data analysis: sets of trees. *Annals of Statistics*, 35(5): 1849–1873. page 391, 392

Warner, F. W. (1971). *Foundations of Differentaible Manifolds and Lie Groups*. Scott, Foresman and Co., Glenview, IL. page 61

Watson, G. N. (1944). *Theory of Bessel Functions*, 2nd edn. Cambridge University Press, Cambridge. page 262

Watson, G. S. (1965). Equatorial distributions on a sphere. *Biometrika*, 52: 193–201. page 227

Watson, G. S. (1983). *Statistics on spheres*. University of Arkansas Lecture Notes in the Mathematical Sciences, Vol. 6. John Wiley & Sons, Inc., New York. page 227, 396

Watson, G. S. (1986). The shape of a random sequence of triangles. *Advances in Applied Probability*, 18: 156–169. page 57, 351

Weber, G. W. and Bookstein, F. L. (2011). *Virtual Anthropology: A guide to a New Interdisciplinary Field*. Springer-Verlag, Vienna. page 396

Wilks, S. S. (1962). *Mathematical Statistics*. John Wiley & Sons, Inc., New York. page 256

Wilson, A. (1995). *Statistical models for shape deformations*. PhD thesis, Duke University, Durham, NC. page 390

Wilson, A. and Johnson, V. (1995). Models for shape deformation. In: *Bayesian Statistics* (eds J. Berger, J. Bernardo, A. P. Dawid, and A. F. M. Smith), pp. 801–808. Oxford University Press, Oxford. page 390

Winkler, G. (1995). *Image Analysis, Random Fields and Dynamic Monte Carlo Methods.* Springer-Verlag, Berlin. page 377

Witkin, A. (1983). Scale-space filtering. In: *Proceedings of the Eighth International Joint Conference on Artificial Intelligence* (ed. A. Bundy), pp. 1019–1022. Kaufman, Los Altos. page 390

Ying-Lie, O., Toet, A., Foster, D., Heijmans, H. J. A. M., and Meer, P. (eds) (1994). *Proceedings of the NATO Conference on Shape in Pictures.* Springer-Verlag, Berlin. page 390, 397

Younes, L. (1998). Computable elastic distances between shapes. *SIAM Journal of Applied Mathematics*, 58(2): 565–586. page 371

Younes, L. (2010). *Shapes and Diffeomorphisms*, Vol. 171 of *Applied Mathematical Sciences.* Springer-Verlag, Berlin. page xx, 314, 390, 397

Younes, L., Michor, P. W., Shah, J., and Mumford, D. (2008a). A metric on shape space with explicit geodesics. *Atti della Accademia Nazionale dei Lincei. Classe di Scienze Fisiche, Matematiche e Naturali. Rendiconti Lincei. Serie IX. Matematica e Applicazioni*, 19(1): 25–57. page 374

Younes, L., Qiu, A., Winslow, R. L., and Miller, M. I. (2008b). Transport of relational structures in groups of diffeomorphisms. *Journal of Mathematical Imaging and Vision*, 32(1): 41–56. page 315

Yu, Y.-Y., Fletcher, P. T., and Awate, S. P. (2014). Hierarchical Bayesian modeling, estimation, and sampling for multigroup shape analysis. In Golland, P., Hata, N., Barillot, C., Hornegger, J., and Howe, R., editors, *Medical Image Computing and Computer-Assisted Intervention MICCAI 2014* (eds P. Golland, N. Hata, C. Barillot, J. Hornegger, and R. Howe), Vol. 8675 of *Lecture Notes in Computer Science*, pp. 9–16. Springer International Publishing, New York. page 349

Yuan, Y., Zhu, H., Styner, M., Gilmore, J. H. and Marron, J. S. (2013). Varying coefficient model for modeling diffusion tensors along white matter tracts. *The Annals of Applied Statistics*, 7: 102–125. page 396

Yuille, A. L. (1991). Deformable templates for face recognition. *Journal of Cognitive Neuroscience*, 3: 59–70. page 5

Zacharias, J. and Knapp, E. W. (2013). Geometry motivated alternative view on local protein backbone structures. *Protein Science*, 22(11): 1669–1674. page 40

Zahn, C. T. and Roskies, R. Z. (1972). Fourier descriptors for plane closed curves. *IEEE Transactions on Computing*, C21: 269–281. page 364

Zelditch, M. L., Swiderski, D. L., and Sheets, H. D. (2012). *Geometric Morphometrics for Biologists: a Primer*, 2nd edn. Academic Press, New York. page xx, 396

Zhang, L., Wahba, G., and Yuan, M. (2016). Distance shrinkage and Euclidean embedding via regularized kernel estimation. *Journal of the Royal Statistical Society: Series B (Statistical Methodology)*. DOI: 10.1111/rssb.12138. page 358

Zhou, D., Dryden, I. L., Koloydenko, A. A., Audenaert, K. M. R., and Bai, L. (2016). Regularisation, interpolation and visualisation of diffusion tensor images using non-Euclidean statistics. *Journal of Applied Statistics*, 43: 943–978. page 150, 396

Zhou, D., Dryden, I. L., Koloydenko, A. A., and Bai, L. (2013). Procrustes analysis for diffusion tensor image processing. *International Journal of Computer Theory and Engineering*, 5: 108–113. page 396

Zhou, R. R., Serban, N., Gebraeel, N., and Müller, H.-G. (2014). A functional time warping approach to modeling and monitoring truncated degradation signals. *Technometrics*, 56(1): 67–77. page 325

Zienkiewicz, O. C. (1971). *The Finite Element Method in Engineering Science*. McGraw-Hill, London. page 305

Ziezold, H. (1977). On expected figures and a strong law of large numbers for random elements in quasi-metric spaces. In: *Transactions of the Seventh Prague Conference on Information Theory, Statistical Decision Functions, Random Processes and of the 1974 European Meeting of Statisticians*, Vol. A, pp. 591–602. Academia: Czechoslovak Academy of Sciences, Prague. page 32, 33, 103, 113, 125, 318

Ziezold, H. (1989). On expected figures in the plane. In: *Proceedings of GEOBILD 89* (eds A. Hübler, W. Nagel, B. D. Ripley, and G. Werner), pp. pp. 105–110. Akademie-Verlag, Berlin. page 146, 318

Ziezold, H. (1994). Mean figures and mean shapes applied to biological figure and shape distributions in the plane. *Biometrical Journal*, 36: 491–510. page 67, 101, 125, 146, 318, 323

Zou, H. and Hastie, T. (2005). Regularization and variable selection via the elastic net. *Journal of the Royal Statistical Society: Series B (Statistical Methodology)*, 67(2): 301–320. page 214

Zusne, L. (1970). *Visual Perception of Forms*. Academic Press, New York. page 395

Index

WILEY SERIES IN PROBABILITY AND STATISTICS

ESTABLISHED BY WALTER A. SHEWHART AND SAMUEL S. WILKS

Editors: *David J. Balding, Noel A. C. Cressie, Garrett M. Fitzmaurice, Geof H. Givens, Harvey Goldstein, Geert Molenberghs, David W. Scott, Adrian F. M. Smith, Ruey S. Tsay, Sanford Weisberg*

Editors Emeriti: *J. Stuart Hunter, Iain M. Johnstone, Joseph B. Kadane, Jozef L. Teugels*

The *Wiley Series in Probability and Statistics* is well established and authoritative. It covers many topics of current research interest in both pure and applied statistics and probability theory. Written by leading statisticians and institutions, the titles span both state-of-the-art developments in the field and classical methods.

Reflecting the wide range of current research in statistics, the series encompasses applied, methodological and theoretical statistics, ranging from applications and new techniques made possible by advances in computerized practice to rigorous treatment of theoretical approaches.

This series provides essential and invaluable reading for all statisticians, whether in academia, industry, government, or research.

† ABRAHAM and LEDOLTER · Statistical Methods for Forecasting
  AGRESTI · Analysis of Ordinal Categorical Data, *Second Edition*
  AGRESTI · An Introduction to Categorical Data Analysis, *Second Edition*
  AGRESTI · Categorical Data Analysis, *Third Edition*
  ALSTON, MENGERSEN, and PETTITT (editors) · Case Studies in Bayesian Statistical
    Modelling and Analysis
  ALTMAN, GILL, and McDONALD · Numerical Issues in Statistical Computing for the
    Social Scientist
  AMARATUNGA and CABRERA · Exploration and Analysis of DNA Microarray and
    Protein Array Data
  AMARATUNGA, CABRERA, and SHKEDY · Exploration and Analysis of DNA
    Microarray and Other High-Dimensional Data, *Second Edition*
  ANDĚL · Mathematics of Chance
  ANDERSON · An Introduction to Multivariate Statistical Analysis, *Third Edition*
* ANDERSON · The Statistical Analysis of Time Series
  ANDERSON, AUQUIER, HAUCK, OAKES, VANDAELE, and WEISBERG · Statistical
    Methods for Comparative Studies
  ANDERSON and LOYNES · The Teaching of Practical Statistics
  ARMITAGE and DAVID (editors) · Advances in Biometry
  ARNOLD, BALAKRISHNAN, and NAGARAJA · Records
* ARTHANARI and DODGE · Mathematical Programming in Statistics
  AUGUSTIN, COOLEN, DE COOMAN, and TROFFAES (editors) · Introduction to
    Imprecise Probabilities
* BAILEY · The Elements of Stochastic Processes with Applications to the Natural
    Sciences
  BAJORSKI · Statistics for Imaging, Optics, and Photonics
  BALAKRISHNAN and KOUTRAS · Runs and Scans with Applications
  BALAKRISHNAN and NG · Precedence-Type Tests and Applications

*Now available in a lower priced paperback edition in the Wiley Classics Library.
†Now available in a lower priced paperback edition in the Wiley–Interscience Paperback Series.

BARNETT · Comparative Statistical Inference, *Third Edition*

BARNETT · Environmental Statistics

BARNETT and LEWIS · Outliers in Statistical Data, *Third Edition*

BARTHOLOMEW, KNOTT, and MOUSTAKI · Latent Variable Models and Factor Analysis: A Unified Approach, *Third Edition*

BARTOSZYNSKI and NIEWIADOMSKA-BUGAJ · Probability and Statistical Inference, *Second Edition*

BASILEVSKY · Statistical Factor Analysis and Related Methods: Theory and Applications

BATES and WATTS · Nonlinear Regression Analysis and Its Applications

BECHHOFER, SANTNER, and GOLDSMAN · Design and Analysis of Experiments for Statistical Selection, Screening, and Multiple Comparisons

BEH and LOMBARDO · Correspondence Analysis: Theory, Practice and New Strategies

BEIRLANT, GOEGEBEUR, SEGERS, TEUGELS, and DE WAAL · Statistics of Extremes: Theory and Applications

BELSLEY · Conditioning Diagnostics: Collinearity and Weak Data in Regression

† BELSLEY, KUH, and WELSCH · Regression Diagnostics: Identifying Influential Data and Sources of Collinearity

BENDAT and PIERSOL · Random Data: Analysis and Measurement Procedures, *Fourth Edition*

BERNARDO and SMITH · Bayesian Theory

BHAT and MILLER · Elements of Applied Stochastic Processes, *Third Edition*

BHATTACHARYA and WAYMIRE · Stochastic Processes with Applications

BIEMER, GROVES, LYBERG, MATHIOWETZ, and SUDMAN · Measurement Errors in Surveys

BILLINGSLEY · Convergence of Probability Measures, *Second Edition*

BILLINGSLEY · Probability and Measure, *Anniversary Edition*

BIRKES and DODGE · Alternative Methods of Regression

BISGAARD and KULAHCI · Time Series Analysis and Forecasting by Example

BISWAS, DATTA, FINE, and SEGAL · Statistical Advances in the Biomedical Sciences: Clinical Trials, Epidemiology, Survival Analysis, and Bioinformatics

BLISCHKE and MURTHY (editors) · Case Studies in Reliability and Maintenance

BLISCHKE and MURTHY · Reliability: Modeling, Prediction, and Optimization

BLOOMFIELD · Fourier Analysis of Time Series: An Introduction, *Second Edition*

BOLLEN · Structural Equations with Latent Variables

BOLLEN and CURRAN · Latent Curve Models: A Structural Equation Perspective

BONNINI, CORAIN, MAROZZI, and SALMASO · Nonparametric Hypothesis Testing: Rank and Permutation Methods with Applications in R

BOROVKOV · Ergodicity and Stability of Stochastic Processes

BOSQ and BLANKE · Inference and Prediction in Large Dimensions

BOULEAU · Numerical Methods for Stochastic Processes

BOX · Improving Almost Anything, *Revised Edition*

* BOX and DRAPER · Evolutionary Operation: A Statistical Method for Process Improvement

BOX and DRAPER · Response Surfaces, Mixtures, and Ridge Analyses, *Second Edition*

BOX, HUNTER, and HUNTER · Statistics for Experimenters: Design, Innovation, and Discovery, *Second Edition*

BOX, JENKINS, and REINSEL · Time Series Analysis: Forecasting and Control, *Fourth Edition*

BOX, LUCEÑO, and PANIAGUA-QUIÑONES · Statistical Control by Monitoring and Adjustment, *Second Edition*

*Now available in a lower priced paperback edition in the Wiley Classics Library.

†Now available in a lower priced paperback edition in the Wiley–Interscience Paperback Series.

*Now available in a lower priced paperback edition in the Wiley Classics Library.
†Now available in a lower priced paperback edition in the Wiley–Interscience Paperback Series.

*Now available in a lower priced paperback edition in the Wiley Classics Library.
†Now available in a lower priced paperback edition in the Wiley–Interscience Paperback Series.

*Now available in a lower priced paperback edition in the Wiley Classics Library.

†Now available in a lower priced paperback edition in the Wiley–Interscience Paperback Series.

HURD and MIAMEE · Periodically Correlated Random Sequences: Spectral Theory and Practice

HUSKOVA, BERAN, and DUPAC · Collected Works of Jaroslav Hajek—with Commentary

HUZURBAZAR · Flowgraph Models for Multistate Time-to-Event Data

JACKMAN · Bayesian Analysis for the Social Sciences

† JACKSON · A User's Guide to Principle Components

JOHN · Statistical Methods in Engineering and Quality Assurance

JOHNSON · Multivariate Statistical Simulation

JOHNSON and BALAKRISHNAN · Advances in the Theory and Practice of Statistics: A Volume in Honor of Samuel Kotz

JOHNSON, KEMP, and KOTZ · Univariate Discrete Distributions, *Third Edition*

JOHNSON and KOTZ (editors) · Leading Personalities in Statistical Sciences: From the Seventeenth Century to the Present

JOHNSON, KOTZ, and BALAKRISHNAN · Continuous Univariate Distributions, Volume 1, *Second Edition*

JOHNSON, KOTZ, and BALAKRISHNAN · Continuous Univariate Distributions, Volume 2, *Second Edition*

JOHNSON, KOTZ, and BALAKRISHNAN · Discrete Multivariate Distributions

JUDGE, GRIFFITHS, HILL, LÜTKEPOHL, and LEE · The Theory and Practice of Econometrics, *Second Edition*

JUREK and MASON · Operator-Limit Distributions in Probability Theory

KADANE · Bayesian Methods and Ethics in a Clinical Trial Design

KADANE AND SCHUM · A Probabilistic Analysis of the Sacco and Vanzetti Evidence

KALBFLEISCH and PRENTICE · The Statistical Analysis of Failure Time Data, *Second Edition*

KARIYA and KURATA · Generalized Least Squares

KASS and VOS · Geometrical Foundations of Asymptotic Inference

† KAUFMAN and ROUSSEEUW · Finding Groups in Data: An Introduction to Cluster Analysis

KEDEM and FOKIANOS · Regression Models for Time Series Analysis

KENDALL, BARDEN, CARNE, and LE · Shape and Shape Theory

KHURI · Advanced Calculus with Applications in Statistics, *Second Edition*

KHURI, MATHEW, and SINHA · Statistical Tests for Mixed Linear Models

* KISH · Statistical Design for Research

KLEIBER and KOTZ · Statistical Size Distributions in Economics and Actuarial Sciences

KLEMELÄ · Smoothing of Multivariate Data: Density Estimation and Visualization

KLUGMAN, PANJER, and WILLMOT · Loss Models: From Data to Decisions, *Third Edition*

KLUGMAN, PANJER, and WILLMOT · Loss Models: Further Topics

KLUGMAN, PANJER, and WILLMOT · Solutions Manual to Accompany Loss Models: From Data to Decisions, *Third Edition*

KOSKI and NOBLE · Bayesian Networks: An Introduction

KOTZ, BALAKRISHNAN, and JOHNSON · Continuous Multivariate Distributions, Volume 1, *Second Edition*

KOTZ and JOHNSON (editors) · Encyclopedia of Statistical Sciences: Volumes 1 to 9 with Index

KOTZ and JOHNSON (editors) · Encyclopedia of Statistical Sciences: Supplement Volume

KOTZ, READ, and BANKS (editors) · Encyclopedia of Statistical Sciences: Update Volume 1

*Now available in a lower priced paperback edition in the Wiley Classics Library.

†Now available in a lower priced paperback edition in the Wiley–Interscience Paperback Series.

*Now available in a lower priced paperback edition in the Wiley Classics Library.

†Now available in a lower priced paperback edition in the Wiley–Interscience Paperback Series.

McNEIL · Epidemiological Research Methods

MEEKER and ESCOBAR · Statistical Methods for Reliability Data

MEERSCHAERT and SCHEFFLER · Limit Distributions for Sums of Independent Random Vectors: Heavy Tails in Theory and Practice

MENGERSEN, ROBERT, and TITTERINGTON · Mixtures: Estimation and Applications

MICKEY, DUNN, and CLARK · Applied Statistics: Analysis of Variance and Regression, *Third Edition*

* MILLER · Survival Analysis, *Second Edition*

MONTERO, FERNÁNDEZ-AVILÉS and MATEU · Spatial and Spatio-Temporal Geostatistical Modeling and Kriging

MONTGOMERY, JENNINGS, and KULAHCI · Introduction to Time Series Analysis and Forecasting

MONTGOMERY, PECK, and VINING · Introduction to Linear Regression Analysis, *Fifth Edition*

MORGENTHALER and TUKEY · Configural Polysampling: A Route to Practical Robustness

MUIRHEAD · Aspects of Multivariate Statistical Theory

MULLER and STOYAN · Comparison Methods for Stochastic Models and Risks

MURTHY, XIE, and JIANG · Weibull Models

MYERS, MONTGOMERY, and ANDERSON-COOK · Response Surface Methodology: Process and Product Optimization Using Designed Experiments, *Fourth Edition*

MYERS, MONTGOMERY, VINING, and ROBINSON · Generalized Linear Models. With Applications in Engineering and the Sciences, *Second Edition*

NATVIG · Multistate Systems Reliability Theory with Applications

† NELSON · Accelerated Testing, Statistical Models, Test Plans, and Data Analyses

† NELSON · Applied Life Data Analysis

NEWMAN · Biostatistical Methods in Epidemiology

NG, TAIN, and TANG · Dirichlet Theory: Theory, Methods and Applications

OKABE, BOOTS, SUGIHARA, and CHIU · Spatial Tesselations: Concepts and Applications of Voronoi Diagrams, *Second Edition*

OLIVER and SMITH · Influence Diagrams, Belief Nets and Decision Analysis

PALTA · Quantitative Methods in Population Health: Extensions of Ordinary Regressions

PANJER · Operational Risk: Modeling and Analytics

PANKRATZ · Forecasting with Dynamic Regression Models

PANKRATZ · Forecasting with Univariate Box-Jenkins Models: Concepts and Cases

PARDOUX · Markov Processes and Applications: Algorithms, Networks, Genome and Finance

PARMIGIANI and INOUE · Decision Theory: Principles and Approaches

* PARZEN · Modern Probability Theory and Its Applications

PEÑA, TIAO, and TSAY · A Course in Time Series Analysis

PESARIN and SALMASO · Permutation Tests for Complex Data: Applications and Software

PIANTADOSI · Clinical Trials: A Methodologic Perspective, *Second Edition*

POURAHMADI · Foundations of Time Series Analysis and Prediction Theory

POURAHMADI · High-Dimensional Covariance Estimation

POWELL · Approximate Dynamic Programming: Solving the Curses of Dimensionality, *Second Edition*

POWELL and RYZHOV · Optimal Learning

PRESS · Subjective and Objective Bayesian Statistics, *Second Edition*

PRESS and TANUR · The Subjectivity of Scientists and the Bayesian Approach

*Now available in a lower priced paperback edition in the Wiley Classics Library.

†Now available in a lower priced paperback edition in the Wiley–Interscience Paperback Series.

*Now available in a lower priced paperback edition in the Wiley Classics Library.
†Now available in a lower priced paperback edition in the Wiley–Interscience Paperback Series.

† SEBER and WILD · Nonlinear Regression

SENNOTT · Stochastic Dynamic Programming and the Control of Queueing Systems

* SERFLING · Approximation Theorems of Mathematical Statistics

SHAFER and VOVK · Probability and Finance: It's Only a Game!

SHERMAN · Spatial Statistics and Spatio-Temporal Data: Covariance Functions and Directional Properties

SILVAPULLE and SEN · Constrained Statistical Inference: Inequality, Order, and Shape Restrictions

SINGPURWALLA · Reliability and Risk: A Bayesian Perspective

SMALL and McLEISH · Hilbert Space Methods in Probability and Statistical Inference

SRIVASTAVA · Methods of Multivariate Statistics

STAPLETON · Linear Statistical Models, *Second Edition*

STAPLETON · Models for Probability and Statistical Inference: Theory and Applications

STAUDTE and SHEATHER · Robust Estimation and Testing

STOYAN · Counterexamples in Probability, *Second Edition*

STOYAN and STOYAN · Fractals, Random Shapes and Point Fields: Methods of Geometrical Statistics

STREET and BURGESS · The Construction of Optimal Stated Choice Experiments: Theory and Methods

STYAN · The Collected Papers of T. W. Anderson: 1943–1985

SUTTON, ABRAMS, JONES, SHELDON, and SONG · Methods for Meta-Analysis in Medical Research

TAKEZAWA · Introduction to Nonparametric Regression

TAMHANE · Statistical Analysis of Designed Experiments: Theory and Applications

TANAKA · Time Series Analysis: Nonstationary and Noninvertible Distribution Theory

THOMPSON · Empirical Model Building: Data, Models, and Reality, *Second Edition*

THOMPSON · Sampling, *Third Edition*

THOMPSON · Simulation: A Modeler's Approach

THOMPSON and SEBER · Adaptive Sampling

THOMPSON, WILLIAMS, and FINDLAY · Models for Investors in Real World Markets

TIERNEY · LISP-STAT: An Object-Oriented Environment for Statistical Computing and Dynamic Graphics

TROFFAES and DE COOMAN · Lower Previsions

TSAY · Analysis of Financial Time Series, *Third Edition*

TSAY · An Introduction to Analysis of Financial Data with R

TSAY · Multivariate Time Series Analysis: With R and Financial Applications

UPTON and FINGLETON · Spatial Data Analysis by Example, Volume II: Categorical and Directional Data

† VAN BELLE · Statistical Rules of Thumb, *Second Edition*

VAN BELLE, FISHER, HEAGERTY, and LUMLEY · Biostatistics: A Methodology for the Health Sciences, *Second Edition*

VESTRUP · The Theory of Measures and Integration

VIDAKOVIC · Statistical Modeling by Wavelets

VIERTL · Statistical Methods for Fuzzy Data

VINOD and REAGLE · Preparing for the Worst: Incorporating Downside Risk in Stock Market Investments

WALLER and GOTWAY · Applied Spatial Statistics for Public Health Data

WEISBERG · Applied Linear Regression, *Fourth Edition*

WEISBERG · Bias and Causation: Models and Judgment for Valid Comparisons

WELSH · Aspects of Statistical Inference

WESTFALL and YOUNG · Resampling-Based Multiple Testing: Examples and Methods
    for p-Value Adjustment
* WHITTAKER · Graphical Models in Applied Multivariate Statistics
WINKER · Optimization Heuristics in Economics: Applications of Threshold Accepting
WOODWORTH · Biostatistics: A Bayesian Introduction
WOOLSON and CLARKE · Statistical Methods for the Analysis of Biomedical Data,
    Second Edition
WU and HAMADA · Experiments: Planning, Analysis, and Parameter Design
    Optimization, *Second Edition*
WU and ZHANG · Nonparametric Regression Methods for Longitudinal Data Analysis
YAKIR · Extremes in Random Fields
YIN · Clinical Trial Design: Bayesian and Frequentist Adaptive Methods
YOUNG, VALERO-MORA, and FRIENDLY · Visual Statistics: Seeing Data with
    Dynamic Interactive Graphics
ZACKS · Examples and Problems in Mathematical Statistics
ZACKS · Stage-Wise Adaptive Designs
* ZELLNER · An Introduction to Bayesian Inference in Econometrics
ZELTERMAN · Discrete Distributions—Applications in the Health Sciences
ZHOU, OBUCHOWSKI, and McCLISH · Statistical Methods in Diagnostic Medicine,
    Second Edition

*Now available in a lower priced paperback edition in the Wiley Classics Library.
†Now available in a lower priced paperback edition in the Wiley–Interscience Paperback Series.